GEOMETRICAL AND TRIGONOMETRIC OPTICS

In recent years optics has evolved into one of the most flourishing fields in physics. Photonics has found increasing application in products ranging from optical thermometers, camera monitors, and LED lighting, to numerous military applications. This book covers the geometrical aspects of optics, the fundamental level of understanding the technology.

A comprehensive treatment of the subject, the textbook begins with how light is generated, and how fast it travels. The concepts of how materials, such as glass, interact with light, and how various materials affect the velocity of light, are discussed, as well as the ramifications of change in the speed of light. The concept of the index of refraction, and how it is used with Snell's law to produce image forming systems, is developed.

An ideal textbook for advanced undergraduate level courses in geometrical optics, this book will also interest those who wish to learn the concepts and theory of geometrical optics. Each chapter contains worked examples, and there are exercises to reinforce the reader's understanding of the material.

Eustace L. Dereniak is a Professor of Optical Sciences and Electrical and Computer Engineering at the University of Arizona. His research interests are in the areas of detectors for optical radiation, imaging spectrometers, and imaging polarimeters instrument development.

Teresa D. Dereniak received her Bachelor of Science in Mechanical Engineering and a Masters in Business Administration from Cornell University. Her technical experience consists of product development engineering in the biomedical field.

GEOMETRICAL AND TRIGONOMETRIC OPTICS

EUSTACE L. DERENIAK

The University of Arizona

TERESA D. DERENIAK

CAMBRIDGE
UNIVERSITY PRESS

University Printing House, Cambridge CB2 8BS, United Kingdom

One Liberty Plaza, 20th Floor, New York, NY 10006, USA

477 Williamstown Road, Port Melbourne, VIC 3207, Australia

4843/24, 2nd Floor, Ansari Road, Daryaganj, Delhi - 110002, India

79 Anson Road, #06-04/06, Singapore 079906

Cambridge University Press is part of the University of Cambridge.

It furthers the University's mission by disseminating knowledge in the pursuit of education, learning and research at the highest international levels of excellence.

www.cambridge.org
Information on this title: www.cambridge.org/9780521887465

First published 2008

A catalogue record for this publication is available from the British Library

Library of Congress Cataloging in Publication data
Dereniak, Eustace L.
Geometrical and trigonometric optics / Eustace Dereniak, Teresa D. Dereniak.
p. cm.
Includes bibliographical references and index.
ISBN 978-0-521-88746-5
1. Geometrical optics. 2. Optical engineering. I. Dereniak, Teresa D. II. Title.
QC381.D44 2008
535'.32–dc22
2008005640

ISBN 978-0-521-88746-5 Hardback

Contents

Preface

This book is an introduction to geometrical optics and is intended for use in courses on optical engineering. Although several excellent books already exist that cover both physical and geometrical optics, it has been our experience that the purpose of these books is not to teach the fundamentals of geometrical optics, but rather to introduce the subject in order to prepare students for more challenging physical optics courses to follow. In contrast, this book will teach the subject in such a way that it will provide future optical engineers with a solid background and the skills necessary to understand modern computer optical design programs used by lens designers. Furthermore, unlike previous textbooks, this book uses a right-handed coordinate system that lends itself more efficiently to lens design techniques, such as integrating multiple lenses in cascade.

This book is geared toward the professional engineer or student who wants to gain a broad understanding of geometrical optics with the details worked out. It uses a student's prior background in basic high school level algebra, trigonometry, and calculus to build a foundation for the concept of image formation, using linear equations to describe where the image is formed, its size and its classical third-order aberrations, and to teach all fundamental topics necessary to understand complex optics used in optical instruments.

The book has grown out of class notes developed for an introductory course in geometrical optics offered as part of the optical engineering curriculum at the University of Arizona College of Optical Sciences, Tucson, Arizona. Students taking this course had no prior exposure to optics, other than in secondary school. Thus, the organization of the book is designed to follow a one-semester course, covering 12 chapters in 16 weeks. Each chapter contains several worked examples as well as many problems left as exercises for the student.

The authors are grateful to many individuals for their help and encouragement in the writing of this book. The greatest indebtedness is to William L. Wolfe and Robert R. Shannon who, with great amounts of encouragement and many personal contacts, gave one author (Eustace Dereniak) the opportunity of becoming a professor in the College of Optical Sciences at the University of Arizona. This book was originally the brainchild of Margy Green, who initiated it as a student of mine and was to be the coauthor. (After a year of effort, she realized that I was not dedicated to completing this work in a timely manner.) We would like to thank Simon Capelin for his assistance with the publication of this book. We are also extremely grateful to the many students who labored through the geometrical optics course at the University of Arizona College of Optical Sciences and who corrected many errors in the notes. We would like to offer specific thanks to the following individuals for their extra efforts in providing valuable inputs, corrections and additions: Andre Alenin, Andrea Amaro, Erik Arendt, Riley Aumiller, Ruth Burzik, Ken Cardell, Erika Counselor, Julia Craven, Steven Dziuban, Les Foo, Nathan Hagen, Darren Jackson, Michal Kudenov, Jeremiah Lange, Kris LaPorte, Elias Martinez, Tracie Mosciski, Darren Miller, Jonathon Nation, Mike Nofziger, Virginia Pasek, Luz Palomarez, Logan Pastor, Jeff Richie, Louie Rosiek, Jim Scholl, Tim Shih, Katia Shtyrkova, James Tawney, Tomasz Tkaczyk, Dorothy Tinkler, Corrie Vandervlugt, and Emma Walker. Our heartfelt thanks goes to Barbara Grant for her excellent work in reviewing the content of several chapters and for helping to secure the contract with Cambridge University Press that made this book possible. A special note should be made of Tracy Heran (see Figure 3.4) for her laborious and diligent work constructing the figures and putting the equations into the correct Greek format. Another grateful acknowledgement should be given to Trish Pettijohn for typing and proofreading the manuscript while correcting much of our terrible grammar. In any case, the final blame for any errors and omissions belongs to the authors.

1
Light propagation

1.1 Background history

The history of optics is filled with examples of unique uses and situations that are beyond the scope of this book which is intended for the student on a first year optics course. However, a brief review is necessary to show how man has been trying to understand and describe light over the last 2500 years.

The word "optics" originated in a book on visual perception written by Euclid some 2000 years ago. Euclid developed geometrical theories to account for the observation of images by mirrors. Some names that come to mind in the history of optics are Ptolemy, Bacon, Brahe, Kepler, and more recently Newton, Huygens, Fermat, Young, and Einstein.

There is a story by Archimedes (212 BC) that the Greeks defended Syracuse (in modern-day Sicily) from the Roman fleet by reflecting sunlight with the soldiers' shields and burning the ships' sails by focusing the intense heat of the Sun's rays.

Muslims in the thirteenth century were purported to have the ability to create a burning mirror to use for burning cities (in the Holy Land). Roger Bacon, a monk under Pope Clement IV, was motivated by this threat to study optics as a weapon of war. He developed similar devices for the Christian crusaders battling the Muslims.

Ptolemy of Alexandria, a Greek from Egypt (about AD 190), knew that two transparent substances, glass and water, had indices of refraction of 3/2 and 4/3, respectively. These were calculated by casting shadows of objects illuminated by the Sun into water and glass.

Ibn Al-Haytham of Cairo (Khan, 2007) made probably the most precise measurements of the index of refraction in the tenth century, while Europe was still in the Dark Ages. His scientific experiments were the best to that date, and were marveled at until modern times.

Willebrord Snell in the seventeenth century empirically wrote the refraction law that bears his name, but could not explain the relationship, because light was thought to be composed of corpuscular particles. This confusion gave rise to many explanations of why light bent toward the normal of a surface in a denser medium.

Isaac Newton's contributions to the dispersion of light through a prism were some of the greatest and yet worst work in optics. He did this work before the age of 26, and was thrown out of London's Science Academy for his revolutionary approach to optics. He found light to be composed of many colors, and made a prism system to display these colors. The fact that he had incorrectly concluded that white light could not be focused with glass, due to dispersion, set back the development of the achromat doublet by many decades. This delay in the development of the achromat was a consequence of his great dominance in the field of optics, with the widely held assumption that if Newton said it, it must be so!

Modern optics is driven by optical systems modeled after the human eye. In fact, it might turn out that the modeling and copying of the human eye's functions may have been carried too far. In most systems of detection, the optical configuration for forming images and the associated signal processing techniques all mimic the human eye. Although geometrical optical systems typically mimic the eye as an imaging system, other optical systems in nature do not. These include compound eyes, polarization sensing eyes, as well as color sensing in the insect world.

This book covers only the geometrical aspects of optics, which can be thought of as the lowest level in the hierarchy of optics. The assumptions that light travels in a straight line and that all equations are linear will be held throughout this textbook. We will consider optics to be mainly confined to light radiation that is detectable by the human eye (i.e. the visible spectrum as opposed to the entire electromagnetic spectrum).

1.2 Nature of light

What is light? That question can be very difficult to answer. In fact, throughout the ages, optics theory has bounced back and forth between corpuscular quanta and wave models. Present-day scientists' description of light depends on the application on which they are working. The formation of light from heat, e.g. from a fire or an incandescent light bulb, is described as being due to excited atoms, and this can be explained in classical terms.

The atoms which are thermally excited have electrons which are "bumped" into higher energy orbits from which they decay to lower energy orbits. During this process, the electrons release a quantum of radiation with a frequency (v)

of radiation in proportion to the energy (E) released. The atoms of the substance being heated have a large range of discrete energy level orbitals, and the decay from these orbitals to lower energy states releases a continuum of energies. However, the planet model of the atom is not complete, because "strictly speaking," since the electrons use energy when orbiting the nucleus, the atom should eventually collapse.

The relationship between energy and the frequency of a quantum of light emitted is

$$E = hv, \tag{1.1}$$

where h is Planck's constant and v is the frequency of light.

Light propagates into space whenever a charged particle is accelerated or decelerated. A common example is the X-ray machine used in radiology departments, in which a beam of electrons is focused onto an anode with high voltage (50 kV). When the electrons are stopped at the anode, X-rays are emitted at the speed of light.

Light bulbs give off a continuum of visible light because of the many different electron energy levels decaying to a continuum of lower energy levels, thus, energy quanta of many frequencies are emitted. The heated tungsten filament produces white light, which is light that contains all frequencies of light from zero to infinity. However, there is radiation being given off at frequencies above and below that which the human eye can detect. The electromagnetic spectrum has been classically divided up into regions by energy level; however, the exact dividing points are not well defined. The main spectral regions of interest are shown in Table 1.1.

Table 1.1. *Various common names of spectral regions, with the approximate center frequency.*

Wave name	v cycles s^{-1}
Gamma rays	$\sim 3\ (10^{24})$
X-rays	$3\ (10^{16})$
Ultraviolet	$8\ (10^{14})$
Visible	$6\ (10^{14})$
Infrared	$3\ (10^{12})$
Microwave	$3\ (10^{11})$
UHF	$3\ (10^{8})$
VHF	$3\ (10^{8})$
FM	10^{8}
AM	10^{6}
Audio	10^{4}

The visible spectrum, to which we humans respond, is between 4 (10^{14}) and 7.5 (10^{14}) Hz. Monochromatic light, such as that from a laser, has a center frequency with a very narrow bandwidth. For example, a HeNe laser has a center frequency of 4.74 (10^{14}) Hz, while a laser diode (InGaAs) has a frequency of 4.47 (10^{14}) Hz.

Example 1.1

What is the energy of a photon from a laser diode of frequency 4.47(10^{14}) Hz?

$$E = h\upsilon = 6.6\left(10^{-34}\right) \times 4.47\left(10^{14}\right) \text{joules}$$

$$E = 2.95\left(10^{-19}\right) \text{joules} = \frac{2.95\left(10^{-19}\right) \text{joules}}{1.6\left(10^{-19}\right) \text{coulombs}} = 1.7 \text{ electron volts.}$$

The previous discussion assumes light to be made up of particles or quanta of energy ($h\upsilon$). An alternative approach is to consider light as an electromagnetic (EM) wave. From many physics observations, it is concluded that whenever an electric charge is accelerated, a wave is emitted (similar to our photon model). The waves that are formed consist of electric and magnetic fields that propagate at the speed of light.

James Maxwell logically coined the term "electromagnetic waves." An EM wave is a self-propagating wave consisting of electric and magnetic fields fluctuating together. Maxwell developed equations describing these EM waves, and derived the wave equation, which is an expression that describes their propagation.

Maxwell's equations also predicted how fast these waves would move, i.e. their velocity. He found that the velocity is dependent on two constants of the medium (permittivity and permeability), and that the velocity itself is also a constant – a revolutionary conclusion. Maxwell discovered that light, in fact all electromagnetic radiation, produces an electrical field that travels at a constant velocity, in air, of about 3 (10^{8}) m s^{-1}. A changing electric field (**E**) induces a changing magnetic field (**H**), as shown in Figure 1.1.

Most waves encountered in nature, e.g. water waves, propagate in a medium; however, EM waves can also propagate in a vacuum. The EM wave keeps itself going through its own internal mechanism, so once launched, the EM wave no longer depends on its source, the accelerated charge, and propagates in a straight line in a homogeneous medium. It propagates on its own, and carries some characteristics of the source which generated it.

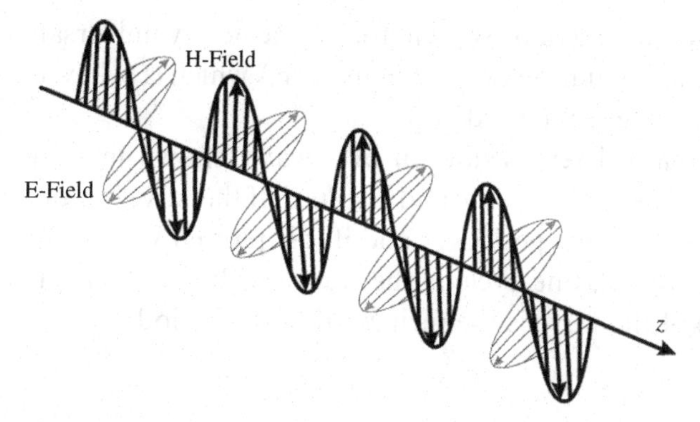

Figure 1.1 A diagram of an EM wave.

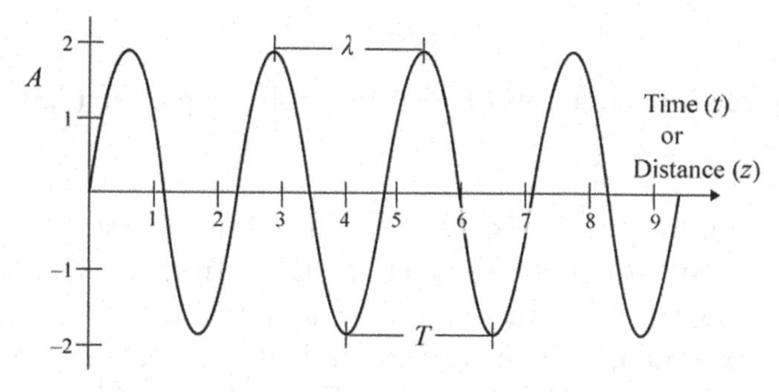

Figure 1.2 A sinusoidal EM wave plotted in time or distance.

Mathematically, we represent light as a sinusoidal electric field propagating through time (t) and space (z) as

$$|\mathbf{E}(z, t)| = A \sin(kz - ct), \tag{1.2}$$

where A is the amplitude, k is the wave number, z is the axial distance, c is the speed of light and t is the time. Figure 1.2 shows a sine wave plotted from Equation (1.2) in time (t) or distance (z). Pick a position in space (z-fixed), and watch the light wave as it passes this position. The wave would be modulated in a sinusoidal way, as shown in time (t).

For a fixed position in space (z constant), the amplitude of the light wave varies sinusoidally with t. For an instantaneous time t (a snapshot in time), the light wave's intensity would be sinusoidal over space (i.e. in the z variable). Either variable is correctly modeled as a sine wave.

Recall that the frequency is constant for a given monochromatic light source. However, the velocity of the wave may change as we propagate through different media. The speed of light in a vacuum (free space) is approximately $3\,(10^8)\ \mathrm{m\,s^{-1}}$

or 186 287 miles per second. We will follow the nearly universal convention of representing the vacuum velocity of light as "*c*," which is believed to come from the Latin word *celeritas* (speed).

The mathematical representation of a wave, shown in Equation (1.2), is sinusoidal in two variables, time and distance. If the wave is plotted versus time (t), one cycle is the time period T, and if it is plotted versus distance (z), also shown in Figure 1.2, one cycle is the wavelength (λ) of that EM wave. The frequency (v) of the wave is the reciprocal of the period:

$$v = 1/T. \tag{1.3}$$

The velocity of the EM wave, $3\ (10^8)\ \mathrm{m\,s^{-1}}$, is the distance it travels in one period (λ) divided by the time it takes to move one period (T), so

$$c = \lambda/T, \tag{1.4}$$

which can be rewritten in terms of frequency using Equation (1.3),

$$c = \lambda v. \tag{1.5}$$

The velocity of light in free space is considered the fastest velocity known. Sunlight takes about 8 minutes to reach the Earth from the sun. Light could travel between Los Angeles and New York about 62 times in a second. In a homogeneous medium, light travels in straight lines called rays. This ray concept is a fundamental description of light, albeit one which oversimplifies what is really propagating.

Example 1.2

Find the velocity of light for a laser diode that has a frequency of $4.47(10^{14})$ Hz and a wavelength of 670 nm (red).

$$v = 4.47\left(10^{14}\right)\text{ hertz;}$$
$$\lambda = 670\,\text{nm (red)},$$
$$c = 4.47\left(10^{14}\right)(670)\left(10^{-9}\right) \approx 3\left(10^{8}\right)\mathrm{m\,s^{-1}}.$$

The visible spectrum of light lies between 400 nm and 700 nm in wavelength. The convention is to define that spectrum in terms of wavelength; however, the description in terms of frequency should be used. This is because the frequency of light does not change once generated.

Very crudely, we can divide the visible region of the electromagnetic spectrum into three parts: red, yellow, and blue (600–700 nm, 500–600 nm, and

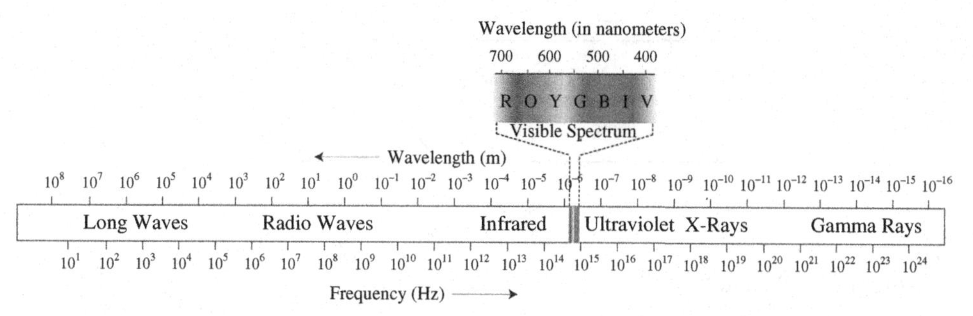

Figure 1.3 The EM spectrum.

400–500 nm, respectively) to represent human color sensitivity. (We could explain this sensitivity, as per Charles Darwin, by noting the need for our ancestors to know when the bananas were ripe.)

In this visible range, red light has the longest wavelength (lowest frequency) of 700 nm or a frequency of 4.3 (10^{14}) Hz. Red is the least energetic region of the visible spectrum. If we assume an atom has a diameter of about 0.65 nm, then the red wavelength is about the length of 1000 atoms laid side by side. Blue/violet on the other hand, which has the shortest wavelength (400 nm) and is the most energetic light wave to which the eye responds, has a wavelength corresponding to about 600 atoms.

When all the wavelengths of light are present for the entire spectrum (400–700 nm), white light is observed by humans. The wavelengths between 400 and 700 nm each form one of the colors red, orange, yellow, green, blue, indigo, and violet, making rainbows visible in the sky to humans. The visible spectrum is a very small part of the electromagnetic spectrum as seen in Figure 1.3.

A key point here, used throughout the book, is that these EM waves are modeled geometrically as straight lines, thus producing rays for the study of geometrical optics. The field of geometrical optics manipulates these rays to form images and illuminations or to transfer information.

A wave is produced by an accelerating charge, such as an electron in an atom changing energy levels. This gives a quantum of energy, $h\upsilon$, so very simply, one would conclude that each EM wave would have an energy of $h\upsilon$. Here are the three main points of our conception of EM waves so far:

- Atoms give off photons of energy ($h\upsilon_i$).
- Each wave can be thought of as an EM ray with energy of $h\upsilon_i$. This is wrong, but conceptually acceptable at this point.
- Many photons (N_i) give off $\Sigma_i N_i h\upsilon_i$ energy. Conceptually, this is sufficient for a very preliminary observation, but again is wrong. We can think of light as traveling in waves, with each wave having the energy $Nh\upsilon$, where N is the number of photons in the wave.

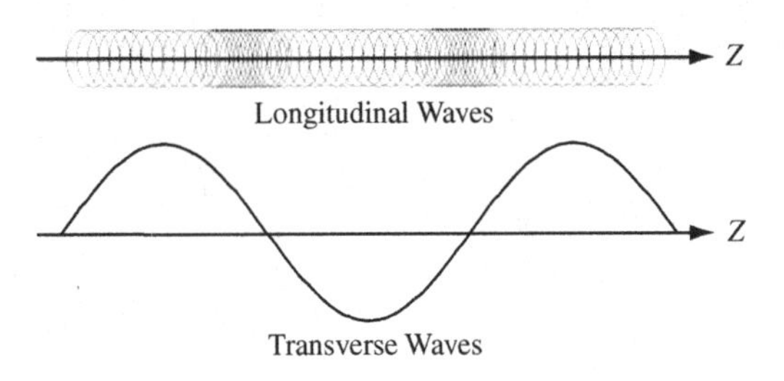

Figure 1.4 Wave types.

There are two types of waves: transverse waves, also known as EM waves, which have been discussed in this chapter, and longitudinal waves. A pictorial representation of these wave types is shown in Figure 1.4

Longitudinal waves need a medium in which to propagate, and without such a medium, their energy is lost. Sound waves are an example of longitudinal waves. Note that sound waves, seismic waves, and other kinds of waves that require matter in which to propagate travel much slower than the speed of light.

In the case of EM waves, the energy is contained in the electric and magnetic fields, which can exist in a vacuum. In fact, they propagate fastest in a vacuum. In other media, the velocity is less than the speed of light in a vacuum because the atoms making up the material are excited and relaxed, slowing the fields. The energy propagates perpendicular to the $\mathbf{E} \times \mathbf{H}$ field direction, as shown in Figure 1.1.

1.3 Wavefronts and rays

Geometrical optics represents the EM wave as a vector pointing in the direction of propagation: a straight line representation, called a ray. This model is somewhat misleading and incorrect, but for the most part, the ray model may be used in the context of geometrical optics to produce useful results.

There are two types of radiation sources in geometrical optics: point and extended. A point source, such as for starlight, may be thought of as a source from which rays emanate in all directions. See Figure 1.5. The rays are actually propagating into 4π steradians, or into three-dimensional space. (Solid geometry will be discussed later.) The ray is simply the path followed by a single photon of light, or an imaginary line drawn in the direction the wave is traveling.

In a homogeneous isotropic medium, the ray paths are straight lines which have varying amplitudes of both the **E**- and **H**-fields, as shown in Figure 1.1. If we connect all the peak values of the EM wave that are a distance of 100 peaks

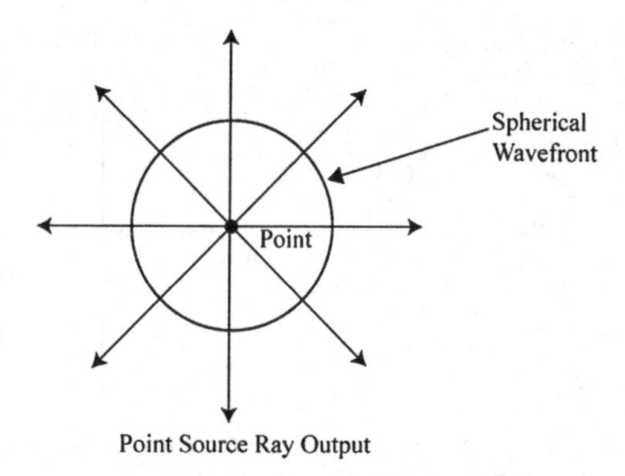

Figure 1.5 Rays propagating from a point source of radiation.

from the source, we would produce a spherical surface (see Figure 1.5). The points on this spherical surface would be at equal distances from the point source, and these distances would equal the radius of the sphere. The rays are the radii of the spherical wavefronts, and are perpendicular to these wavefronts. One may think of the wavefronts as being at each crest or trough of the EM wave which is emanating from a point source. The expression for a spherical wave is:

$$\frac{A}{r} e^{i\varphi} e^{i(\mathbf{k} \bullet \mathbf{r} - \omega t)}, \tag{1.6}$$

where $k = 2\pi/\lambda$ is the wave number, φ is the phase, $\omega = 2\pi/T$, and r is the distance from the point source to the wavefront.

At a boundary of two homogeneous media (such as air and glass), the ray direction changes suddenly, but the ray remains a straight line in each medium. However, if the medium were not homogeneous, e.g. it is like our atmosphere in which the density changes with altitude, the ray would bend continuously.

Even if the rays are changing direction, the wavefront is always perpendicular to the ray. A combination of the rays or the sum of several rays forms a beam of light such as a search light, which is represented by many rays.

A wavefront is, therefore, a set of points with equal phase located at regular intervals from the source of light. Phase is the relationship of the sinusoidal period of the EM wave. A wave emanates from a point source in all directions as a spherical wavefront, centered at the source, as shown in Figure 1.5. It is important to note that the optical path length relative to the source is constant over the wavefront.

As this EM wave propagates (at 186 282 miles per second), the spherical wavefront becomes a plane surface at large distances. Thus, in this case, we have a series of plane waves (see Figure 1.6).

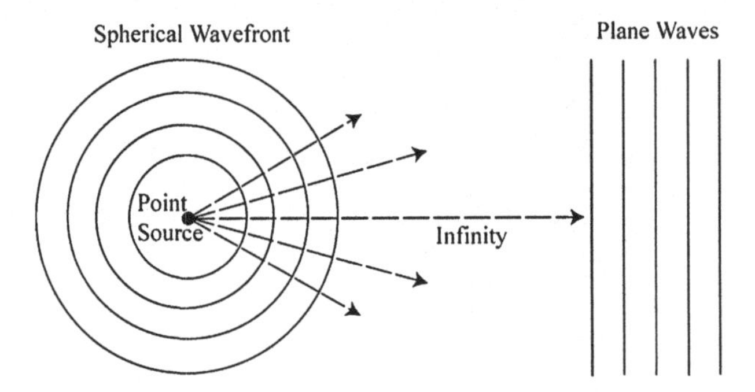

Figure 1.6 Spherical wavefronts emanating from a point source which become plane waves as the radius becomes infinite.

1.4 Index of refraction

When a charged particle is accelerated it emits EM radiation. EM radiation is described according to Maxwell's equations (Maxwell, 1865), which are beyond the scope of this book. If this radiation has the correct energy, it will be in the visible spectrum (light as we know it). These light waves follow descriptions derived by Maxwell in his equations of EM light waves for a time-varying field (electric or magnetic) (Born and Wolf, 1959). The result, after some minor manipulation, is the wave equation for EM light waves in a charge-free homogeneous medium:

$$\nabla^2\mathbf{E} - \mu_\mathrm{m}\,\varepsilon_\mathrm{m}\frac{\partial\mathbf{E}}{\partial t} = 0, \tag{1.7}$$

where $\mathbf{E}$ is the time varying electric field, μ_m is the permeability of the medium, and ε_m is the permittivity of the medium

The corresponding speed of light in the medium is

$$v_\mathrm{m} = \frac{1}{\sqrt{\mu_\mathrm{m}\,\varepsilon_\mathrm{m}}}. \tag{1.8}$$

For the case of free space (vacuum), the permeability and permittivity are well known:

$$\mu_0 = 4\pi(10^{-7})\,\mathrm{N\,s^2\,C^{-2}},$$

$$\varepsilon_0 = 8.85(10^{-12})\,\mathrm{C^2\,N^{-1}m^{-2}},$$

which gives (using Equation (1.8)) the speed of light (c) as $2.99792458 \times 10^8\,\mathrm{m\,s^{-1}}$ in free space. The speed of light is most often approximated to

$$c \approx 3(10^8)\,\mathrm{m\,s^{-1}}. \tag{1.9}$$

The velocity of light in a medium (as given in Equation (1.8)) is related to the index of refraction, or the refractive index. The refractive index of a material is the factor by which EM radiation is slowed down (relative to the velocity in a vacuum) when it travels inside the material. For a general material, the index is given for the relative permittivity (ε_r) and relative permeability (μ_r) by

$$n = \sqrt{\varepsilon_r \mu_r}, \tag{1.10}$$

where

$$\varepsilon_r = \varepsilon_m / \varepsilon_0, \quad \mu_r = \mu_m / \mu_0.$$

So if v is the phase velocity of radiation of a specific frequency in a medium, the refractive index, by substitution, is given by

$$n = c / v_m. \tag{1.11}$$

This number is typically bigger than 1: the denser the material, the more the light is slowed down.

The velocity can be expressed as either phase or group velocity. The phase velocity is defined as the rate at which the crests of the waveform propagate; or the rate at which the phase of the waveform is moving. The group velocity is the rate at which the envelope of the waveform is propagating; i.e. the rate of variation of the amplitude of the waveform. It is the group velocity that (almost always) represents the rate at which information (and energy) may be transmitted by the wave. For example, the velocity at which a pulse of light travels down an optical fiber.

In a medium, the EM wave creates a disturbance of the electrons which is proportional to the permittivity of the medium. This oscillation of the electrons causes a new electromagnetic wave, which is slightly out of phase with the original wave. The resulting two waves at the same frequency interact to produce a new wave with a shorter wavelength, thus causing a slower velocity.

For non-magnetic materials, the permeability (μ_m) is approximately equal to that of free space, which has a permeability of μ_0, so the square of the index of refraction, from Equation (1.10), is equal to the relative permittivity or dielectric constant (K_0) of the material:

$$n^2 = K_0. \tag{1.12}$$

Strictly speaking, the parameter used to describe the interaction of the EM field, or light wave, with matter should be a complex index of refraction:

$$\bar{n} = n - iK_e, \tag{1.13}$$

where n can also be called the index of refraction and K_e is called the extinction coefficient. In a dielectric material such as glass, none of the light is absorbed, and therefore $K_e = 0$.

In 1967, a Russian scientist by the name of Veselago proposed that materials with both negative permeability and negative permittivity would produce a negative index of refraction (Vesalago, 1968). This has been proven to be true in the microwave region but not, as yet, in the visible optical region.

The refractive index of materials found in nature is positive; however, man-made materials can be engineered to have a negative index. Materials that have a negative index of refraction are called metamaterials. Metamaterials exhibit a negative index as a result of negative permeability and negative permittivity. At wavelengths much smaller than the free-space wavelength, $\lambda \ll \lambda_0$, ε and μ are independent of each other. Therefore, the meta-material can have a negative ε or μ. Due to a negative index, by Snell's law the light inside the medium would make a negative angle with the surface normal of the medium. Thus, a negative index causes negative refraction. Such a concept was first suggested by Veselago in the 1960s, but only became a reality with the development of meta-materials such as metallic nanowires and photonic crystals in 1968. Photonic crystals are engineered to have a large positive index of refraction.

1.5 Optical path length (*OPL*) and reduced thickness

In a homogenous medium, light travels in straight lines called rays, as previously discussed; however, if the medium changes abruptly (e.g. air to glass), the ray's velocity also changes. The frequency of light is not affected by crossing into a new medium or material, but the velocity (distance/time) is always reduced ($v_m < c$). The velocity in a medium is related to the speed of light in a vacuum by a factor called the refractive index, or index of refraction. The speed of light is greatest in a vacuum or free space, as discussed in Section 1.4. The ratio of the velocity of light in a vacuum to the velocity of light in matter is the refractive index, as shown in Figure 1.7.

The refractive index is always greater than 1 for any material. Since the frequency of light is fixed, once a ray is launched, the velocity in matter is slower than the fastest velocity, c. Rearranging Equation (1.11) for velocity in matter:

$$v_m = c/n. \tag{1.14}$$

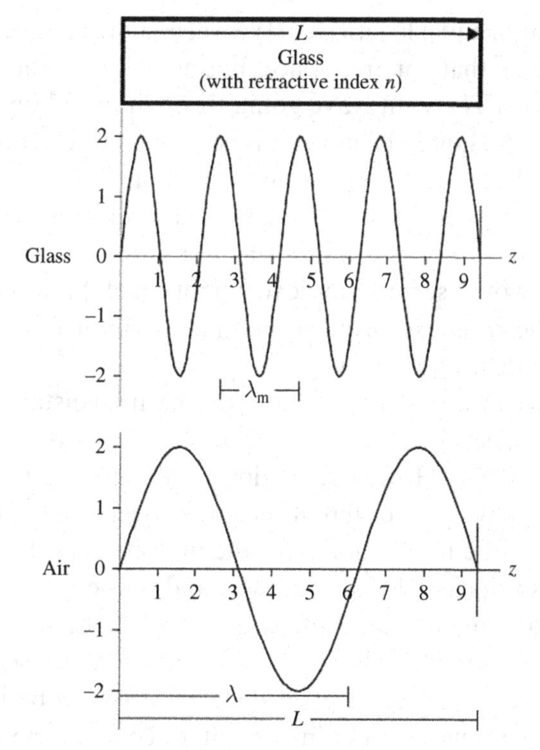

Figure 1.7 Path length is longer for given thickness of glass; light seems to be traveling through more periods.

Recalling our discussion of sinusoidal waves/rays, the velocity in the medium can be written in terms of wavelength as

$$v_{\mathrm{m}} = \upsilon\lambda_{\mathrm{m}} = c/n, \tag{1.15}$$

where λ_{m} is the wavelength of light in matter and υ is the frequency. Substituting Equation (1.5) for the free space velocity of light c into Equation (1.15):

$$\upsilon\lambda_{\mathrm{m}} = \upsilon\lambda/n \tag{1.16}$$

and solving for the wavelength in the medium:

$$\lambda_{\mathrm{m}} = \lambda/n. \tag{1.17}$$

Thus, the wavelength of light in a medium is shorter than the wavelength of that same light in a vacuum (free space).

This wavelength difference is important in considering optical path length (*OPL*). The light in a medium passes through more periods of the sinusoidal model than it could in free space. Shown in Figure 1.7 is the sinusoidal path for a ray passing through a glass medium compared with a ray of equal frequency passing through air of equal length (L). (We will assume that the refractive index of air is the same as that for free space, but actually the difference is 1.0080 versus 1, as defined.) The light wave going through a medium traverses more periods of light in the same distance. This concept is often confusing, because light with a wavelength of 1 μm passing through air will have its wavelength changed to 500 nm as it passes through a glass medium with an index of refraction equal to 2 ($n = 2$). Since the eye responds to wavelengths of 400–700 nm, this would seem to indicate (erroneously) that the color of light had changed in the glass. However, recall that the frequency does not change, and color is dependent on frequency.

If a distance L separates two buildings, the measured distance has nothing to do with the medium between the buildings. If it is filled with water, the distance between the two is still L. However, the time it takes for light to travel between the buildings is different for different media between the buildings. The time difference is due to the interaction with the molecules in the medium, which impede the light's velocity, slowing it down, and thus causing the light to take more time to traverse the same physical distance. Light is absorbed and reemitted at the same velocity via electron–electron transitions within the medium. Therefore, a new concept of distance needs to be used to account for this delay in the time of flight in the water. This new optical path length (OPL) takes into account the slower velocity within the material, and is the product of distance and refractive index:

$$OPL = nL. \tag{1.18}$$

Thus, light passing through matter seems to traverse a longer distance than light propagating in free space.

The effect of this increased path length is very easily demonstrated by an observer viewing a fish in a tank of water. In reality, the fish is at a distance L from the eye as shown in Figure 1.8, but it appears to be in a plane at the dashed line. The light rays from the fish are refracted as they emerge from the water to a larger angle, so they appear to originate from the dashed line shown in Figure 1.8. The details of this refraction effect will be derived in Section 2.6 for this reduced thickness. The two distances, the optical path distance in air (nL) and in water ($n'L'$), must be equal:

$$nL = n'L'. \tag{1.19}$$

L' is called the reduced thickness. The larger the index, the smaller the reduced distance. Using $n = 1$ for air and n' for water yields

$$L' = L/n'. \tag{1.20}$$

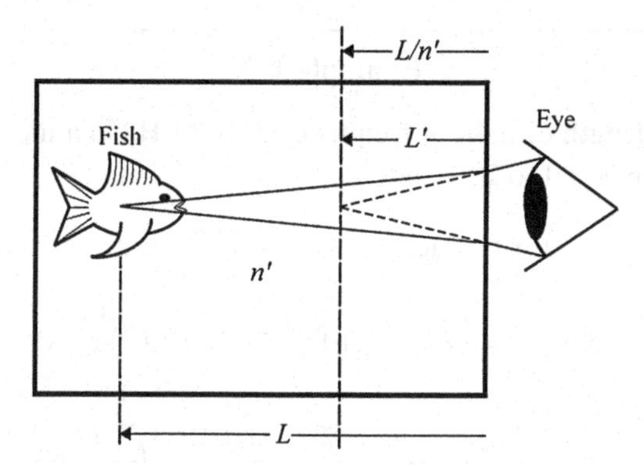

Figure 1.8 The fish appears to be at L/n'. The distance L' is called the reduced thickness or the equivalent air thickness.

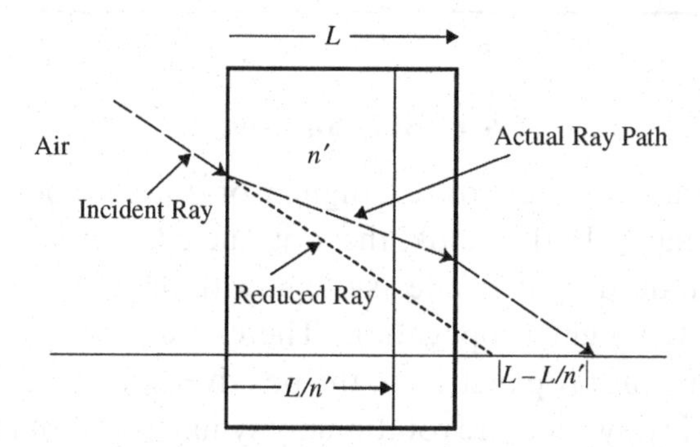

Figure 1.9 Optical effects of a plane parallel plate.

Note that the actual nominal distance L, the optical path length (OPL), and the reduced distance (L/n') all have units of length (e.g. millimeters, feet or meters).

The OPL in air is equivalent to the OPL in the medium (reduced thickness) which can be conceptualized as the equivalent amount of phase change for a ray if it were traveling through air. The number of periods that occur is an indicator of the OPL. In a medium, the wavelength is smaller, so more periods are present for a given distance, and, since the observation is calibrated in air, an object looks closer.

If a window is in an optical path, as shown in Figure 1.9, the equivalent thickness (reduced thickness L/n') can be used instead of the window thickness L; thus, providing a means of drawing straight lines for rays through the media. Reduced distance is often used in optical layout for ease of drawing as opposed to ray tracing the actual ray path.

Example 1.3

What is the wavelength of light of frequency $3(10^{14})$ Hz in a medium with the index of refraction equal to 2?

time $(T) = 1/3(10^{-14})$ or frequency$(v) = 3(10^{14})$ if index of refraction $= 2$:

$$\text{velocity} = c/n = 3(10^8)/2 = 1.5(10^8)\ \text{m/s}$$

$$\text{wavelength}: v_m = \frac{\lambda_m}{T} = \frac{c}{n} \Rightarrow \lambda_m = \frac{cT}{n} = \frac{3(10^8)\left[(1/3)(10^{-14})\right]}{2} = 0.5\ \mu\text{m}.$$

In free space the wavelength would be 1.0 μm:

$$\lambda = cT = c/v$$

1.6 Coordinate system

To honor Descartes, a system of rectangular or oblique coordinates is called Cartesian coordinates. In this book, the x, y, and z Cartesian coordinates will be used in the standard right-handed system, with the z axis being the average direction of positive light propagation. The coordinate system is shown in Figure 1.10, with light propagation in the positive z direction.

Since most optical systems are rotationally symmetric about the z axis, polar coordinates are often used to replace the x–y plane coordinates. The orientation of the polar coordinates and the x–y plane is shown in Figure 1.11.

The relationships of polar coordinates (ρ and θ) to x and y coordinates are:

$$\begin{aligned} x &= \rho \sin\theta, \\ y &= \rho \cos\theta. \end{aligned} \tag{1.21}$$

Due to rotational symmetry, the x and y units of length are equal. In addition, a plane containing the z axis and the x axis has the same values and geometry

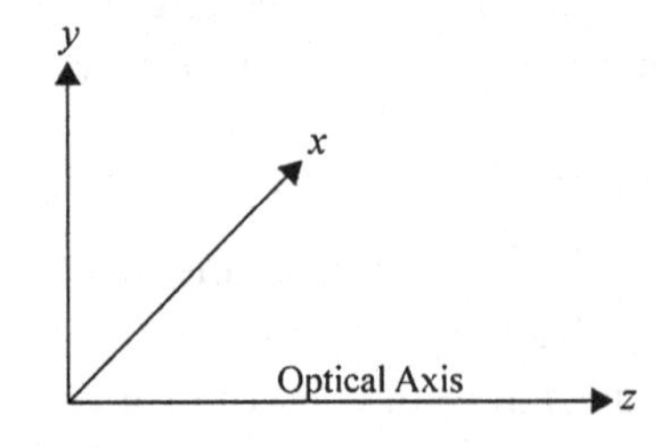

Figure 1.10 Right-hand Cartesian coordinate system.

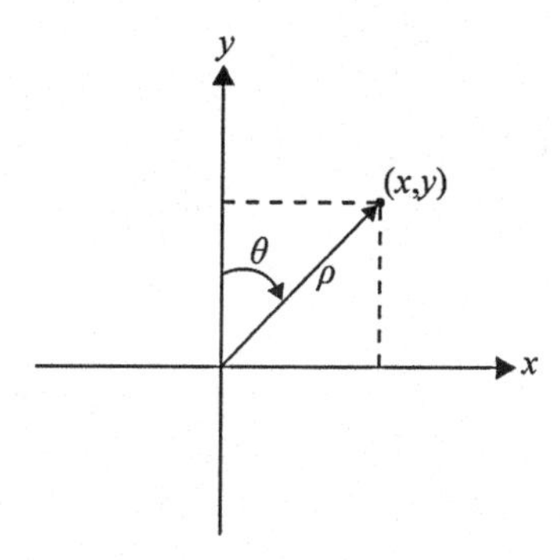

Figure 1.11 Polar coordinates in the x–y plane.

as one containing the z axis and the y axis. So if one plots a y–z plane, it is equivalent to any plane at an angle θ containing the z axis.

1.7 Solid angle

Radiation emitted from a point source can take on any direction in a hemisphere centered on the source. To characterize the radiation fully in geometrical optics, the concept of solid geometry and, more specifically, solid angles, must be developed. Linear angles, although conceptually simple and much more common, are not able to transmit the energy of light. It is useful to use a solid angle subtended by a surface that is viewed from a point or vertex of a cone, as shown in Figure 1.12. The solid angle, Ω, that an object subtends from a point is a measure of how large that object appears at that point in three-dimensional space.

The solid angle is a cone generated by a line that passes through the vertex and a point on a surface which is enclosed as the line moves to contour the surface. The size of the angle is measured in steradians (sr), and is defined, in the differential limit, as the surface area intercepted by the cone in an imaginary sphere, from the center of that sphere, divided by the square of the sphere's radius. A square degree, which seems more logical, is not an SI unit, but it can be used as a solid angle unit. The unit would be denoted as "sq. deg." or "deg^2," but these are not widely used. Steradian is more commonly used. A solid angle is related to the surface area of a sphere in the same way a linear angle (radians) is related to the circumference of a circle. Recall the surface of a sphere is $4\pi r^2$, so dividing by r^2 from our definition, the total solid angle about a point in space is 4π steradians. The USA subtends about 1/4 sr from the center of the Earth, a standard ice cream cone similarly subtends about 1/4 sr.

To determine the expression for a solid angle in a rotationally symmetric geometry about the z axis, consider an element, da, of a surface at a distance r from the vertex of the Cartesian coordinate system. The center of the

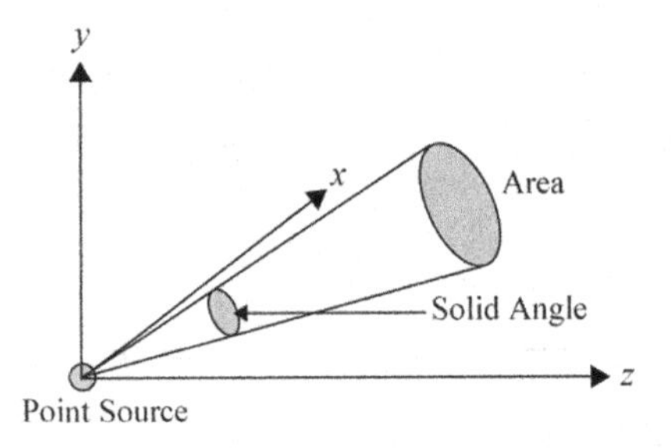

Figure 1.12 Solid angle.

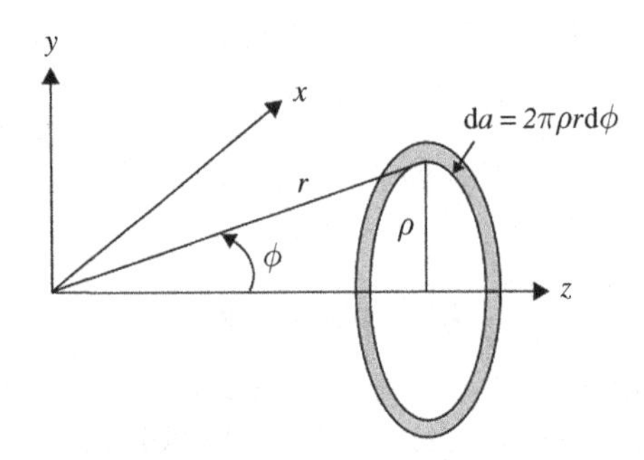

Figure 1.13 Differential solid angle.

differential area projected on the x–y plane is at some angle (ϕ) and distance ρ as shown in Figure 1.13.

Using the definition of a solid angle, the differential area (da) shown in Figure 1.13, and a spherical coordinate system gives:

$$\mathrm{d}a = 2\pi\rho r\mathrm{d}\,\phi = 2\pi r \sin\phi\, r\, \mathrm{d}\phi, \tag{1.22}$$

$$\mathrm{d}a = 2\pi r^2 \sin\phi\, \mathrm{d}\phi. \tag{1.23}$$

Using the definition of a solid angle related to da,

$$\mathrm{d}\Omega = \frac{\mathrm{d}a}{r^2} = 2\pi \sin\phi\, \mathrm{d}\phi. \tag{1.24}$$

Finding the total solid angle in terms of a surface now becomes an integration exercise. We will assume rotational symmetry about the z axis, as shown in Figure 1.14, to find the solid angle of a right circular cone with the surface being a spherical cap area. If ϕ is the cone half angle of a right circular cone, as shown in Figure 1.14, the integral can be set up and solved as:

$$\Omega = \int \mathrm{d}\,\Omega = 2\pi \int_0^{\phi_{\max}} \sin\phi\, \mathrm{d}\phi \tag{1.25}$$

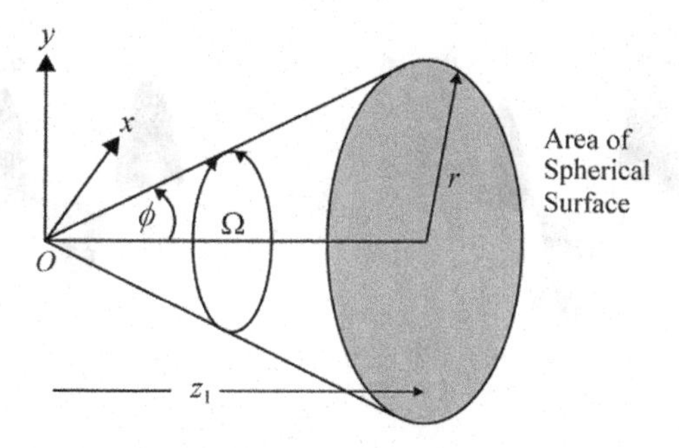

Figure 1.14 Spherical geometry for a solid angle.

$$\Omega = 2\pi \int_0^{\phi_{max}} \sin\phi \, d\phi = 2\pi[-\cos\phi]_0^{\phi_{max}} \quad (1.26)$$

$$= 2\pi[1 - \cos\phi_{max}]. \quad (1.27)$$

Equation (1.27) is the solid angle subtended for a cone half angle of ϕ_{max}. For a 1sr solid angle, the corresponding linear cone half angle is about 32.5°. For small angle approximations, a simple relationship can be used to relate the cone half angle to the solid angle with very good accuracy. A trigonometric approximation for small angle is

$$\cos\phi \approx 1 - \phi^2/2. \quad (1.28)$$

Substituting Equation (1.28) into Equation (1.27), gives

$$\Omega \approx \pi\phi^2. \quad (1.29)$$

This approximation is good to within 1% for cone half angles less than 20°, and is good to within 0.1% for cone half angles less than 6°.

Example 1.4

How many steradians are there in a hemisphere?

$$\phi = 90°,$$
$$\Omega = 2\pi(1 - \cos 90°) = 2\pi \text{ steradians}.$$

1.8 Polarization

A light wave has its electric and magnetic fields perpendicular to the direction of propagation (z), as shown in Figure 1.1, for monochromatic light. The

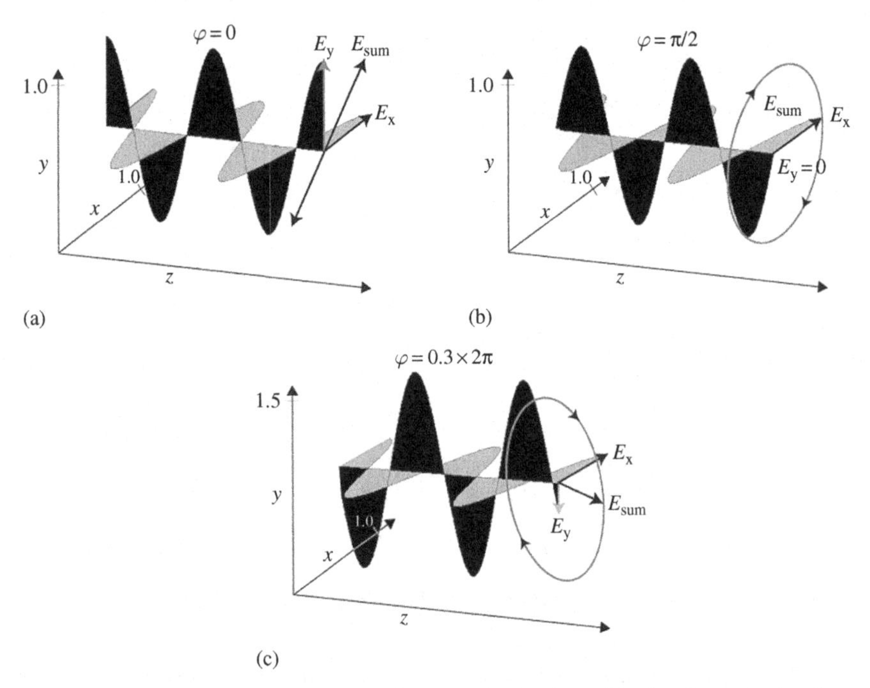

Figure 1.15 Polarization states: (a) linear polarization; (b) circular polarization; (c) elliptical polarization.

polarization of a plane wave is described by the electric field vector, while the magnetic field is ignored, since this is always perpendicular to it and in phase with it. Since the plane wave electric vector is in the x–y plane, it can be broken down into two components perpendicular to the direction of light travel. The electric field propagating along the z axis is the sum of two copropagating orthogonal waves. One electric field oscillates along the x axis and the other along the y axis. For this simple wave, in which the amplitude varies in a sinusoidal manner, the components have the same frequency. However, the x- and y-components may have different amplitudes and different phases. If we pick a position on the z axis and observe the electric vector as time passes, we observe the description of polarization. The tip of the vector sweeps out shapes in time on this fixed x–y plane location, and this shape describes the state of polarization. The observer must be looking in the negative z direction to describe the state of polarization.

Consider the simple case of two orthogonal components in phase for a given frequency, as shown in Figure 1.15(a). The strengths of the two components are equal or are related by a constant ratio, so the electric field (the vector sum of two orthogonal components) can always be represented by a fixed line vector in the x–y plane, as shown. The electric field is a line on this fixed x–y

plane, and the light is linearly polarized. The slope of this line, indicating the composite electric field, does not change with time.

The second case is where the orthogonal electric components have exactly the same amplitude, but they are 90° out of phase. One component is zero when the other component is a maximum or minimum, as shown in Figure 1.15(b). The x-component can lead or lag behind the y-component to create this condition. In this case, the electric vector in the plane will be formed by summing the two components, and a circle will be mapped by the tip of the electric vector. This is called circular polarization. The direction of rotation depends on the phase between the two components, right circular or left circular polarization, depending on which direction the electric vector rotates.

Where the two components are not in phase (φ) and do not have the same amplitude, elliptical polarization is created. The sum of electric vectors will trace out an ellipse on the x–y plane. There is a special case in which the phase difference is 90°, but different amplitudes for the vertical and horizontal fields still give an elliptical polarization. This is shown in Figure 1.15(c).

Birefringent materials can introduce these phase differences in the x and y directions. This effect can occur only if the structure of the material is anisotropic. Birefringent materials (e.g. calcite) have polarization-dependent indices of refraction: i.e. the x- and y-components of the electric field experience different velocities as the wave passes through the medium. These indices are called the ordinary (n_o) and extraordinary (n_e) refractive indices. In Figure 1.15(b) the x-component leads the y-component, so the y-component has the fast velocity (lower refractive index), and the x-component has the slow velocity (higher refractive index).

The difference in optical path length can be represented by the phase difference where one wavelength is equal to 2π radians. The phase delay introduced between the electric field in the x and y directions for a given thickness (t) of birefringent material is given by

$$\varphi = \frac{2\pi(n_e - n_o)t}{\lambda}. \tag{1.30}$$

As stated earlier, light is often produced by a large number of individual radiators producing a continuum of independent waves, which is termed incoherent radiation. There is no single frequency, but a spectrum of frequencies in the electric fields. This type of incoherent light can also be polarized. It just implies that the electric fields of all these components are acting as previously discussed. In this case, the terms "partially polarized" and "degree

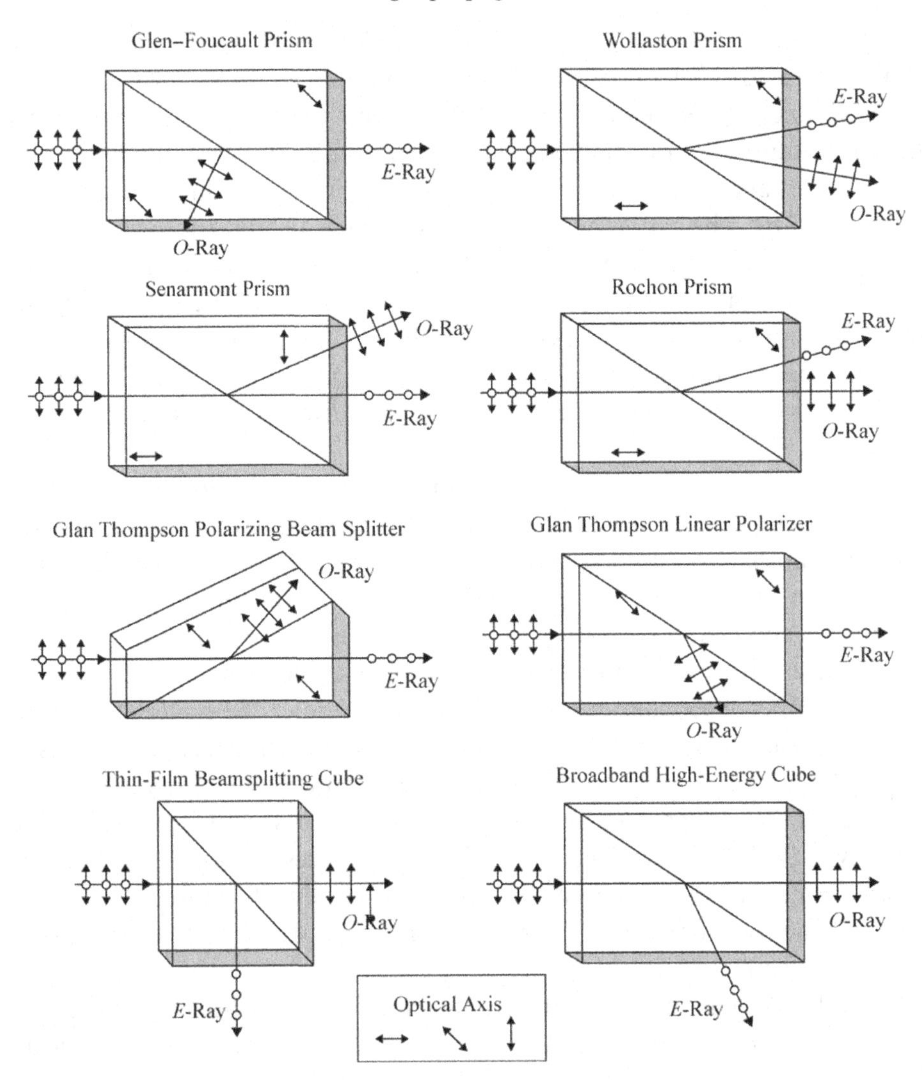

Figure 1.16 Polarizing prisms.

of polarization" are often used to describe the state of polarization. Some examples of how polarized light is produced are shown in Figure 1.16.

Problems

1.1 A fish is 1 ft below the surface of the water ($n = 4/3$). At 2 ft above the water, a fisherman is viewing this fish. What is the optical path length from the fish to the fisherman's eye?

1.2 An eagle's eye can detect 15 photons per second of light at 440 nm. How much radiant power (watts) can the eagle detect, or what is the threshold power?

1.3 The human eye can see 550 nm radiation (light). What is the energy (electron volts) of the light at that wavelength? Determine the frequency and the period of this light.

1.4 The threshold sensitivity of the eye is 100 photons per second. The eye is most sensitive at 550 nm. Determine the threshold power (watts) the eye can detect.

1.5 If the eye responds to light from 400 to 700 nm, what is the energy (electronvolts) of light at each end of the spectrum?

1.6 What is the wavelength of:
(a) The EM radiation of a HeNe laser, which has a frequency (υ) of $4.7\,(10^{14})$ Hz?
(b) Your personal computer, operating at 233 MHz?

1.7 What is the speed of light in the following media? What is the corresponding wavelength of a HeNe laser propagating in each medium?
(a) Water ($n = 1.3\bar{3}$).
(b) Glass ($n = 1.5$).
(c) ZnSe ($n = 3.5$).
(d) Diamond ($n = 2.426$).
(e) Methyl methacrylate ($n = 1.49166$).

1.8 A laser pointer ($\upsilon = 4.47\,(10^{14})$ Hz) is propagated through equal (20 cm) distances of air ($n = 1$) and glass ($n = 1.517$), as illustrated below. What is the optical path difference between the two beams at A' and B'?

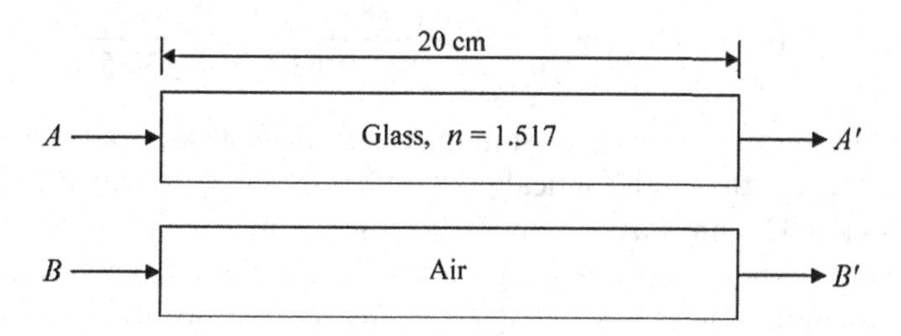

1.9 If the refractive index for a piece of optical glass is 1.516, calculate the speed of light in the glass.

1.10 If the distance of the Moon from the Earth is $3.840\,(10^5)$ km, how long will it take laser light to travel from the Earth to the Moon and back again?

1.11 How long does it take light from the Sun to reach the Earth? Assume the distance of the Earth from the Sun to be $1.50\,(10^8)$ km.

1.12 A beam of light passes through a block of glass 50 cm thick, then through water for a distance of 100 cm, and finally through another block of glass 25 cm thick. If the refractive index of both pieces of glass is 1.5250, and the refractive index of water is 1.3330, find the total optical path length.

1.13 A water tank is 50 cm long inside and has glass ends that are each 2.0 cm thick. If the refractive index of water is 1.3330, and that of glass is 1.5, find the overall optical path length.

1.14 A picture is viewed through a plane parallel plate of glass with $n = 1.5$. The image appears 2 mm closer to the observer than without the glass present. How thick is the glass plate?

1.15 A laser beam is split into two parallel beams that travel a distance of 20 cm. In the path of one beam, a 2 cm transparent glass plate is placed, causing an optical path difference of 1 cm. What is the refractive index of the glass?

1.16 A hawk can detect a standard spherical cow carcass (1 m sphere) at 60 miles with 1000 photons of 500 nm wavelength light at his eye.

(a) What is this angle that the hawk can see in radians? In degrees? In arc minutes? In arc seconds?

(b) What is the energy (joules) of this light?

1.17 A source of light has a frequency of $9\,(10^{14})$ Hz.

(a) What is its wavelength in glass with index of refraction of 1.5?

(b) What is its temporal period?

(c) In what spectral region is it located in air, e.g., visible, infrared, etc.?

1.18 The index of refraction of a calcite ($CaCO_3$) glass depends on the electronic vector orientations (ordinary and extraordinary). The equations for the two refractive indices versus wavelength (λ) are (λ in micrometers):

$$n_o^2 - 1 = \frac{0.8559\lambda^2}{\lambda^2 - 0.003457} + \frac{0.8391\lambda^2}{\lambda^2 - 0.019881} + \frac{0.0009\lambda^2}{\lambda^2 - 0.038809} + \frac{0.6845\lambda^2}{\lambda^2 - 49.07003},$$

$$n_e^2 - 1 = \frac{1.0856\lambda^2}{\lambda^2 - 0.006236} + \frac{0.0988\lambda^2}{\lambda^2 - 0.020164} + \frac{0.317\lambda^2}{\lambda^2 - 131.515}.$$

(a) For a 13 mm thick (z dimension) piece of calcite glass, what is the difference in optical path length (optical path difference, or OPD) for the two orientations at 632.8 nm wavelength (HeNe laser in air)?

(b) What thickness is required to get a quarter wave difference ($\lambda/4$) for 1000 nm wavelength of light between the ordinary and extraordinary rays?

1.19 A fish is frozen in ice in Alaska, and appears to be 10 inches inside the ice. How far beneath the surface does the fish actually lie?

1.20 What is the index of refraction if the velocity of light is $2\,(10^8)$ m s^{-1}?

1.21 The straightest road in the USA is the Simi Highway, Route 28, in Michigan. A driver, seated in a car, is viewing a deer crossing the road. The eyes of the driver are about 3 ft above the ground, and the deer is about 3 ft in height. If the radius of the Earth is 6400 km:

(a) How far ahead (maximum) of the car can the driver just see the very top of the deer?

(b) If the deer is 5 ft long, what angle does the deer subtend from the driver?

1.22 One (1) nautical mile $= 6076$ feet, which is also equal to 1 arcminute ($1'$) subtended from the center of the Earth. Knowing these facts, what is the diameter of the Earth?

1.23 For a straight line going through two points on a x–y coordinate system as shown below, find the equation of the line in the $y = mx + b$ form.

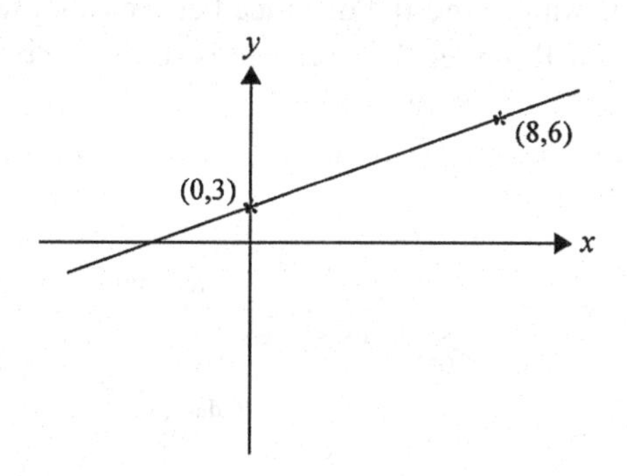

1.24 How much time does it take blue light ($\lambda = 400$ nm) to travel 80 km in the media below? (This is the distance for fiber optic repeater stations.) Assume its velocity in air equals $3\,(10^8)$ m s^{-1}.
(a) In fiber, $n = 1.6$.
(b) In air, $n = 1$.
(c) In fiber, $n = 1.61$.
(d) In fiber, $n = 1.59$.

1.25 The waves on the ocean can be approximated by a sinusoidal surface as: surface $= A\ (1 + \cos\ (2\pi v))$. If there is 3 ft between crests of the waves and the waves are 1 ft high in height, how much volume (ft^3) of water is there in a single wave that is 10 feet long?

1.26 A satellite travels around the Earth at 7.5 km s^{-1} at an altitude of 320 nautical miles.
(a) How long does the satellite take to make one revolution around the Earth?
(b) How long does it take the satellite to travel between LA and NY (3000 miles)?
(c) How much energy is expended if this satellite (mass = 1kg) crashes into a huge asteroid (mass asteroid $\gg$ mass satellite)?

1.27 If the visible spectrum of light (400–700 nm) is divided into three regions, each region containing 1 megajoule of energy, how many photons, on average, does each region contain?

1.28 If light has a frequency of $4.2\,(10^{14})$ Hz, what is the wavelength in the following glasses:
(a) borosilicate ($n = 1.5$);
(b) gallium arsenide ($n = 2.2$);
(c) germanium ($n = 4$)?

1.29 What is the speed of light in a glass of refractive index (n):
(a) $n = 1.5$;
(b) $n = 2.2$;
(c) $n = 4$;
(d) $n = 10^5$?

1.30 An observer in the crow's nest of a ship (A) sees the crow's nest of another ship (B) at the horizon, due to the curvature of the Earth of radius R. The crow's nests are 50 ft above the water, and the distance between the two ships is 17.5 miles. The layout of the Earth under this scenario is shown below.
(a) What is the radius of the earth (ft)?
(b) What is the angular arc (in radians) that these two ships subtend from the center of the Earth?

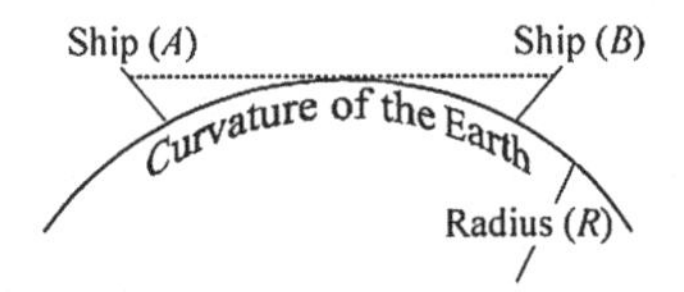

1.31 The city of Tucson is at the latitude 32° N, and the Earth's axis of rotation points to the North Star. Draw a diagram to show the angle at which an observer sees the North Star from the horizon in Tucson.

1.32 A corduroy gravel road has a very interesting spatial pattern. The peak-to-peak distances of the crests are 13 in due to the size of automobile tires. The depth is 1.3 in.
(a) How much dirt is removed for one dip (groove) on a 16 ft wide road?
(b) What is the spatial frequency of the pattern (cycles per meter)?
(c) What is the temporal frequency the driver experiences (cycles per minute) if a car is going at 25 mph with 16 in diameter tires?

1.33 Manhole covers are round. Very seldom do you see them square or rectangular. Why?

1.34 Why does light propagate in a vacuum but sound waves do not?

1.35 What is the energy (joules) for the following numbers of photons at various wavelengths:
(a) one photon at a wavelength of 632.8 nm;
(b) 10^9 photons at a wavelength of 1 μm;
(c) 10^{10} photons at a wavelength of 10 μm;
(d) 1000 photons at a wavelength of 500 nm?

1.36 What is the energy (joules) of a single photon at the following wavelengths:
(a) 400 nm;
(b) 500 nm;
(c) 600 nm;
(d) 700 nm;
(e) 5 μm;
(f) 10 μm.

1.37 What is the velocity of light (m s^{-1}) in the following glasses:
(a) N-BK7, $n = 1.517$;
(b) GaAs, $n = 2.4$;
(c) silicon, $n = 3.4$;
(d) germanium, $n = 4$?

1.38 A bundle of rays is split into two beams, one propagating through a length of ice ($n = 1.33$) and the other through an equal length of plastic ($n = 1.4$). If the light has a wavelength of 632.8 nm, what length is necessary to produce one-half wavelength difference between the two beams after transmission through the media?

1.39 A fish is in a 30 cm diameter tank filled with water. If the fish is at the surface on the far side of the tank, what is the optical path length if the tank has glass ($n = 1.5$) walls that are 8 mm thick?

1.40 An eagle is traditionally very sharp eyed, in that it can detect a rabbit (1 ft) at 10 miles. What is the angular resolution of an eagle's eye.

1.41 The Alabaster Gypsum Plant is seen from a ship at sea level in Lake Huron, MI. The building's horizontal length subtends an angle of 1.5° from a point 2 miles away. The building is 25 meters high (known from the crane used to dump gypsum on the ships).
(a) What is the building's horizontal length in meters?
(b) What is the furthest distance away that a ship can be in order for the top of the building to be visible at sea level.
Hint: Take into account the Earth's curvature only (radius = 3964 miles).

1.42 Green light has a wavelength of 475 nm. What is the frequency (υ) of green light?

1.43 The Sun is 93 million miles away. How long does it take sunlight to reach the Earth?

1.44 The diameters of the Sun, Earth, and Moon are 864000, 7927, and 2160 miles, respectively. The mean distance of the Sun from the Earth is $93\,(10^6)$ miles, the mean distance of the Moon from the Earth is 238 857 miles, and the period of the Moon is 27.3 days.
(a) Find the diameter of the umbra of the Moon on the Earth.
(b) Find the diameter of the umbra of the Earth on the Moon.
(c) What is the duration (seconds) of a total lunar eclipse?
(d) What is the duration (seconds) of a total solar eclipse?

1.45 The green photon has a wavelength of 480 nm. What is the energy (joules) of a single photon?

1.46 What is the speed of light in a medium with an index of refraction of 1.5?

1.47 A swimmer and a man-eating shark are 20 meters apart (in water). From the shark's point of view, how far away does the swimmer appear to be?

1.48 For an angle (θ) of 1.5°:
(a) What is the sin of θ?
(b) What is the tan of θ?
(c) How many radians is 1.5°?
(d) Compare the values in (a), (b), and (c).

1.49 For light with a frequency of $4\,(10^{14})$ hertz, what is its wavelength in:
(a) air;
(b) glass ($n = 1.5$);
(c) ZnSe ($n = 3.1$)?

1.50 If a medium has the following permeability and permittivity

$$\mu_{\mathrm{m}} = 4\pi(10^{-7})\,\mathrm{N\,s^2\,C^{-2}}$$
$$\varepsilon_{\mathrm{m}} = 2.5(10^{-11})\,\mathrm{C^2\,N^{-1}\,m^{-2}}$$

(a) What is the speed of light in the medium?
(b) What is the dielectric constant for the medium?

1.51 A non-magnetic medium has a dielectric constant of 9. What is its index of refraction?

1.52 For a cone with a 4 in base diameter and a height of 10 in:
(a) What is the exact solid angle as seen from the vertex?
(b) If $F/\#$ is defined as the height to base diameter ratio (2.5), derive an expression for the approximate solid angle in terms of $F/\#$.

1.53 Find the largest solid angle for which the approximation of solid angle has less than 10% error:

$$\text{Approximate:}\ \Omega = \pi a^2;$$
$$\text{Exact:}\qquad \Omega = 2\pi(1 - \cos\alpha).$$

1.54 For a linear angle of $\pm 20^\circ$ (full field of view):
(a) What is the approximate solid angle?
(b) What is the exact value of the solid angle?
(c) What is the error introduced by the approximation?

1.55 For a linear angle of 20° (full field of view):
(a) What is the approximate solid angle?
(b) What is the exact value of the solid angle?
(c) What is the error introduced by the approximation?

1.56 Reading a newspaper through a plane parallel plate of glass ($n = 1.5$), the print appears to be 10 mm closer to the reader than without the glass. How thick is the glass plate?

1.57 A ray of light in air is incident on the polished surface of a block of glass at an angle of 10°.
(a) If the refractive index of the glass is 1.5160, find the angle of refraction to four significant figures.
(b) Assuming the sines of the angles in Snell's law can be replaced by the angles themselves (in radians), what would be the angle of refraction (in degrees)?
(c) Find the percentage error for the $\sin\alpha = \alpha$ approximation in part (b).

Bibliography

Born, M. and Wolf, E. (1959). *Principles of Optics*, sixth edn. Cambridge: Cambridge University Press.
Guimond, S. and Elmore, D. (2004). Polarizing views. *OE Magazine*, **4**, No. 5, 26.
Hecht, E. (1998). *Optics*, third edn. Reading, MA: Addison-Wesley.

Hood, J. M. (1981). *The History of Optics, OSA/SPIE Workshop*. San Diego: San Diego State University.
Jenkins, F. A. and White, H. E. (1976). *Fundamentals of Optics*, fourth edn. New York: McGraw-Hill.
Khan, S. A. (2007) Arab origins of the discovery of the refraction of light, *OPH*, October.
Maxwell, J. C. (1865). A dynamic theory of the electromagnetic field. *Philosophical Transactions of the Royal Society of London*, **155**, 459–512.
Meyer-Arendt, J. R. (1989). *Introduction to Classical and Modern Optics*, third edn. Englewood Cliffs, NJ: Prentice-Hall.
Sears, F. W. (1958). *Optics*, third edn. London: Addison-Wesley.
Veselago, V. G. (1968). The electrodynamics of substances with simultaneously negative values of ε and μ. *Soviet Physics Uspekhi*, **10**, 509–514.

2

Reflections and refractions at optical surfaces

2.1 Rays

In geometrical optics, light is assumed to travel in a definite direction from a source, exhibiting a behavior known as rectilinear propagation. One useful way to think of light is to imagine it traveling in a very narrow beam, which is to say that light can be modeled as a ray. The emergent ray can be thought of as being a very narrow line. A ray of light is like a mathematically infinite thin line, but a light ray has direction while a line does not. The rays of geometrical optics are perpendicular to the wavefronts, and indicate the most probable path of the photons of quantum optics. A wavefront is an undulation of energy which propagates from one point to another and, as it travels, it carries electromagnetic energy. The wavefront is conceptualized as a surface with a fixed phase across it, and periodic oscillations (sinusoid) in the direction of propagation, with each wavefront having its own wavelength, frequency, amplitude, phase, and polarization.

The point source in Figure 2.1 illustrates the relationship between wavefronts and rays. The source emits energy in a spherical shell in all directions (4π sr), propagating at the speed of light. A wavefront is the locus of points on rays that have the same optical path length from the point source (same number of periods). The wavefront is perpendicular to the rays. At an infinite distance, the spherical wavefront turns into a plane wavefront (at that point the radius equals infinity).

A ray geometrically describes the path of electromagnetic radiation as it is emitted from a source and travels through an optical system. Rays are a simplified way of thinking about wavefronts: they have direction but no phase, so one ray cannot interact with another. Rays, like waves, have a source and a direction, and are drawn in optical diagrams as lines with arrows indicating the direction. Modeling rays as lines makes it easier, in many cases, to see how the wave behaves and to make calculations for its propagation through

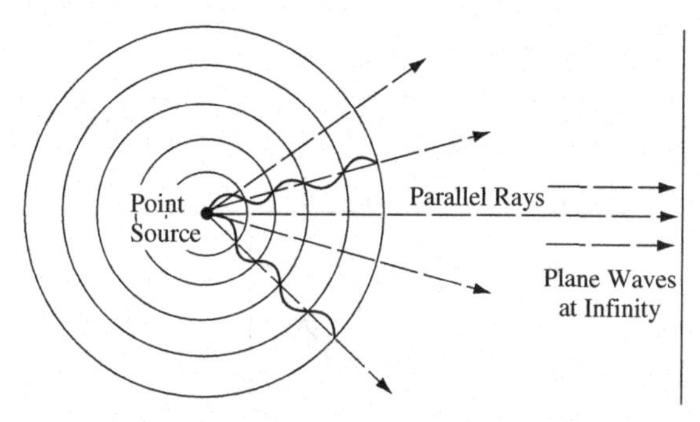

Figure 2.1 Relationship between rays and wavefronts from a point source.

various media. In an isotropic, homogeneous medium, rays propagate as straight lines.

The ray model is used extensively in the field of geometrical optics. For example, the pinhole camera is used in geometrical optics as a classical illustration of how rays form an image. This ray model is convenient in practice, since the laws governing the paths of rays are linear equations. Although it is a good model, there are many optical phenomena that rays cannot explain. For instance, if all rays converged at a single point, there would be an infinite concentration of energy at that point, but this does not, in fact, happen, as we will see when we discuss blur and diffraction. Thus, geometrical optics, by itself, cannot paint the entire picture of optical phenomena.

2.2 Fermat's principle

In its simplest form, Fermat's principle states that light rays of a given frequency traverse the path between two given points in the least amount of time. When two points in space are connected by a ray, the ray represents the shortest time path. The most obvious example of this is the passage of light through a homogeneous medium in which the speed of light doesn't change with position. In this case, the path of shortest time is equivalent to the shortest distance between the points, which is a straight line. Thus, Fermat's principle is consistent with light traveling in a straight line in a homogeneous medium.

To illustrate how the path of a ray from one point in a medium to another point in a dissimilar medium obeys the principle of minimum time of flight, consider the situation of a farmer trying to herd cows back to the barn. If the cows are in the forest, the herd's motion will be slow compared to their speed over a cleared flat field. What path taken through the forest and clear field would get the cows to the barn in the shortest amount of time? This illustration

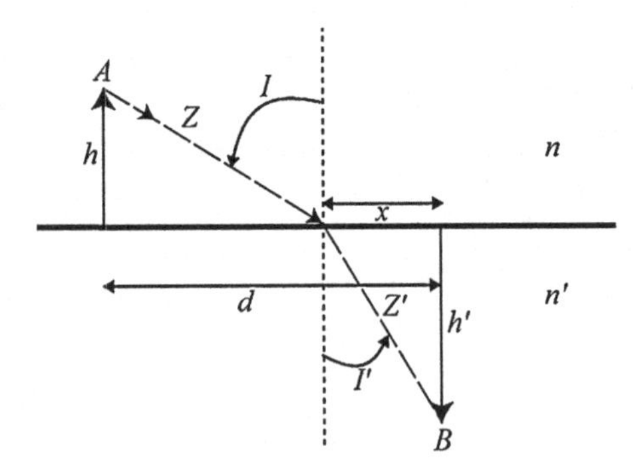

Figure 2.2 Light path for propagation from point A in a medium with index n to point B in a medium with index n'.

is analogous to that of a light ray traveling through two different media. Light rays, as shown in Figure 2.2, travel from point A (source) in one medium with index n, to point B (receiver) in another medium with index n', in the minimum amount of time.

The total time (t_{T}) from point A to point B is the sum of the distance traveled in each medium divided by the velocity in that medium.

$$t_{\mathrm{T}} = \frac{Z}{v_n} + \frac{Z'}{v_{n'}}. \tag{2.1}$$

The velocities of light in the respective media are:

$$v_n = \frac{c}{n}, \qquad v_{n'} = \frac{c}{n'}. \tag{2.2}$$

Therefore, substituting into the total time of travel:

$$t_{\mathrm{T}} = \frac{nZ}{c} + \frac{n'Z'}{c}. \tag{2.3}$$

Recall that the optical path length (OPL) from A to B or the equivalent path traveling at the speed of light (c) is ct_{T}, which is equal to the sum of the products of the refractive index and the distance through the corresponding medium. Rewriting Equation (2.3):

$$nZ + n'Z' = ct_{\mathrm{T}} = OPL. \tag{2.4}$$

This can be interpreted as the equivalent distance light travels in free space:

$$OPL = nZ + n'Z'. \tag{2.5}$$

The distances Z and Z' can be expressed in terms of distances in the Pythagorean Theorem as shown in Figure 2.2:

$$Z^2 = h^2 + (d - x)^2, \tag{2.6a}$$

$$Z'^2 = h'^2 + x^2. \tag{2.6b}$$

Substituting for the Z and Z' in Equation (2.5),

$$OPL = n(h^2 + (d - x)^2)^{1/2} + n'(h'^2 + x^2)^{1/2}. \tag{2.7}$$

To find the extremum (minimum) path or the shortest OPL, take the differential with respect to x and set it equal to zero:

$$\begin{aligned}\frac{\mathrm{d}(OPL)}{\mathrm{d}x} &= \frac{n/2}{(h^2 + (d - x)^2)^{1/2}}(-2d + 2x) + \frac{n'/2}{(h'^2 + x^2)^{1/2}}(2x) \\ &= 0 \text{ (minimum)}.\end{aligned} \tag{2.8}$$

Rearranging and isolating terms,

$$n \underbrace{\left[\frac{d - x}{(h^2 + (d - x)^2)^{1/2}}\right]}_{\sin I} = n' \underbrace{\left[\frac{x}{(h'^2 + x^2)^{1/2}}\right]}_{\sin I'}. \tag{2.9}$$

Thus, from Figure 2.2, it can be seen that

$$n \sin I = n' \sin I'. \tag{2.10}$$

The change in velocity of light at a surface causes refraction of the light ray. This refraction was discovered by an English scientist, Willebrord Snell, and is thus called Snell's law. It was also known by Ptolemy (1600 BC), but was not then formally developed mathematically.

Note that the reflection case (to be discussed later) illustrates a point about Fermat's principle: the minimum time may actually be a local, rather than a global, minimum. The global minimum distance from A to B is still just a straight line between the two points! In fact, light starting from point A will reach point B by both routes – the direct route and the reflected route. We will discover that the reflected route forces the angle of incidence to equal the angle of reflection.

Situations exist in which the actual path taken by a light ray may represent a maximum time or even one of many possible paths, all requiring equal time. As an example of the latter case, consider light propagating from one focus to the other focus inside an ellipsoidal mirror, along any of an infinite number of possible paths. Since an ellipse is the set of all points whose combined distances from the two foci remain constant, all paths are indeed of equal time. The actual path taken by a light ray in its propagation between two given points is determined by the minimum time.

2.3 Snell's law

Refraction is the bending of the path of a light wave as it passes across the boundary separating two media. Refraction is caused by the change in speed experienced by a ray/wave when it enters a different medium. Light can either refract towards the surface normal (by slowing down while crossing the boundary) or away from the surface normal (by speeding up while crossing the boundary). The larger the difference between the dielectric constants or permittivities of the media, the more that light refracts.

2.3.1 Experimental verification of Snell's law

To illustrate this point, consider a hemisphere of glass. Suppose that a laser beam is directed toward the flat side of the hemisphere at the exact center as shown in Figure 2.3.

The angle of incidence (I) can be measured at the point of incidence relative to the normal of the surface. Since the light is passing from a medium in which it travels fast into one in which it travels more slowly, this ray will refract, bending towards the normal at angle I'. Once the light ray enters the glass, it travels in a straight line until it reaches the second glass/air boundary. At the second boundary, the light ray is traveling along the normal to the curved surface. The ray does not refract upon exiting, since the angle of incidence is 0° for this second surface; therefore, the ray of laser light exits at the same angle as the refracted ray of light entered the glass hemisphere at the first boundary. Setting up an experiment like this can provide two angles to be measured and recorded. The angle of incidence (I) of the laser beam can be changed, and refracted angle (I') measurements can be recorded. This process can be repeated until a complete data set of values has been collected. The data in Table 2.1 are a representative set of results for such an experiment.

The data in Table 2.1 do not reveal a linear relationship between the angle of incidence and the angle of refraction, i.e. a doubling of the angle of incidence

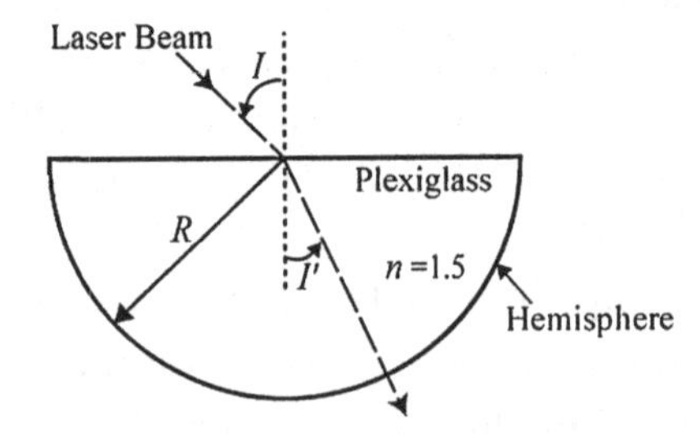

Figure 2.3 Hemisphere of glass with ray incident at the center of curvature.

Table 2.1. *Measurements of angle due to refraction of ray*

Angle of incidence, I (degrees)	Angle of refraction, I' (degrees)
00.0	00.0
05.0	03.3
10.0	06.7
15.0	09.9
20.0	13.2
25.0	16.4
30.0	19.5
35.0	22.5
40.0	25.4
45.0	28.1
50.0	30.7
55.0	33.1
60.0	35.3
65.0	37.2
70.0	38.8
75.0	40.1
80.0	41.0
85.0	41.6

from 30° to 60° does not result in a doubling of the angle of refraction (from 19.5° to 35.3°). Thus, a plot of this data set would not yield a straight line. If, however, the sine of the angle of incidence and the sine of the angle of refraction were plotted, as in Figure 2.4, the plot would be a straight line. Therefore, a linear relationship exists between the sines of the angles.

From the graph in Figure 2.4, an equation can be written relating the angle of incidence (I) and the angle of refraction (I') for light passing from air ($n = 1$) into Plexiglas ($n' = 1.5$):

$$\sin I = 1.5 \sin I'. \tag{2.11}$$

The equation above is an experimentally verified example of Snell's law, which was derived mathematically in Section 2.2. Observe that the slope of this line is 1.5, the value of the index of refraction for Plexiglas. If the hemisphere of glass were replaced by a hemispherical dish of water, the constant of proportionality would be 4/3, the index of refraction for water. The same pattern would result for light traveling from air into any material. Experimentally, it is found that for a ray of light traveling from air ($n = 1$) into some material of refractive index n', a general equation for Snell's law can be written as:

$$\sin I = n' \sin I'. \tag{2.12}$$

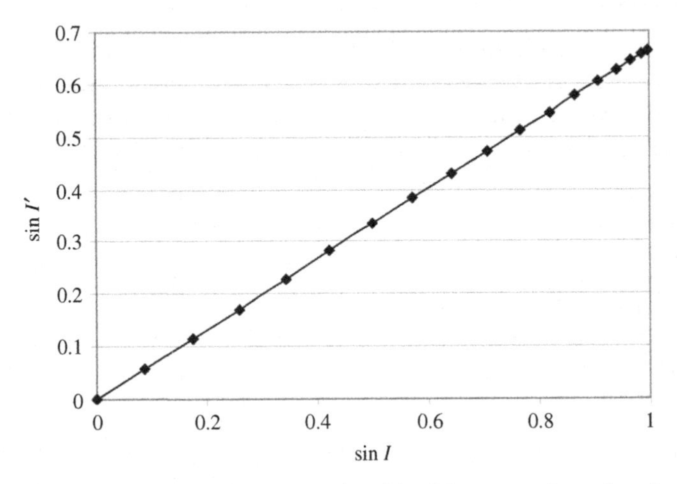

Figure 2.4 Plot of the sine of the angle of incidence against the sine of the angle of refraction.

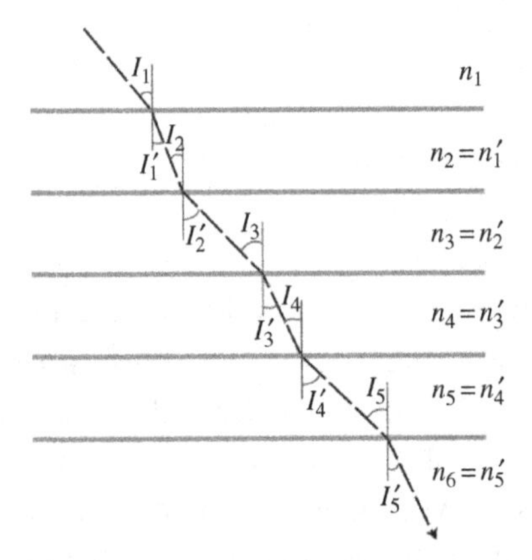

Figure 2.5 Multi-layer stack refracting a single ray.

2.3.2 *Multilayer stack of glass*

A consequence of Snell's law is that, for a given ray propagation, the product $n \sin I$ is a constant; therefore, a multi-layer stack, such as an interference filter, may be analyzed easily, as illustrated in Figure 2.5. If one wants to know the output ray angle (I') for medium 6 and the input ray angle (I_1) at the top of the stack, the application of Snell's law once, instead of applying it five times, gives the same answer as the chain rule:

$$n_1 \sin I_1 = n_1' \sin I_1' = n_2 \sin I_2 = n_3 \sin I_3 = n_4 \sin I_4 = n_5 \sin I_5 = n_6 \sin I_5'$$

$$\therefore n_1 \sin I_1 = n_6 \sin I_5'. \tag{2.13}$$

Therefore, to find the angle at which a ray exits a stack, it is necessary to apply Snell's law only in its final form in Equation (2.13), and not at each interface.

2.4 Reflection versus refraction at an interface

The incident ray crossing an interface in Figure 2.6 is transmitted into the second medium as well as reflected back into the initial medium, as shown. Snell's law applies to both of these secondary rays:

$$n \sin I_{\mathrm{i}} = n' \sin I' = -n \sin I_{\mathrm{r}}. \tag{2.14}$$

By Snell's law, the angle of incidence (I_{i}) and the angle of reflection (I_{r}) are equal in magnitude but opposite in sign. This change in sign is accounted for by a negative index of refraction due to the direction of light propagation.

2.4.1 Fresnel reflectance and transmittance equations

The incident radiant power is equal to the reflected radiant power plus the transmitted radiant power, assuming no scattering:

$$E_{\mathrm{i}} = E_{\mathrm{r}} + E_{\mathrm{t}}.$$

Normalizing with respect to E_{i} to get the relative reflectance (ρ) and transmittance (τ):

$$1 = \rho + \tau. \tag{2.15}$$

The Fresnel reflectance, ρ, is expressed for small angles of incidence ($\sin I = I$ with an error of less than 5%) as a function of the index of refraction:

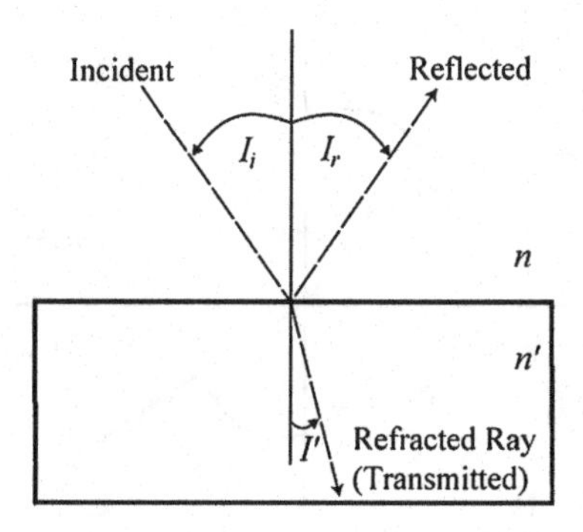

Figure 2.6 Refraction and reflection of an incident ray.

$$\rho = \left[\frac{n' - n}{n' + n}\right]^2. \tag{2.16}$$

If we use regular visible glass with a refractive index of 1.5 as an example, the reflectance is

$$\rho = \left[\frac{1.5 - 1}{1.5 + 1}\right]^2 = 0.04. \tag{2.17}$$

The corresponding Fresnel transmittance for normal incidence is

$$\tau = \left[\frac{4n'n}{(n' + n)^2}\right]. \tag{2.18}$$

For regular visible glass with $n = 1.5$:

$$\tau = \left[\frac{(4)(1.5)(1)}{2.5^2}\right] = 0.96. \tag{2.19}$$

Therefore the sum of the reflected and the transmitted light is equal to 100% of the incident light.

2.4.2 *Total internal reflection (TIR)*

A ray propagating from a denser to a less dense medium is limited in possible incidence angles. For example, a ray going from glass to air (velocity speeds up) has a limited acceptance angle at which the ray will transmit into the air, since light velocity can only increase to its highest value, c. As shown in Figure 2.7, a ray incident at angle I is refracted at a larger angle in the less dense medium (faster velocity). Thus, when the refracted ray angle (I') is larger than 90°, the ray will no longer enter the less dense (faster velocity) medium (n').

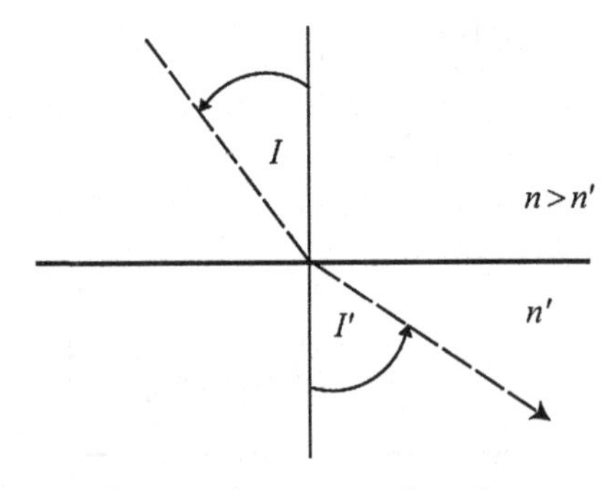

Figure 2.7 Ray emerges from a denser to a less dense medium.

Since the velocity of light can only increase up to the limit c (the speed of light in air), the angle of refraction of a light ray traveling into a less dense medium is limited to a maximum value. If that angle is exceeded, the ray of light is totally reflected. This angle is called the critical angle, and for angles greater than this value, total internal reflection (TIR) occurs. Therefore, the critical angle can be calculated as:

$$I' = \pi/2 \text{ or } 90^\circ, \tag{2.20}$$

$$\sin I_c = n'/n, \tag{2.21}$$

where I_c is the critical angle. At all incident angles larger than this angle, reflection occurs with 100% efficiency. If and only if the ray is emerging from some medium into air ($n' = 1$), Equation (2.21) may be rewritten as

$$\sin I_c = 1/n. \tag{2.22}$$

This important phenomenon often governs the use of deviating prisms which depend on TIR for their operation. In the case of TIR, 100% of the energy is reflected, thus producing a perfect mirror.

2.5 Handedness/parity

An optical image in a simple camera system is rotated 180° and has odd-handedness with respect to the direction of light propagation. Handedness or parity is the orientation of the image relative to the standard right-handed person, or the original orientation of the object. When interpreting the image handedness, one is looking back toward the source ($-z$ direction). However, a film or other recording medium is observed from the opposite direction (the $+z$ direction), so the hard copy of the image is rotated 180°. The image can be rotated by optical systems at various angles, but is typically at fixed 30°, 45°, 60°, 90° or 180° orientations as shown in Figure 2.8. Images with odd parity are shown in Figure 2.9.

Figure 2.8 Right-handed image with various rotation angles.

Figure 2.9 Left-handed image with various rotation angles.

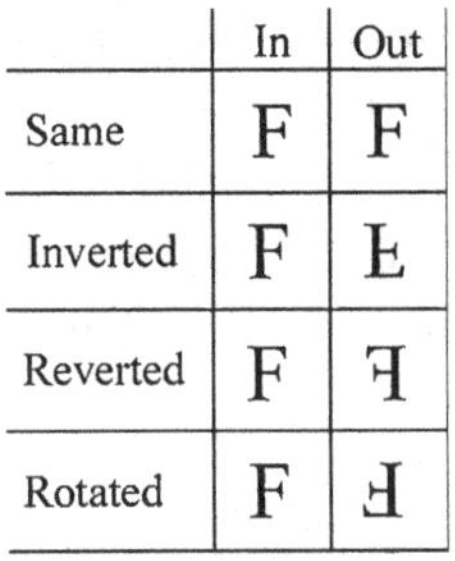

	In	Out
Same	F	F
Inverted	F	Ɫ
Reverted	F	ꟻ
Rotated	F	Ⅎ

Figure 2.10 The image looking into the optical system with the letter F as the input and the corresponding effective image of F via inversion, reversion, or rotation.

Odd-parity images are often referred to as *left-handed*, with a given angle orientation included in the description. The mirror is the simplest optical device to demonstrate handedness or parity. The virtual image formed by a plane mirror has odd-handedness or parity (a left-handed image of yourself) due to one reflection. The rule for handedness is given by (-1^m), where m is the number of reflections in the system: -1 indicates odd-handedness and $+1$ even-handedness. A classic example is the barber shop mirror effect where two plane mirrors are placed in parallel facing each other with the observer sitting in between. The observer's multiple images at the various locations are either odd or even in the series of images seen on successive mirrors.

The image parity and orientation are described by four terms often accepted by optical engineers as describing an image: same, inverted, reverted, and rotated (by a number of degrees) for a relationship between the object and image, as shown in Figure 2.10. Note that rotation is an inversion and a reversion with no handedness change.

For the inverted case, the image is flipped (odd) about the x–z plane, resulting in an odd image. The reverted image is flipped (odd) about the y–z plane, and also results in an odd-handed image. For the case of the rotated image (even), it is effectively inverted and reverted to provide rotation.

2.6 Plane parallel plate (PPP) and reduced thickness

The plane parallel plate (PPP) is the simplest optical element, and yet it produces a couple of interesting optical effects. First, it shifts the image laterally if the PPP is tilted. Second, it shifts the image longitudinally for a convergent wave, producing a change in the optical path length which is a function of angle. This is seen in everyday life, albeit subtly, as one looks through a window. The glass causes an apparent distance change which most observers do not recognize; however, if one looks into a fish tank, the fish appear closer than they really are, due to the refractive index of water being greater than that of the air.

Rays of light traversing a PPP, as shown in Figure 2.11, bend and re-emerge parallel. We do not see the rainbow effect, since the superposition of all colors is still parallel after two refractions, obeying Snell's law. Therefore, we still see the light as white or the original color, but the bundle is shifted laterally as shown in Figure 2.11.

Two things that happen to rays traversing a PPP:

(1) Rays are displaced without changing divergence angles.
(2) The optical path length (*OPL*) traversed by the rays increases.

A PPP does not cause the rays to diverge or converge, as demonstrated in Figure 2.11. The light must be monochromatic or collimated white light in order for these effects to be achieved. To find the amount of shift for a ray traversing a PPP, consider the geometry shown in Figure 2.12:

$$\sin(\theta - \theta') = d/\ell, \qquad \cos\theta' = t/\ell. \tag{2.23}$$

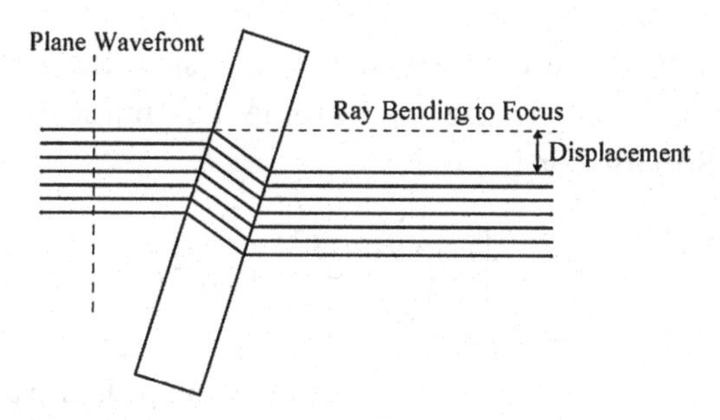

Figure 2.11 Tilted PPP.

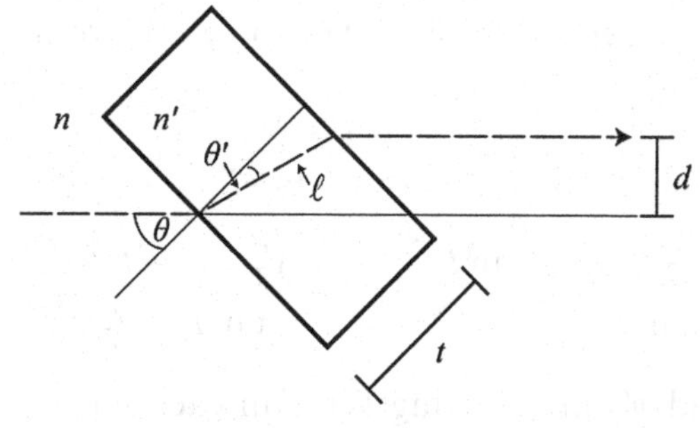

Figure 2.12 Shift of a ray due to a PPP at angle of incidence, θ.

Rearranging and using the trigonometric identity $\sin(a-b) = \sin a \cos b - \cos a \sin b$:

$$\sin\theta\cos\theta' - \cos\theta\sin\theta' = \frac{d}{\ell} = \frac{d}{\ell}\frac{t}{t} = \frac{d}{t}\cos\theta', \tag{2.24}$$

$$\sin\theta - \sin\theta'\left(\frac{\cos\theta}{\cos\theta'}\right) = \frac{d}{t}.$$

By Snell's Law: $\sin\theta' = n/n' \sin\theta$:

$$\sin\theta - \frac{n}{n'}\sin\theta\left(\frac{\cos\theta}{\cos\theta'}\right) = \frac{d}{t}.$$

Solving for d:

$$d = t\sin\theta\left(1 - \frac{n\cos\theta}{n'\cos\theta'}\right). \tag{2.25}$$

This is the equation for the displacement of the parallel ray. For small angles, the assumption can be made that $\cos\theta \cong \cos\theta' = 1$, and that $\sin\theta = \theta$:

$$d = t\theta(1 - n/n'). \tag{2.26}$$

This displacement distance can be used to measure the index of refraction for an unknown material. If the measurement is made in the air, $n = 1$ in Equation (2.25). Solving for the index, n':

$$n'^2 = \left[\frac{t\sin\theta\cos\theta}{t\sin\theta - d}\right]^2 + \sin^2\theta. \tag{2.27}$$

Therefore, by measuring the thickness (t), deviation (d), and angle of incidence (θ), the index of refraction can be calculated.

The longitudinal distance shift along an optical axis ($\Delta z'$) for a converging bundle of rays can be derived for a PPP in air ($n = 1$) from Figure 2.13. From geometry:

$$\tan I = \frac{y - y_2}{t} \quad \text{and} \quad \tan I' = \frac{y - y_2'}{t}$$

$$y_2 = y - t\tan I \qquad y_2' = y - t\tan I'$$

$$\tan I = y_2/z \qquad \tan I = y_2'/z'.$$

Substituting y_2 and y'_2 and solving for y in each equation, the initial ray height is:

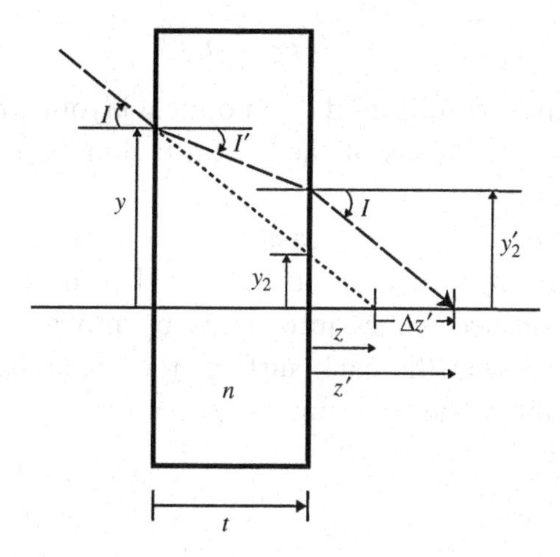

Figure 2.13 Longitudinal shift of an on-axis point with a PPP.

$$y = z\tan I + t\tan I, \qquad y = z'\tan I + t\tan I'.$$

Equating expressions for y:

$$z\tan I + t\tan I = z'\tan I + t\tan I'.$$

Collecting terms related to tan I and tan I':

$$(z' - z - t)\tan I = -t\tan I'.$$

Rearranging terms:

$$z' - z - t = -t\frac{\tan I'}{\tan I} = -t\frac{\sin I' \cos I}{\cos I' \sin I}.$$

Solving for the shift along the optical axis $(z' - z)$:

$$z' - z = t - \frac{t\ \cos I}{n\cos\ I'}. \tag{2.28}$$

Therefore, the shift in the z or longitudinal location at which a converging ray crosses the axis, as shown in Figure 2.12, is

$$\Delta z' = t - \frac{t\ \cos I}{n\cos I'}. \tag{2.29}$$

This means that the image location is moved further (positive direction) down the z-axis with the introduction of a PPP of thickness t and index n. If we make the small angle approximation for paraxial rays, and both I and I' are small, the shift is:

$$\Delta z' \approx t - t/n. \tag{2.30}$$

If the reduced thickness (t/n) is used in an optical layout which has a converging bundle of rays, the effects of the shift are eliminated in the layout, as shown in Figure 2.14.

The reduced thickness of a PPP is easily measured by focusing with a linear tracking microscope on the front surface and then measuring the distance at which the rear surface comes into focus by moving t/n. As shown in Figure 2.15, while viewing the back surface, the apparent position you are looking at is really above the back surface by about 1/3 the thickness.

$$\tan\theta_2 = \frac{x}{t} = \frac{\sin\theta_2}{\cos\theta_2}, \quad \tan\theta_1 = \frac{x}{y} = \frac{\sin\theta_1}{\cos\theta_1}. \tag{2.31}$$

In fact, using the geometry of Figure 2.15, where θ_1 and θ_2 are related by Snell's law, solving for x and then y in Equation (2.31) gives an expression for the apparent depth as viewed through the microscope (the reduced thickness). This reduced thickness value is the exact reduced thickness in Equation (2.23):

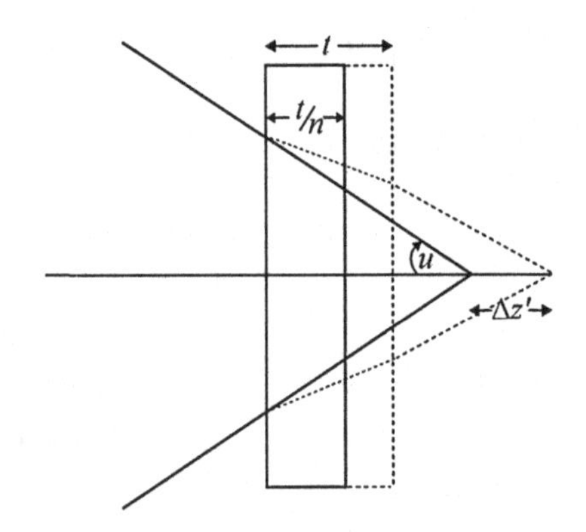

Figure 2.14 Ray trace using a PPP represented in reduced thickness (t/n).

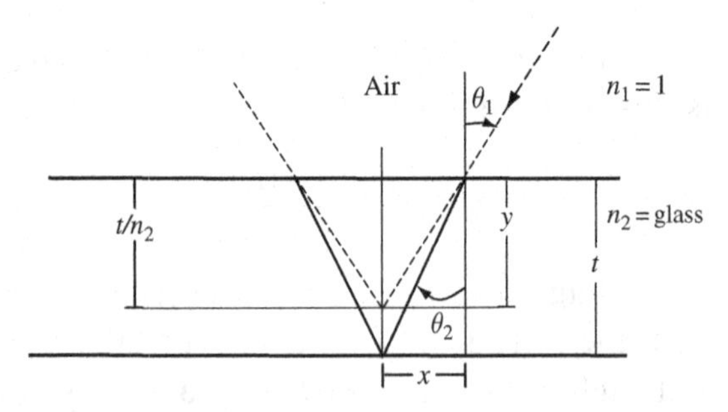

Figure 2.15 Measuring reduced thickness.

$$x = \frac{t \sin \theta_2}{\cos \theta_2} = \frac{y \sin \theta_1}{\cos \theta_1}, \tag{2.32}$$

$$y = t \frac{n_1 \cos \theta_1}{n_2 \cos \theta_2}. \tag{2.33}$$

If we assume that the glass is in air ($n_1 = 1$) and the angles are small, (i.e. 10° or less), Equation (2.33) reduces to

$$y = t/n_2. \tag{2.34}$$

The measurement gives the reduced thickness directly from the distance the microscope moved.

Problems

2.1 How thick must a glass ($n = 1.5$) PPP tilted at 45° be in order to displace the beam by 0.25 in?

2.2 For small angles, Snell's law, $n \sin I = n' \sin I'$, may be approximated by $nI = n'I'$. Find the largest angle I for which the approximation gives a percentage error less than or equal to 10% in I'. Assume $n' = 1.5$, $n = 1$.

$$\% \text{ error} = \frac{I_{\text{apprx}} - I_{\text{exact}}}{I_{\text{exact}}} \times 100\%.$$

2.3 An empty pail (cylinder) resting on its circular base is 80 cm in diameter and 60 cm deep. An observer looking into this pail has a line of sight beginning at the outer top edge to a point on the opposite bottom edge. When the pail is filled with a liquid, the observer, looking from the same direction, now sees the center of the pail instead of the opposite edge (see diagram below). What is the refractive index of the liquid?

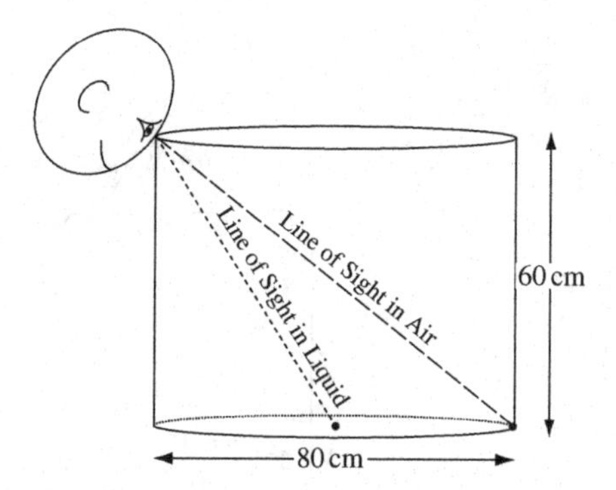

2.4 A quarter wave retardance is to be put in one path for a HeNe laser at 632.8 nm (in air) using two glass slides, one in each path. One slide is 1 mm thick with a refractive index of 1.5, while the second slide has a refractive index of 1.7. How thick must the second slide be to give an optical path difference of $\lambda/4$ between the two paths of light?

2.5 A light ray made up of red and blue light, entering a glass ball, is refracted as it enters the sphere and also as it leaves. As shown below, which light is deviated the most, red or blue? Make a sketch ($n_{\text{blue}} > n_{\text{red}}$).

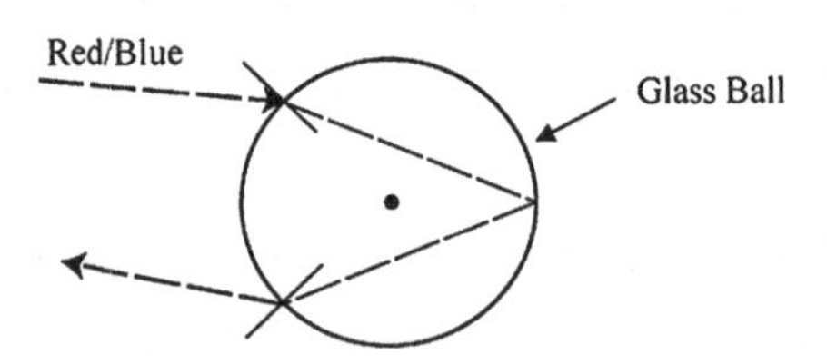

2.6 Show that a corner mirror reflects incident light through 180° regardless of the angle (α) between the incident ray and the mirror normal. Determine the handedness.

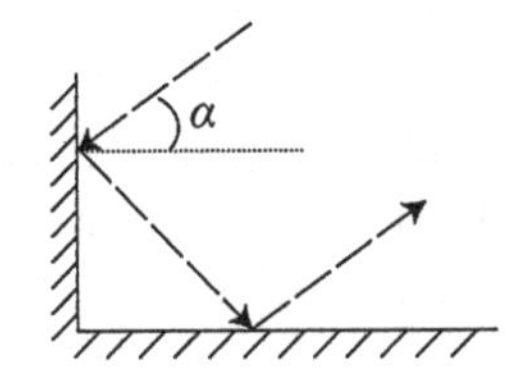

2.7 Imagine you are a fish (abducted by aliens while still in the egg and transported to a galaxy far, far away) in the bottom of a lake filled with an unknown liquid. You are observing a bird (call him Louie) at an angle of $\alpha = 32°$ (as shown, the angle of refraction at the lake surface for this viewing angle is $\beta = 49°$).

(a) What it the refractive index of the liquid in which you are swimming?

(b) Can you see another bird (Louie's mate, Pinkie) at an angle of $\gamma = 47°$? Why or why not?

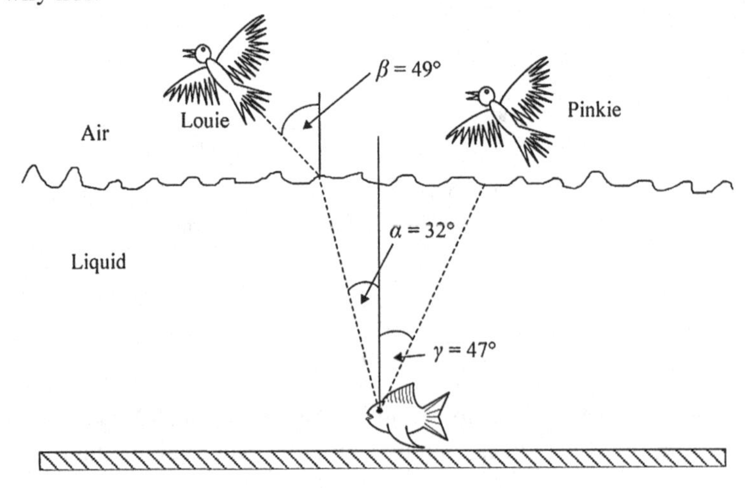

2.8 A ray in air is incident at 45° on a four-level multi-layer stack of glass that is above water as shown below.
(a) Does TIR occur? And if so, at which surface?
(b) If TIR does not occur, at what angle (I') will the ray enter the water?

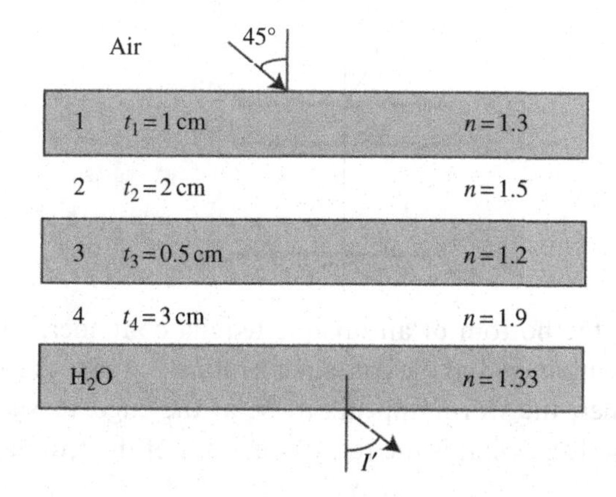

2.9 What is the Fresnel reflection of a ray from a glass surface of index equal to 4 ($n = 4$)? Assume the ray is in air at normal incidence. What is the Fresnel transmittance?

2.10 A PPP thickness is measured to be 2 ± 0.01 cm. The deviation (d) of the ray was 0.5 ± 0.05 mm for an angle of incidence (I) of 5°. What is the refractive index of the glass?

2.11 A scuba diver is looking through water into a glass that is 1 in thick, and finally into air where a crowd of people is forming. What is the maximum angle he needs to view in the water in order to view the entire scene in air?

2.12 If a PPP of glass ($n = 1.9$) is tilted at 45° to a laser beam (HeNe – 632.8 nm), how thick must the glass be (in inches) to get a displacement of 0.4 in the beam direction?

2.13 Plot the angle of refraction and the angle of reflection from a single refracting surface where the ray is going from a more dense material into air for the following media:
(a) H_2O;
(b) a medium with $n = 1.517$;
(c) diamond ($n = 2.426$).

2.14 Plot the critical angle as a function of refractive index for glasses ranging in refractive index from 1.35 to 3. Assume that the ray is heading into air ($n = 1$).

2.15 A diver shines a flashlight upwards from beneath the water at a 42.5° angle to the vertical. Does the beam of light leave the water?

2.16 What are the angles of reflection and refraction for a ray that is incident at an angle of 45° onto glass ($n = 1.5$)?

2.17 A layer of oil of unknown refractive index is floating on top of a layer of carbon disulfide ($n = 1.63$). If a ray of light forms an incidence angle of 60° with the oil, what is the angle of refraction after it hits the carbon disulfide?

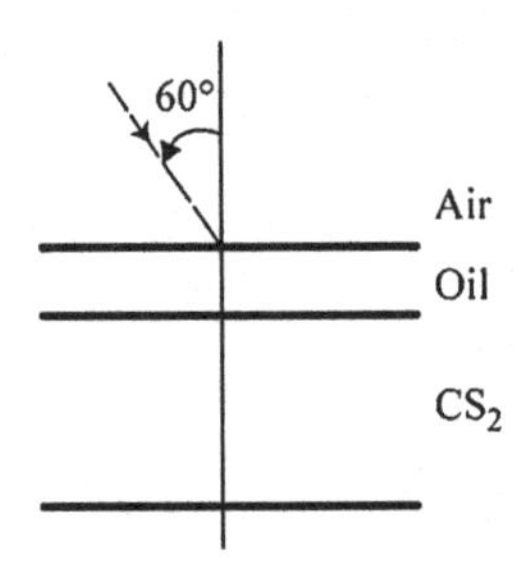

2.18 A worm is at the bottom of an opaque tequila container. The worm is at the center of the container, and the container's radius is 4 cm. As viewed from the top of the container, the worm appears to be at the edge, as shown below. If the liquid is 15 cm deep, what is the refractive index of the tequila?

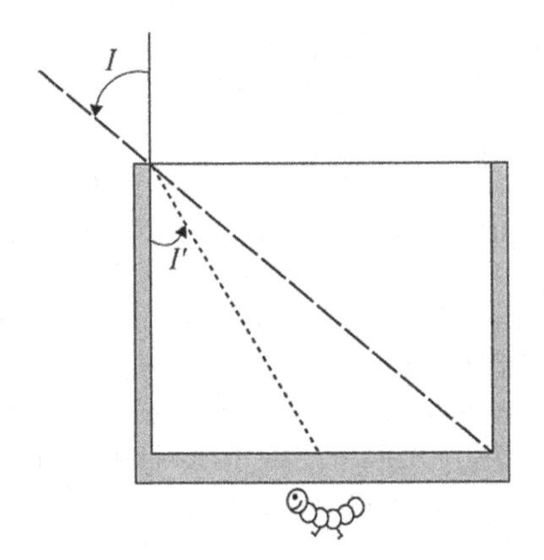

Bibliography

Hecht, E. (1998). *Optics*, third edn. Reading, MA: Addison-Wesley.

Hopkins, R., Hanau, R., Osenberg, H., *et al.* (1962). *Military Standardized Handbook 141 (MIL HDBK-141)*. US Government Printing Office.

Klein, M. V. and Furtak, T. E. (1986). *Optics*, second edn. New York: Wiley.

Longhurst, R. S. (1967). *Geometrical and Physical Optics*, second edn. New York: Wiley.

Smith, W. J. (2000). *Modern Optical Engineering*, third edn. New York: McGraw-Hill.

3
Image formation

Image formation has many meanings to various groups or individuals; however, in geometrical optics its definition is very clear in that it refers to the formation of a light pattern to replicate a scene. The light (radiant power) pattern formed by the optical phenomenon resembles the scene or object, and is called an image. In geometrical optics, an image-forming optical system creates a radiant pattern in two dimensions that resembles the scene that a human eye would perceive as the object. There are two general classes of images in geometrical optics: those formed by lenses and those formed by projections. In present day cameras, lenses are by far the most common means of obtaining an image.

3.1 Pinhole camera

One example of a projection system is the pinhole camera, also referred to as *camera obscura* (Latin for "dark chamber"), which uses a tiny pinhole to collect light without the use of a lens. Figure 3.1 illustrates this simple concept. You may recall, as a child, sitting inside a box while viewing an image projected through a pinhole onto the inside wall. The light from an object passes through a small aperture along a ray, to form an image on a surface. This image may either be projected onto a translucent screen for viewing through the camera, or onto an opaque surface for viewing in reflection. Pinhole cameras require much longer exposure times than conventional cameras because the aperture, which must be tiny in order to produce a reasonably clear image, is much smaller in diameter than a typical lens, and collects much less radiant power.

It is thought that Renaissance artists used pinhole cameras to assist with their painting. One advantages of using such a camera is that it creates an image with the correct perspective, thus greatly increasing the realism of a

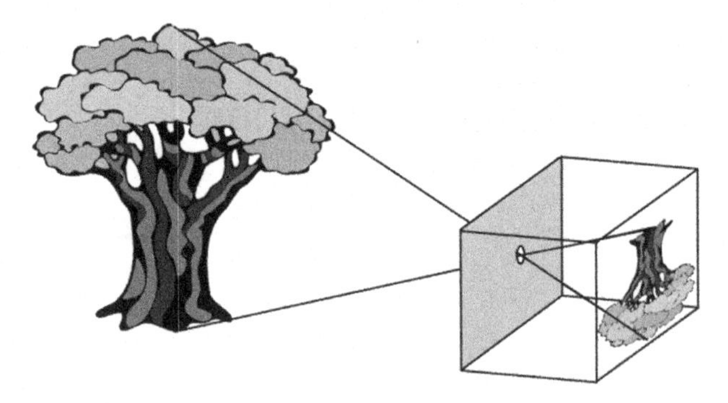

Figure 3.1 Pinhole camera.

painting created by copying such an image onto a canvas (Falco and Hockney, 2000). Painters such as Johannes Vermeer in the seventeenth century were known for their magnificent attention to detail and exact perspectives. It has been widely speculated that they made use of pinhole cameras, but the extent to which this technique was used by artists of the period remains a mystery. Even though the image projected by this simple pinhole apparatus is always upside down, it is possible to use a mirror, as illustrated in Figure 3.2, to project this image right side up and ready for painting.

Image sharpness increases with a smaller pinhole, while light collection, which is proportional to the area of the pinhole, decreases. Practical cameras use a lens rather than a pinhole because this allows a larger aperture, yielding a brighter image. It should be noted that the size of the pinhole is independent of the wavelength of light.

Pinhole cameras are usually handmade by the photographer for a particular purpose. In its simplest form, the photographic pinhole camera consists of a light tight box with a pinhole in one end. The layout of the design is shown in Figure 3.3.

The design of the pinhole camera requires the correct hole size for the distance to the observation plane of the image. The exact relationships are too complex to be developed here, so only an overview of the approach and the techniques used to find the optimum dimensions will be given. The pinhole must be perfectly round (to minimize any diffraction effects or irregularities), and mounted in an extremely thin surface to avoid tunneling or multiple reflections of the light.

A method of calculating the ideal pinhole size is first to define the overall demagnification desired. For example, if you wish to project the image of a 6 ft tall man standing 12 ft away (range) onto a 6 in piece of film, you would choose a length for your camera that is proportional to the projected image. In this case the ideal camera length would be 12 in. Thus in Figure 3.3 the distance (t) would be 12 in. In the ideal design the diffraction spot (Δy) on the observation

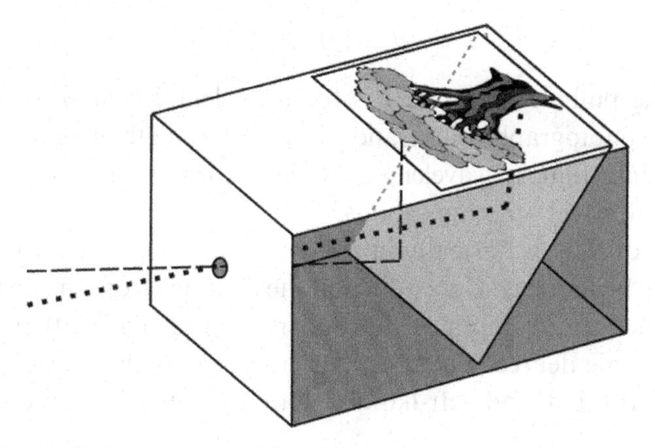

Figure 3.2 Pinhole camera with mirror to correct orientation onto a transparency.

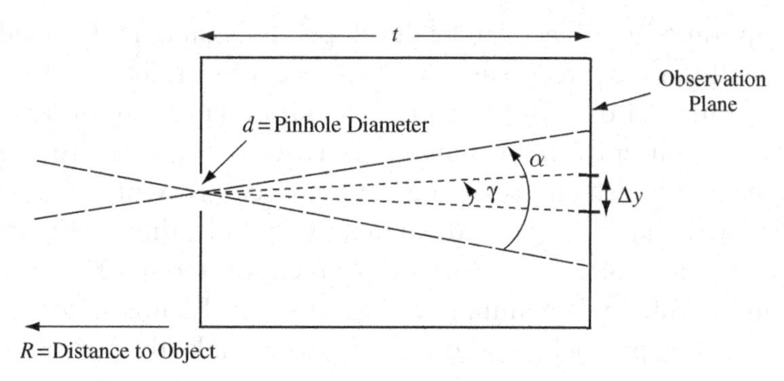

Figure 3.3 Pinhole camera layout.

plane would have the same diameter as the pinhole (d). The angular blur (γ) can be expressed as:

$$\begin{aligned} \Delta y &= \gamma t \\ \gamma &= 2.44\lambda/d \\ \Delta y &= 2.44\lambda t/d. \end{aligned} \tag{3.1}$$

Setting Δy equal to d causes the pinhole diameter to be:

$$d^2 = 2.44\lambda t. \tag{3.2}$$

A method of calculating the optimal pinhole diameter was first devised by Jozef Petzval. The formula was improved upon by Lord Rayleigh, giving the form used today:

$$d = 1.9\sqrt{t\lambda}, \tag{3.3}$$

where d is the pinhole diameter, t is the focal length (the distance from the pinhole to the photographic film), and λ is the wavelength of light. For standard black and white film, a wavelength of light corresponding to yellow-green (550 nm) should yield optimum results.

The depth of field is basically infinite, but this does not mean everything will definitely be in focus. Depending on the distance from the aperture to the film plane, the infinite depth of field means everything is either in or out of focus to the same degree. The image orientation, as shown in Figure 3.1, is inverted, or rotated and left-handed looking into the direction of light propagation.

3.2 Object representation

The description of an object can be developed by assuming it is made up of dots, or small point sources, each emitting into 2π steradians. The example shown in Figure 3.4 of a halftone photograph of Tracy, an optics student, illustrates this concept. The two images in Figure 3.4 are recorded with the halftone technique, which uses a regularly spaced array of dots, similar to that used in newspaper images and dot matrix style printing. In Figure 3.4(a), the image was produced using 40 dots per inch, or about 10000 dots in the image, and the fidelity is maintained. The image in Figure 3.4(b) was made using only 13 dots per inch, or about 1000 dots to make the portrait. Thus, if we consider each ray to be a photon, how many photons are needed to

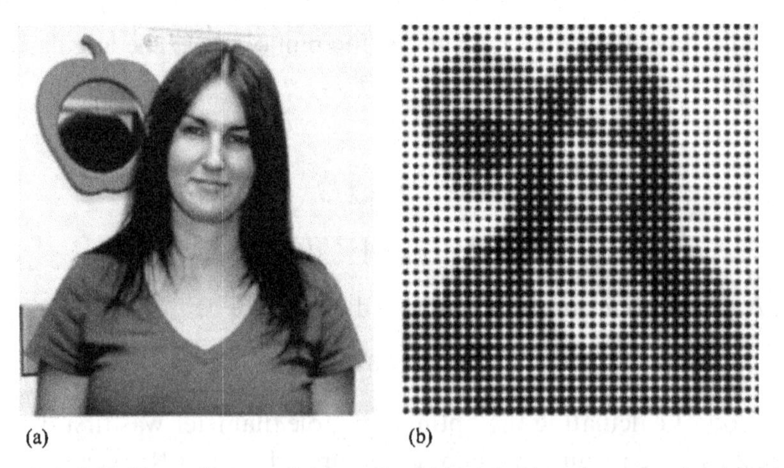

Figure 3.4 Halftone images: (a) standard black/white photograph with high resolution; (b) halftone image using point sources with lower fidelity.

identify an object of a particular size? Most objects are extended sources, meaning they are made up of many point sources. In this context, we will be discussing point sources that form an image. An alternative approach is to decompose the object into sine and cosine functions of various spatial frequencies. This approach will be discussed later, after Fourier mathematical techniques are introduced.

The point source approach to describing an object can provide insight into the formation of an image by recalling the spherical wavefront that is emitted from this point source. The spherical wavefront is radiating in 4π steradians, but the optical imaging system only collects a small solid angle of this source. Therefore, we know intuitively that the point source cannot be faithfully reproduced in the image plane, since we have lost the part of the wavefront that lies outside of the collecting region. As a result, this point source in the image plane will not only be spread out (blurred) but will also lose intensity. Recall that the point source has some number (N) of photons emitted into 4π steradians, so its units of intensity are photon per steradian.

The optical system is assumed to be rotationally symmetric, which means that the optical axis can be thought of as the center of rotation. In the case of the pinhole camera, the optical axis passes through the center of the pinhole, and for lenses, it is the center axis. This requirement of symmetry has a very valuable consequence: i.e. every plane containing the optical axis (z axis) has the same cross-section and contains values in x or y that are identical. Therefore, the y–z plane cross-section is the same as the x–z plane cross-section. Here the y–z plane will be called the meridional plane and the x–z plane the sagittal plane.

This symmetry constraint reduces a three-dimensional system (x, y, z), to two dimensions (y, z), as shown in Figure 3.5. This assumption does not require the object to be rotationally symmetric, but the geometrical coverage will be a circle mapped out on the object.

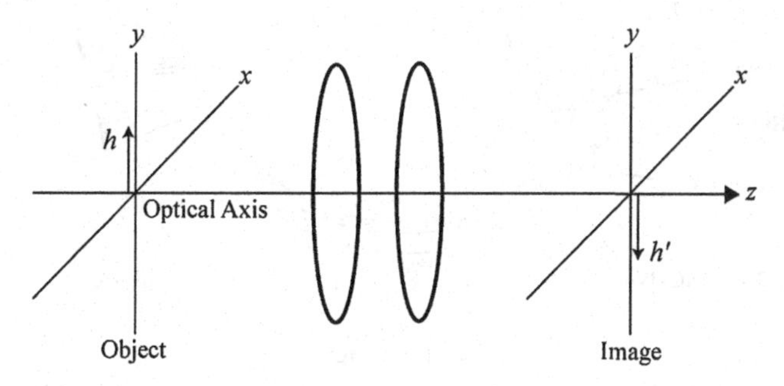

Figure 3.5 Rotationally symmetric optical system.

3.3 Lenses

The use of lenses to form an image provides a means of collecting more energy from the object on scene. In the case of a pinhole camera, a very small number of rays are collected from each point in the object. A lens, however, can cover a much larger area, so the image will be much brighter.

Lens systems that form images, such as camera systems, mimic the human eye. The human eye, shown in Figure 3.6, forms an inverted image on the retina, and has an angular resolution of 1 arcminute.

The various surface curvatures of the eye follow the Arizona eye model given in Table 3.1. Conic refers to the shape of the curvature, such as elliptical, spherical, or hyperbolic. The conics in Table 3.1 between −1 and 0 describe elliptical surfaces, and the conics less than −1 are hyperbolic.

Most lens systems use a layout similar to the human eye to form an image. However, there are instances where this is not the case. For example, in the medical field, an image is often formed indirectly, via the use of a CAT scanner or other techniques. In the insect world, the eye may not form an image at all, but provides other information about a scene, such as color, the polarization of light, or velocity and/or movement of an object.

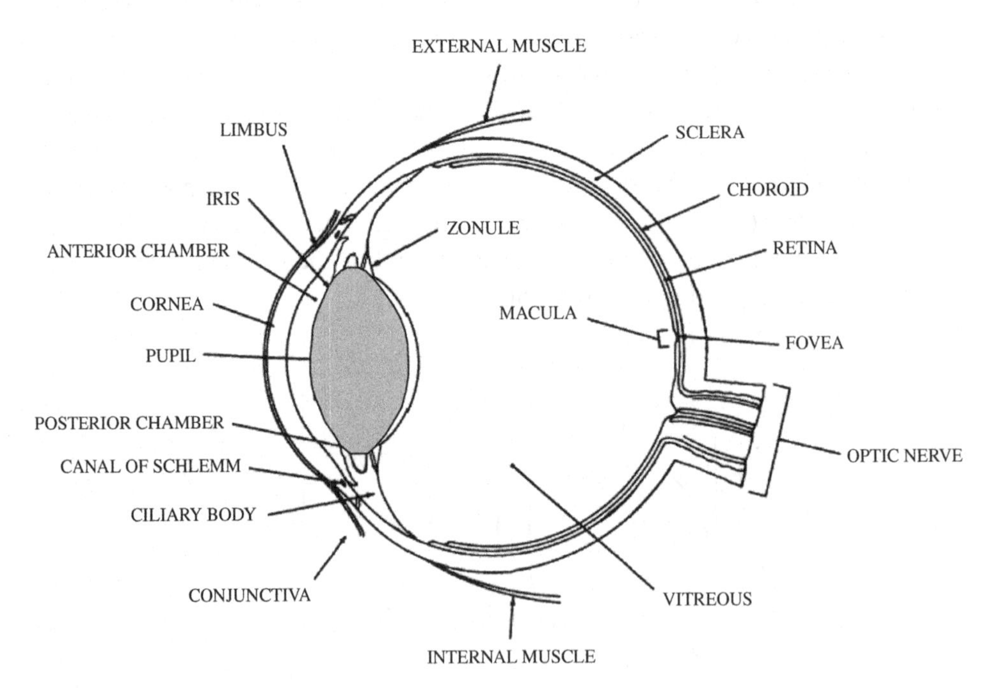

Figure 3.6 Human eye (Hopkins, *et al.*, 1962, p. 4-2).

Table 3.1. *Arizona eye model*

	Radius (mm)	Conic	Thickness (mm)	n_d	n_F	n_C
Anterior cornea	7.80	−0.25	0.5500	1.3771	1.3807	1.37405
Posterior cornea	6.50	−0.25	3.0500	1.3374	1.3422	1.33540
Anterior lens	11.03	−4.30	4.0000	1.4200	1.42625	1.41750
Posterior lens	−5.72	−1.17	16.6423	1.3360	1.34070	1.33410
Retina	−13.40					

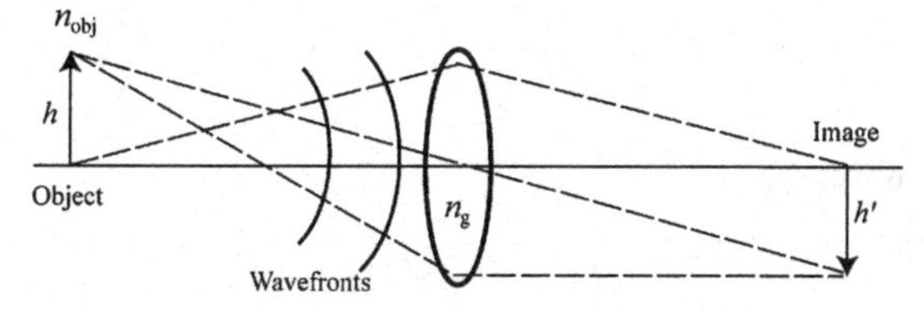

Figure 3.7 Image–object layout using a positive lens.

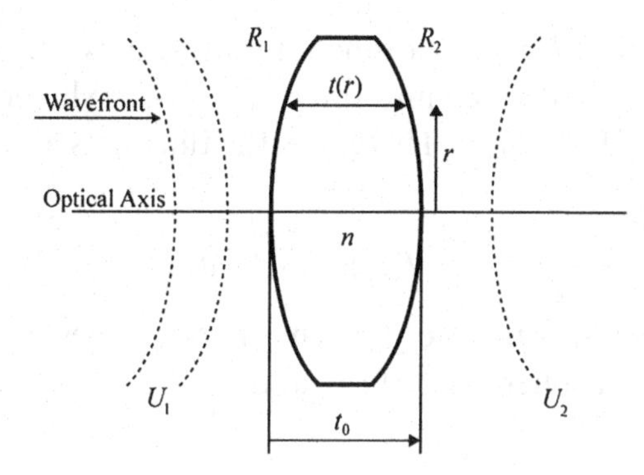

Figure 3.8 Lens layout.

As with the pinhole camera, the image formed by the lens in Figure 3.7 is rotated upside down and has odd parity. The lens acts upon the wavefront from an object. The lens is transparent to the wavelength of light and has an index of refraction (n) greater than 1. The optical surfaces are typically spherical surfaces, assuming symmetry in the lens, as shown in Figure 3.8. The approaching wavefront refracts at the lens at different times, depending on its distance from the optical axis (r). The delays of the various regions of the wavefront are proportional to the thickness of the lens at each radial zone (r), as shown in Figure 3.8. Recall Equation (1.6) for a spherical wave, where the phase (φ) changes the results as the wavefront passes through the lens. This phase change is:

$$\begin{aligned} k\varphi(r) &= k(t_0 - t(r)) + nkt(r) \\ &= kt_0 + k(n-1)t(r), \end{aligned} \tag{3.4}$$

where $k(t_0 - t(r))$ is the phase delay caused by the free space region and it is assumed that the lens is surrounded by air ($n = 1$). The wavefront's velocity in the lens is slower than in air, so the section of the wavefront not in the glass will overtake the section that is in the glass. The emerging wavefront is given by

$$U_2 = U_0 e^{ik(\varphi_0 + \varphi(r))}, \tag{3.5}$$

where the input wavefront was

$$U_1 = U_0 e^{ik\varphi_0}, \tag{3.6}$$

so the output wavefront is

$$U_2 = U_1 e^{ik\varphi(r)}. \tag{3.7}$$

Since the phase change $k\varphi\ (r)$ is a function of thickness, t, at a given zone of radius r, the lens can be divided into three sections, as shown in Figure 3.9, in order to find the thickness, $t(r)$. Therefore, the thickness as a function of zone radius, r, is:

$$t(r) = t_1(r) + t_2 + t_3(r), \tag{3.8}$$

where t_2 is the "edge thickness" of the lens. The thicknesses $t_1(r)$ and $t_3(r)$ are related to the sag of a spherical surface, and can be expressed as

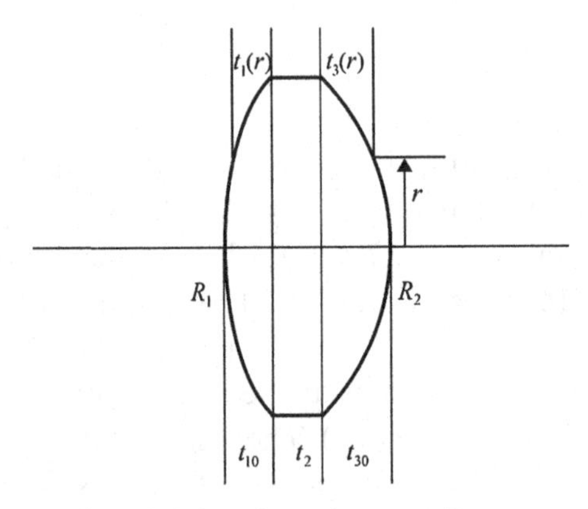

Figure 3.9 Dividing the thick lens into three sections.

$$t_1(r) = t_{10} - \left[R_1 - \left(R_1^2 - r^2\right)^{1/2}\right], \tag{3.9}$$

$$t_3(r) = t_{30} + \left[R_2 - \left(R_2^2 - r^2\right)^{1/2}\right]. \tag{3.10}$$

Rewriting the equations:

$$t_1(r) = t_{10} - R_1\left[1 - \left(1 - \frac{r^2}{R_1^2}\right)^{1/2}\right], \tag{3.11}$$

$$t_3(r) = t_{30} + R_2\left[1 - \left(1 - \frac{r^2}{R_2^2}\right)^{1/2}\right]. \tag{3.12}$$

Assuming the lens radius (r) is small compared with the surface radii (R_1 and R_2), a Taylor series approximation can be made for the square root parts of Equations (3.11) and (3.12), and using the first two terms of that expansion gives

$$t_1(r) = t_{10} - \frac{r^2}{2R_1}, \tag{3.13}$$

$$t_3(r) = t_{30} + \frac{r^2}{2R_2}. \tag{3.14}$$

Now Equation (3.8) becomes

$$t(r) = t_0 - \frac{r^2}{2}\left(\frac{1}{R_1} - \frac{1}{R_2}\right), \tag{3.15}$$

so the phase term, Equation (3.4), becomes

$$k\varphi(r) = kt_0 + k(n-1)\left[t_0 - \frac{r^2}{2}\left(\frac{1}{R_1} - \frac{1}{R_2}\right)\right] \tag{3.16}$$

$$= knt_0 - k\frac{r^2}{2}\left[(n-1)\left(\frac{1}{R_1} - \frac{1}{R_2}\right)\right] \tag{3.17}$$

$$= knt_0 - k\frac{r^2}{2f^*}, \tag{3.18}$$

where we define

$$\frac{1}{f^*} \equiv (n-1)\left(\frac{1}{R_1} - \frac{1}{R_2}\right). \tag{3.19}$$

f^* is the focal length of a thin lens equivalent, which we will discuss later in greater detail. The quadratic phase factor is negative, producing a converging spherical wavefront, as shown in Figure 3.10:

$$\sim e^{-ikr^2/2f}. \tag{3.20}$$

Since the wavefront is truncated due to the finite size of the lens, the point source (p) will not be an exact point image, but will be blurred or smeared.

3.4 Image types

In geometrical optics, a lens can produce a real image or a virtual image. In this context, an image means a two-dimensional pattern in the x–y plane of a three-dimensional scene. If a lens produces a converging wavefront, as shown in Figure 3.10, the image is a real image. The rays converge to a point for each point in the object.

If a screen or paper were placed at the position, I, for the lens setup shown in Figure 3.11, an image would be present on the paper for an object located at O. There would be optical radiation present at the location of the image.

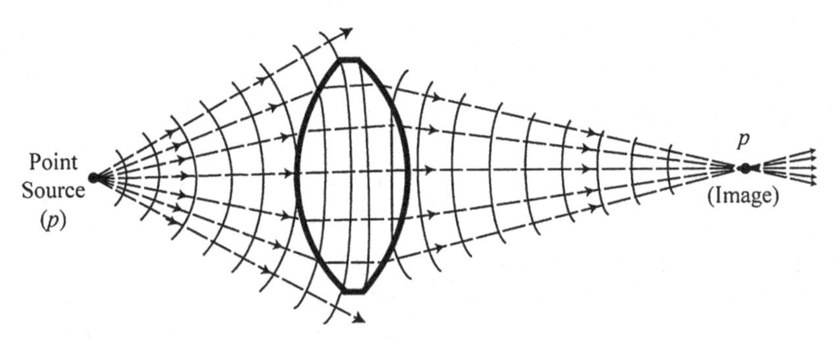

Figure 3.10 Lens effect on wave fronts.

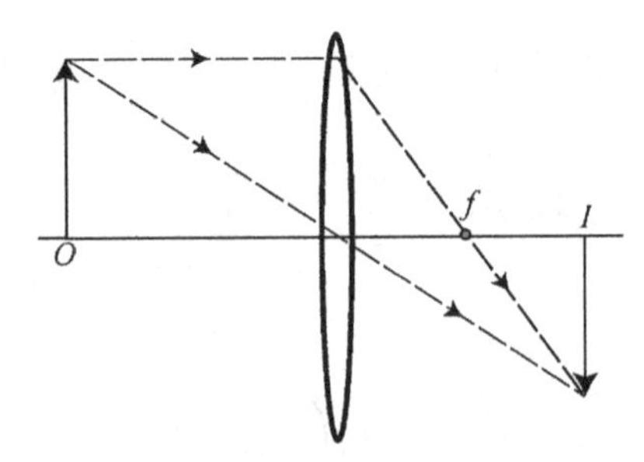

Figure 3.11 Positive lens forming a converging wavefront.

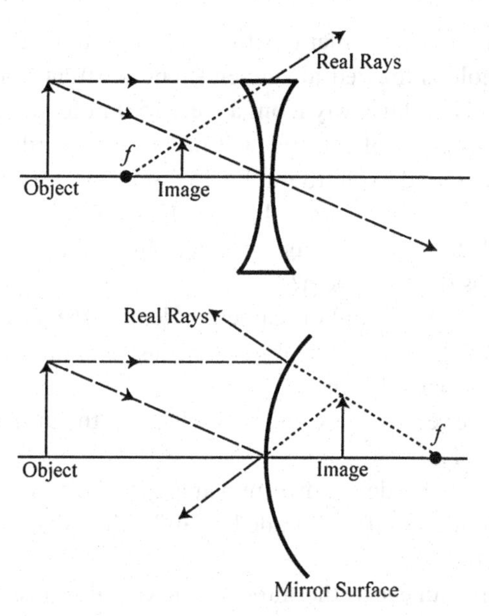

Figure 3.12 Negative lens forming a virtual image.

A virtual image is a representation of an actual object formed by a diverging wavefront, which seems to originate from a virtual image. The rays associated with this wavefront do not cross in real space, only in virtual space. An image would not be produced on a screen placed at the virtual image location. The rays diverge, as shown in Figure 3.12, for a negative lens.

Problems

3.1 The optimum design of a pinhole camera occurs when the hole diameter (d) is equal to: $2.44\lambda l = d^2$. It is required to cover a full field angle of $\pm 15°$. Design the camera (i.e. $l=$? ; $d=$?) for a wavelength of 700 nm.

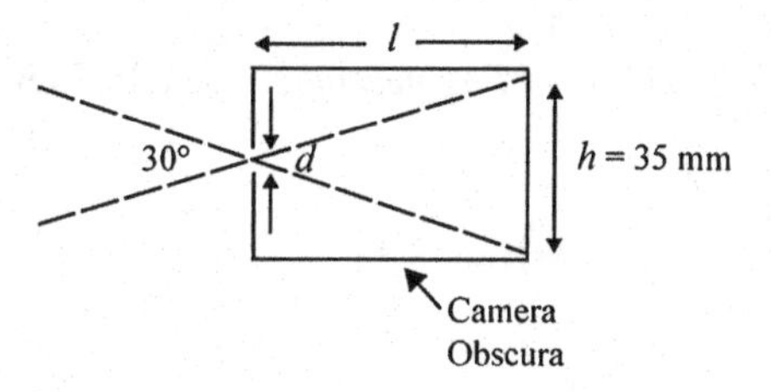

3.2 A pinhole camera produces a 10 cm high image of a tree. This same setup produces a 4 cm high image of a 6 ft tall person standing 5 ft in front of the tree. To make the person's image height 10 cm (equal to the tree), the camera needs to be moved 10 ft forward. How tall is the tree?

3.3 A boy makes a pinhole camera with the dimensions 20 cm × 20 cm × 40 cm (x, y, z). A pinhole is located at one end, and a 10 cm × 10 cm film is placed at the other end. How far away from a tree, 15.0 m high, should the boy place his camera so the image of the tree will just fit on the film?

3.4 A pinhole camera is used to photograph a 15 ft statue of David, located 30 ft away. The image in the pinhole camera is 4 in tall. How long is the camera?

3.5 What is the handedness of the image produced by a pinhole camera when viewing the image from inside the camera?

3.6 Model the human eye as a pinhole camera of length 16 mm with a retina of 8 mm diameter. What is the size of an object that just fills the retina at a distance of 25 cm (standard viewing distance)?

3.7 In Figure 3.4(a) we see the back of Tracy's head in the mirror. Is that a real or a virtual image? Explain.

3.8 For the Arizona eye model, shown in Table 3.1, what is the value of the focal length of the anterior lens if surrounded by air? (Hint: Take the lens out of the eye and put it in air.)

3.9 For the typical human eye at the standard viewing distance of 25 cm, what is the spacing of the dots in a halftone picture such that they can just be resolved?

Bibliography

Baigrie, B. S. (2000). The scientific life of the camera obscura. *Optics and Photonics News*, **11**, 18.

Falco, C. M. and Hockney, D. (2000). Optical insights into Renaissance art. *Optics and Photonics News*, **11**, 52.

Hecht, E. (1998). *Optics*, third edn. Reading, MA: Addison-Wesley.

Hopkins, R., Hanau, R., Osenberg, H., *et al.* (1962). *Military Standardized Handbook 141 (MIL HDBK-141)*. US Government Printing Office.

Jenkins, F. A. and White, H. E. (1976). *Fundamentals of Optics*, fourth edn. New York: McGraw-Hill.

Meyer-Arendt, J. R. (1972). *Introduction to Classical and Modern Optics*. Englewood Cliffs, NJ: Prentice-Hall.

Sears, F. W., Zemansky, M. W. and Young, H. D. (1976). *University Physics*. Reading, MA: Addison-Wesley.

Smith, W. J. (2000). *Modern Optical Engineering*, third edn. New York: McGraw-Hill.

4

Mirrors and prisms

4.1 Plane mirrors

Probably the most common optical element found in a home is the plane mirror. The fragile thin metal layer that comprises the reflecting surface can be on either the outside or the inside of a thicker (typically 1/8 in) protective glass layer. (This is not always the case. For instance, in some scientific mirrors, the reflecting surface is not protected in order to increase reflectivity, particularly in the infrared spectral region.)

A plane mirror not only bends or changes the path of reflected light rays; it also changes the handedness (parity) of a reflected image. To illustrate examples of right-handed and left-handed images, several images of the letter "R" are shown in Figure 4.1.

An image which undergoes an even number of reflections maintains its right-handedness. However, an odd number of reflections changes the handedness to odd (left-handed). A simple expression to remember is the following:

$$(-1)^m, \tag{4.1}$$

where m is the number of reflections. A result of $+1$ yields right-handedness (even parity), while a result of -1 yields left-handedness (odd parity). The images are vertical in all cases. In addition to handedness and parity, there are other special terms that are used to refer to images that are reflected from a mirror. These terms, illustrated in Figure 4.2, are: reverted, inverted, and rotated.

Note that rotation is an inversion and a reversion with no handedness change. For a single reflecting surface, as shown in Figure 4.3, there are four important observations.

(1) A line connecting the object and its image is perpendicular to the mirror surface and is bisected by the mirror.
(2) The image handedness is changed.

Figure 4.1 The letter "R" showing right- and left-handedness corresponding to even or odd parity respectively.

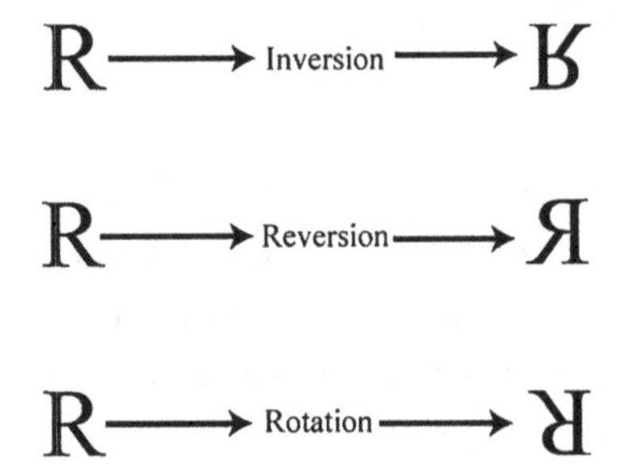

Figure 4.2 The letter "R" indicating images that are inverted, reverted, or rotated through 180°.

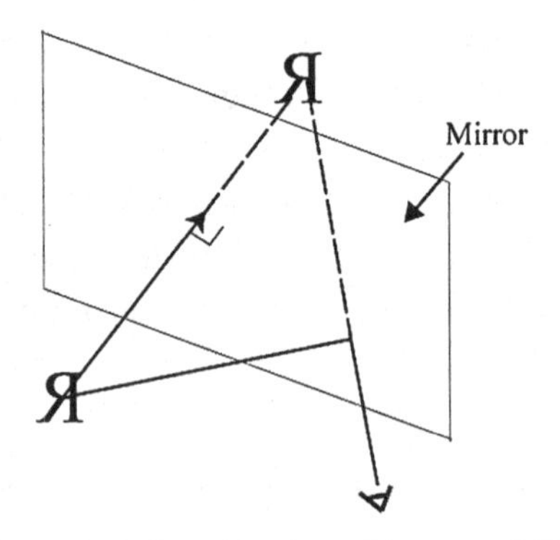

Figure 4.3 Plane mirror forming the virtual image of "R" to the observer.

(3) Any point on the mirror plane is equally distant from the object and its image.
(4) The angles of reflection follow Snell's law.

The letter R is used to illustrate these effects because it is not symmetrical. There are several other letters in the alphabet that have vertical symmetry and horizontal symmetry. These letters should be avoided in observing reflections from mirrors because they will give misleading information in observations or experiments. Consider the letters A, H, I, M, O, T, U, V, W, X, and Y. They all have vertical symmetry, while B, C, D, E, H, I, K, O, and X have horizontal symmetry. Words formed from these letters may also have vertical (TAT, TOT, OTTO, ATOYOTA) or horizontal (BOX, DECODE, HIKE, OXIDE, COOKBOOK) symmetry, which can create problems when evaluating optical systems.

Plane mirrors that are cascaded should be analyzed by considering the sequence of rays propagating from the object being observed. Consider two parallel plane mirrors, as shown in Figure 4.4, separated by a distance, *d*. These mirrors act as a periscope, and displace the image of the object by twice the separation distance, *d*, of the two mirrors. Note that the rays from the virtual image are parallel to the original object rays, and that the image is right-handed.

For two plane mirrors that are tilted toward each other, the intersection of these two surfaces forms a line called the dihedral line (or edge). In a plane perpendicular to the dihedral edge (defined as the principal projection), a projection ray is deviated by twice the angle between the mirrors, or the dihedral angle, θ, as shown in Figure 4.5:

$$\psi = 2\theta. \tag{4.2}$$

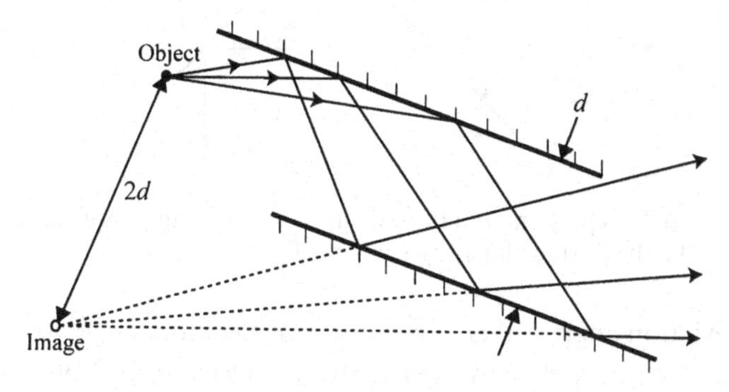

Figure 4.4 Parallel plane mirrors forming a periscope.

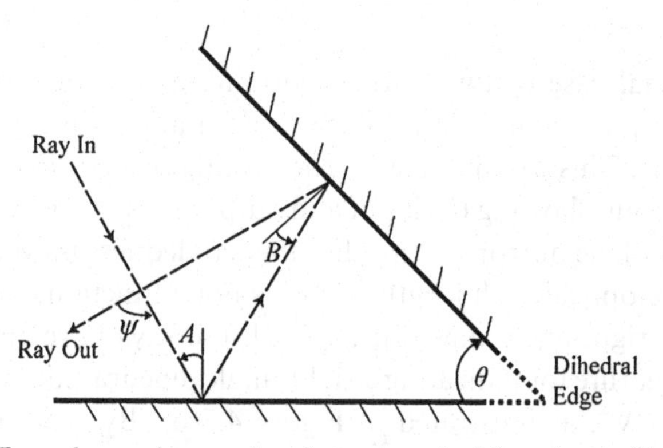

Figure 4.5 Two plane mirrors placed at an angle θ forming a dihedral edge.

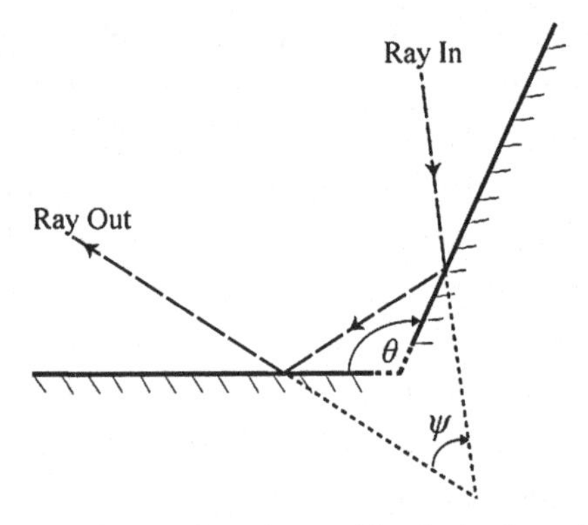

Figure 4.6 Two plane mirrors forming a dihedral edge with an angle greater than 90°.

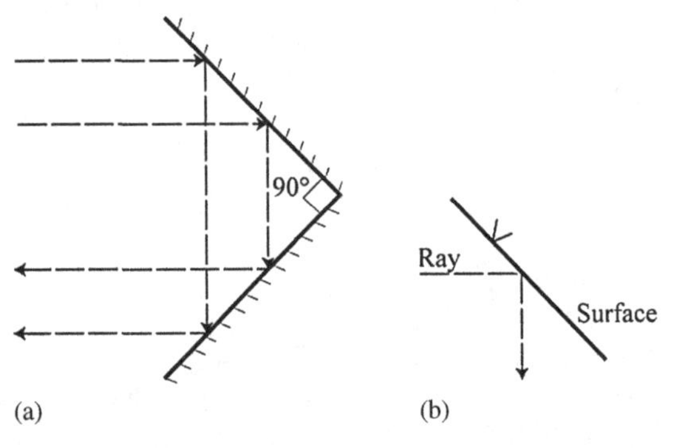

Figure 4.7 (a) Two perpendicular plane mirrors forming a roof, returning the light ray 180°. (b) Surface labeling for a roof.

ψ is the deviation angle and is independent of the input angle for $\theta < 90°$. The input and output rays cross (converge), as shown in Figure 4.5. For dihedral angles greater than 90° ($\theta > 90°$), the input and output rays diverge, as shown in Figure 4.6:

$$\psi = 2\theta - 180°. \tag{4.3}$$

For the special case of two mirrors forming a 90° angle (dihedral angle $\theta = 90°$), both input and output rays are 180°, or antiparallel, in the principal projection plane. This case is called a roof configuration. Figure 4.7(a) illustrates a roof layout, showing the input and output rays. A roof configuration is equivalent to a plane mirror, except that the handedness is even $(-1)^2$.

Note that the optical path length (OPL) is not a function of the position of the ray on the right-angle dihedral edge; all the rays have equal OPL. The typical nomenclature for indicating a right-angle dihedral edge (a roof) is a line with the letter "V", as illustrated in Figure 4.7(b). By using three mutually perpendicular mirrors (called a corner cube, since its construction is a corner), the need to have the input ray in the principal projection plane is eliminated. Any ray input will be reflected out of a corner cube antiparallelly.

4.2 Deviating prisms

Deviating prisms are used to correct the handedness (parity) of an image as well as to change the direction of the ray in the z direction. The deviating prism is probably the most common optical refracting component besides the lens, and it is used in a variety of optical systems. The most common application is in binoculars, where deviating prisms are used to make the image upright (invert). There are several classifications of deviating prisms, and all have different names, none of which are related. Deviating prisms are classified in groups according to their deviation angle: 45°, 90°, or 180°. The various types of designs in each of the groups are shown in Figures 4.8, 4.9, and 4.10. The

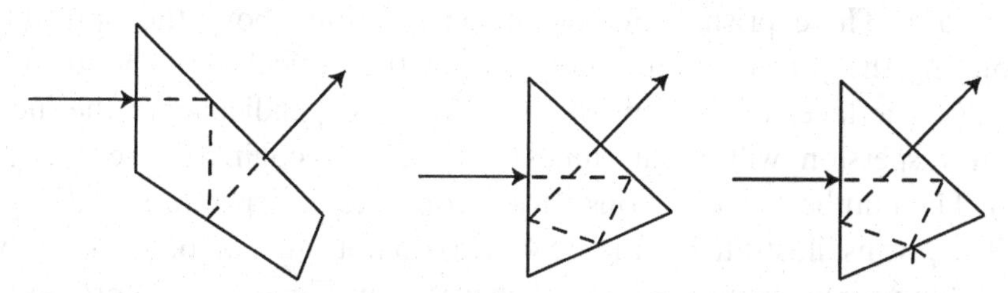

Figure 4.8 45° deviating prisms using a roof in one case, which has two reflections.

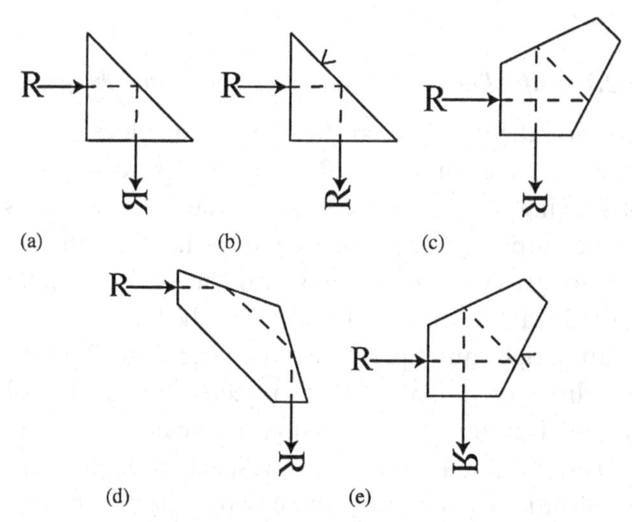

Figure 4.9 90° deviating prisms (looking in the negative z direction): (a) right-angle; (b) Amici; (c) penta; (d) Wollaston; (e) reflex.

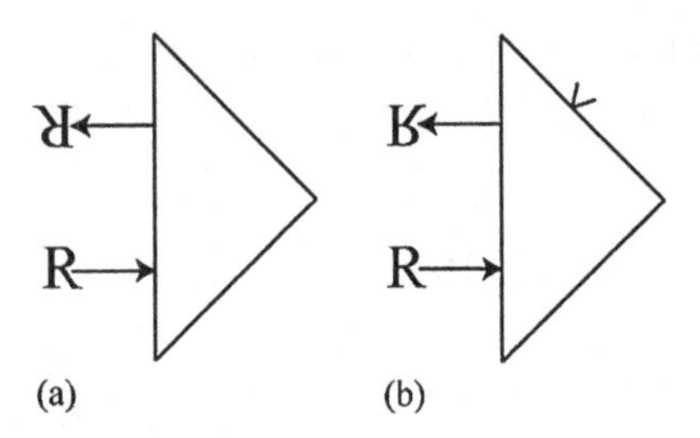

Figure 4.10 180° prisms: (a) Porro prism; (b) corner cube. The corner cube is three orthogonal mirrors (looking in the negative z direction).

180° prism can be replaced with a plane mirror if the required handedness (-1) is acceptable. Many of these produce near 100% reflectivity.

There are also direct view prisms that do not change the direction of the collimated light, but do affect the handedness or orientation (rotation) of the image. These prisms are shown in Figure 4.11(a) and (b) (Shack, private communication). These prisms provide image rotation about the optical axis. On rotating the prism by an angle θ about the optical axis, the image will rotate 2θ. The faces of these prisms should be perpendicular to the incident light or dispersion will occur, unless the light is collimated (e.g. a Dove prism). The number of reflections varies from 1 to 5, depending on the prism type. The prisms illustrated in Figure 4.11(a) do not cause an offset of the beam of light, but provide image rotation. The prisms in Figure 4.11(b) provide even parity for image erecting and a lateral shift in the direction of light propagation, so the optical axis is not collinear.

4.2.1 Unfolding deviating prisms and tunnel diagrams

In order to interpret the virtual image formed by the deviating prisms (sometimes referred to as fixed mirror reflecting prisms) this book follows the convention of viewing the object while facing the negative z axis. The image is not really at the output of the prism, due to the fact that the eye is doing the observing. The image will be illustrated as if the projected image were produced on a frosted glass normal to the emergent beam.

The 90° deviating right-angle prism shown in Figure 4.12 produces an image with odd handedness or parity. In this projected layout, the observation is made from the positive z direction. The angle of incidence is 45° on a reflecting surface, which is itself at a 45° angle with respect to the horizontal; thus, we arrive at a 90° deviation by summing these two angles. If the prism's index of refraction is about 1.5, which is a good guess for visible glass, then TIR occurs, because 45° is greater than the critical angle necessary for TIR.

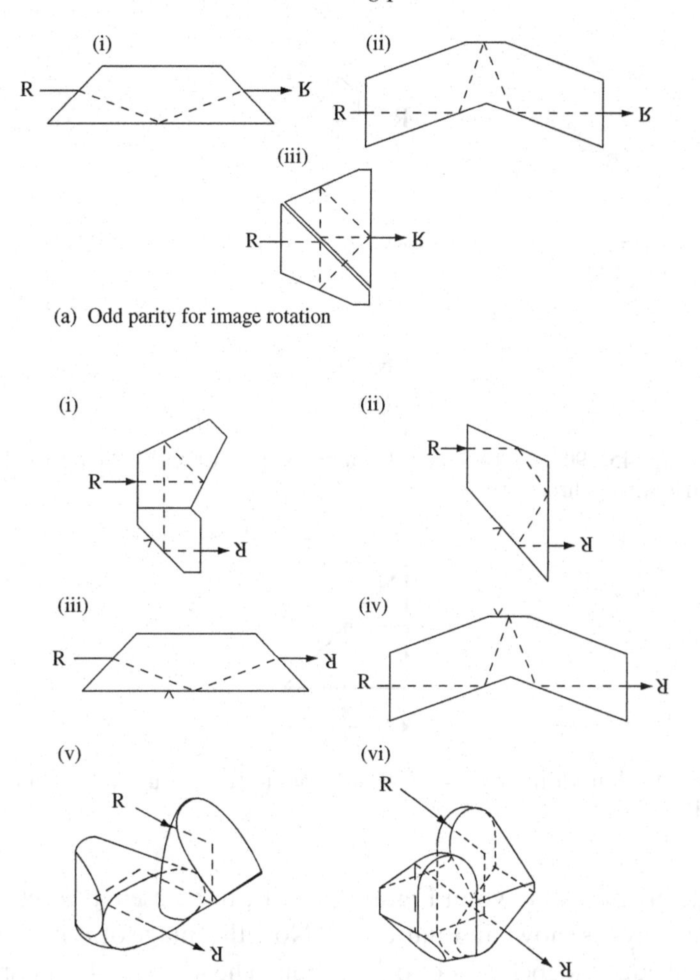

Figure 4.11 Zero angle deviating prisms used to change handedness, with image rotation 2θ for prism rotation θ (looking in the negative z direction): (a) image rotation, (i) dove prism, (ii) reversion, (K) prism, (iii) Pechan prism; (b) image erecting (i) penta-Amici prism, (ii) Leman prism, (iii) erecting dove prism, (iv) Abbe prism, (v) Porro system, (vi) Porro–Abbe system.

The critical angle, as defined in Equation (2.21), is:

$$\sin I_c = 1/1.5$$
$$I_c = 41.8^\circ. \tag{4.4}$$

Therefore, rays hitting a surface at an angle above this value are totally reflected. The right-angle prism shown in Figure 4.9(a) can be used in at least three configurations. In order to bend the ray 90°, this prism can be arranged as a

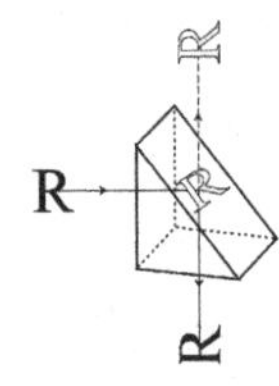

Figure 4.12 45°–90°–45° prism used for 90° deviation and odd parity (looking in the positive z direction).

Figure 4.13 45°–90°–45° prism used as a retro reflector with even parity (looking in the negative z direction).

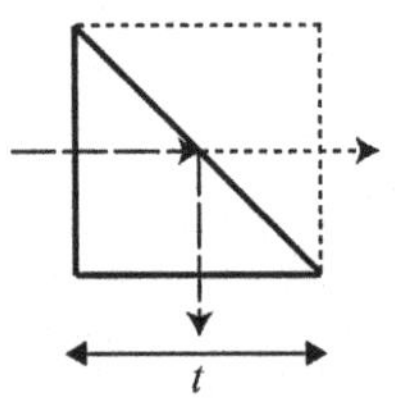

Figure 4.14 Unfolding of a 45°–90°–45° prism to produce a tunnel diagram or PPP.

right-angle prism, as shown in Figure 4.12. It can be used as a retro reflector (180° return ray) as shown in Figure 4.13. Note the image of R is flipped about the dihedral edge independent of orientation. The 45°–90°–45° prism can also be used as a dove prism with no deviation, as shown in Figure 4.11(a)(i).

If we propose that we can unfold the prism around the surface providing the reflection, as shown in Figure 4.14, then we may think of the prism as a PPP, such as those discussed in Section 2.6. This conceptual transformation changes the OPL as well as the direction of energy flow, and results in a tunnel diagram. Tunnel diagrams are shown for penta, Pechan, and Wollaston prisms in Figure 4.15. The unfolded layout, or tunnel diagram, is a PPP which can then be evaluated, as discussed in Chapter 2.

4.2.2 Applications of deviating prisms

One of the most interesting applications of deviating prisms is their use in high resolution color cameras, such as those found in recording studios. Figure 4.16 shows the optical layout of such a camera, which uses dichroic color filters

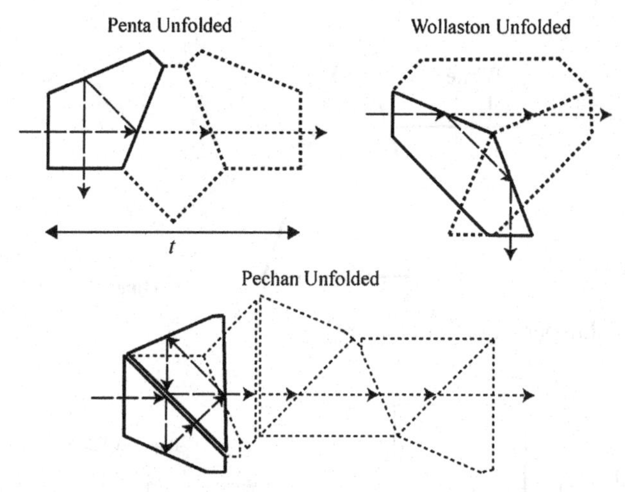

Figure 4.15 Unfolding of a penta, a Wollaston and a Pechan prism to produce a tunnel diagram.

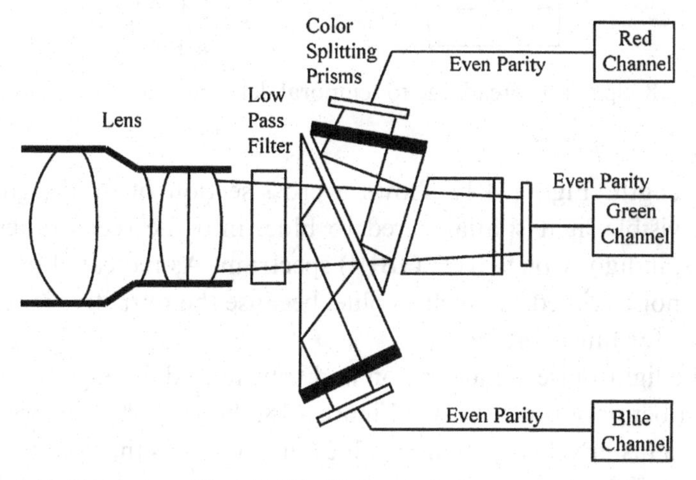

Figure 4.16 Diagram of an RGB high resolution camera illustrating the application of deviating prisms.

between the prisms to produce the necessary color separation into the three channels (red, green, blue (RGB)) needed for modern RGB television signal transmission and display.

4.3 Dispersing prisms

A dispersing prism is a device used to break up light (disperse it in an angular spread) into the spectral colors that make up that light. The various colors have different velocities within media, because the index of refraction varies

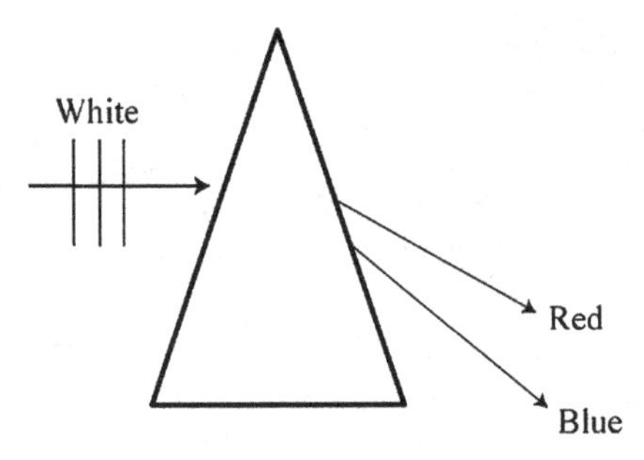

Figure 4.17 Dispersing triangular prism.

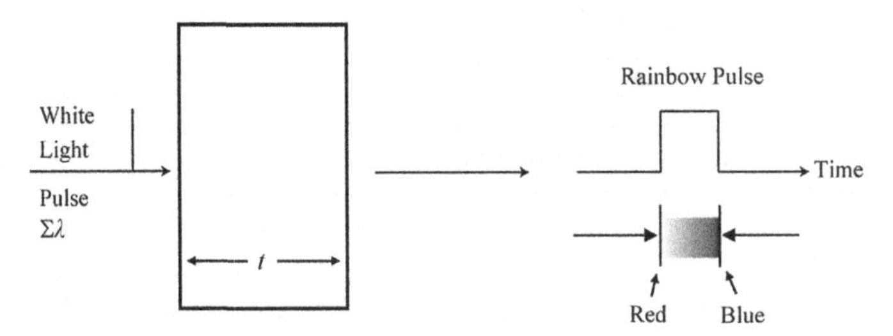

Figure 4.18 Spectral spread due to temporal delay in colors.

with wavelength. Figure 4.17 shows a cross-section of a triangular prism dispersing visible light spatially, red to blue, into the red, orange, yellow, green, blue, indigo, violet (ROYGBIV) spectrum. As indicated in the figure, red light is not deviated as much as blue, because the refractive index is lower for red than for blue light.

If a white light pulse is incident on and transmitted through long pieces of glass (e.g., fiber optic), as shown in Figure 4.18, the red light will move through the glass at a faster velocity than the blue light. Thus, as the light transverses a distance, t, the red wavelength will appear first at the output, and the blue part of the pulse will appear later, yielding a rainbow pulse over time, as shown in the figure. This smear in the pulse, over time, is due to the refractive properties of the glass. This phenomenon is a concern in fiber optic communications where pulse width is important.

Classically dispersing prisms spread the spectrum angularly, as shown in Figure 4.17 for the triangular prism type. However, there are other dispersing prism geometries, such as the double Amici dispersing prism (Figure 4.19(a)), the Pellin–Broca dispersing prism (Figure 4.19(b)), and the Abbe dispersing prism (Figure 4.19(c)). The important difference between dispersing prisms and deviating prisms is that in dispersing prisms the incident ray on the first surface is not at 90°. For all deviating prisms, the input rays are perpendicular to

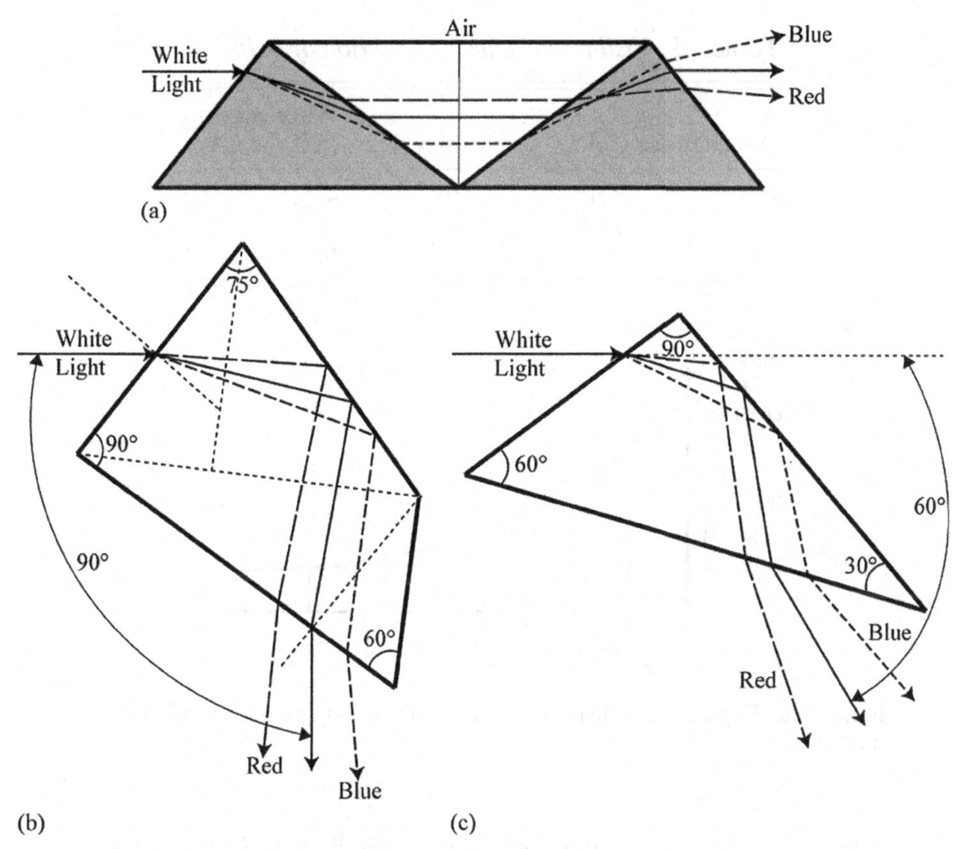

Figure 4.19 Other dispersing prism types: (a) double Amici dispersing prism; (b) Pellin–Broca dispersing prism; (c) Abbe dispersing prism.

the surface, so from Snell's Law the rays for different wavelengths do not spread angularly. The double Amici prism, however, is a direct view spectrum device, so no deviation is introduced. The Abbe dispersing prism has the geometry of a 30°–60°–90° right triangle. In this case, the deviation is nominally 60°, with a spread in spectrum around this value. The Pellin–Broca prism produces a deviation of 90° as well as an angular spectrum. As outlined in dashed lines, the Pellin–Broca prism can be thought of as being made up of two 30°–60°–90° prisms and a 45°–90°–45° prism; however this is just one solid single glass substrate.

4.3.1 Refractive index variation with wavelength

The variation of the index of refraction with wavelength or frequency in the visible spectrum is defined at three specific frequencies or wavelengths, shown in Table 4.1, using the subscript designation of F, d, and C.

Table 4.1. *Visible spectrum description of light*

Designation	Color	Frequency (Hz)	Wavelength (nm vacuum)
F	Blue	$6.172(10^{14})$	486.1
d	Yellow	$5.106(10^{14})$	587.6
C	Red	$4.570(10^{14})$	656.3

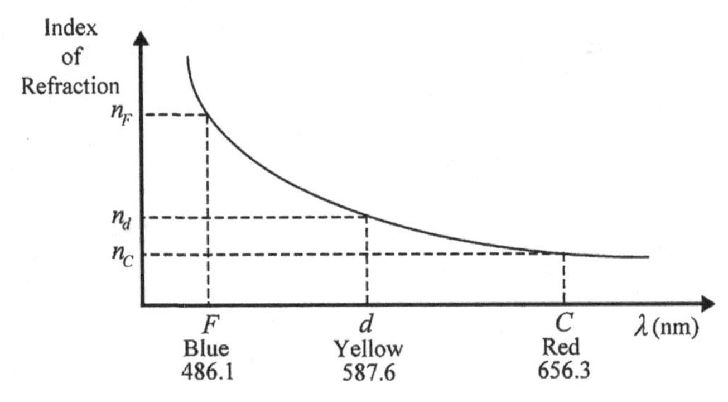

Figure 4.20 Typical glass dispersion curve of the indices of refraction versus wavelength.

The dispersion of a glass is determined by the indices at the three designated wavelengths (*F*, *d*, and *C*). A plot of the refractive index versus wavelength is shown in Figure 4.20. Recall that the frequency of light is constant, and the wavelength depends on the medium in which the light is propagating. These wavelengths are assumed to be in a vacuum or in air (in this text we will not differentiate between the two). The *F* wavelength is for the hydrogen "*F*" emission line. The *d* wavelength is for the helium "*d*" emission line, and the *C* wavelength is for the hydrogen "*C*" emission line. The sodium (Na) emission lines, which are very dominant, are not used because they are a doublet; thus they are ambiguous and not precise. Further discussion on the refractive indices of various glasses will be given in Section 4.4, where it will be shown that each glass type has its own unique indices of refraction for these specified wavelengths of light.

4.3.2 Abbe number ($V^{\#}$)

The Abbe number is a quantitative measure of the average slope of the dispersion curve (refractive index versus wavelength curve; see Figure 4.20).

This slope describes the dispersive characteristics of the individual glass. A flat curve or low slope means that the glass does not disperse the light from red to blue as much as a glass with a larger slope. A steep slope indicates a higher dispersion; thus, the glass spreads the light with a larger angle between blue and red. The Abbe number, sometimes called the glass factor, characterizes this effect and is defined as

$$V^{\#} = \frac{n_d - 1}{n_F - n_C}. \tag{4.5}$$

If the difference in refractive index at the F wavelength and the C wavelength, $n_F - n_C$, is a small value, the Abbe number is large, which indicates a small dispersion and a shallow slope. Conversely, a large dispersion glass has a low Abbe number. The range of values for the Abbe number is from about 20 to 90, with higher values meaning a lower dispersion of the spectrum. Glasses with Abbe numbers ($V^{\#}$) greater than 55 ($V^{\#} > 55$) are classified as crown glasses, and glasses with Abbe numbers less than 50 are called flints.

4.3.3 Deviation for triangular prism

A wedge-shaped piece of glass causes incident light rays to deviate at an angle, as shown in Figure 4.21. The ray deviation (δ_p) is shown for one value of refractive index. Since the refractive index is a function of wavelength, as the wavelength changes so does the deviation angle (δ_p). As a result, the refraction causes the different wavelengths to deviate by different angular amounts; thus, producing a spectrum.

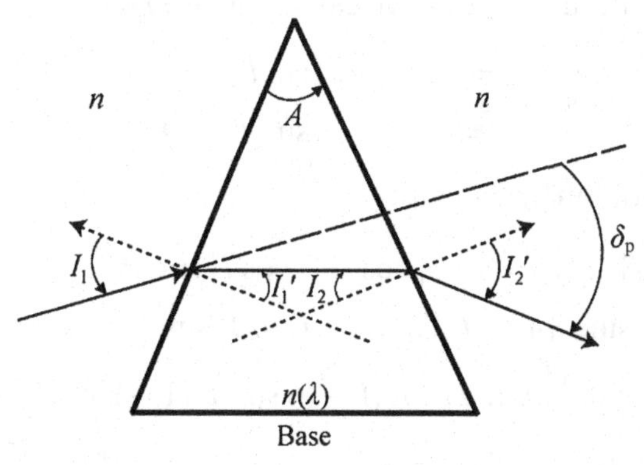

Figure 4.21 Prism deviation of rays.

White light is a superposition of different wavelengths, and the prism causes an angular separation of the incident white light into colors. There are three general observations that can be made about a prism.

(1) The refracted ray bends toward the base.
(2) Shorter wavelengths bend more than longer wavelengths ($n_F > n_C$).
(3) The physical length of the base affects the spread of colors.

In Figure 4.21, let the wedge angle of the prism be called the apex angle, A. The incident light ray at an angle I_1, relative to the normal of side 1 will emerge from side 2 with an angle I_2', relative to its normal (normals are shown as dotted lines). The total angular deflection or ray deviation from its original direction is δ_p. The two refractions at surfaces 1 and 2 follow Snell's law, such that if the prism is in air ($n = 1$):

$$\sin I_1 = n(\lambda)\sin I_1', \tag{4.6}$$

$$n(\lambda)\sin I_2 = \sin I_2', \tag{4.7}$$

where $n(\lambda)$ is the index of refraction as a function of wavelength. The ray deviation, δ_p, caused by the prism, is then given by

$$\begin{aligned} -\delta_\mathrm{p} &= I_1 - I_1' + (-I_2' + I_2) \\ \delta_\mathrm{p} &= -I_1 + I_1' - I_2 + I_2'. \end{aligned} \tag{4.8}$$

From the geometry in Figure 4.21:

$$A = I_1' - I_2. \tag{4.9}$$

Therefore, the ray deviation via substitution is

$$\delta_\mathrm{p} = A - I_1 + I_2'. \tag{4.10}$$

Ideally, the ray deviation should be found in terms of A, I_1, and $n(\lambda)$, and it follows from Equation (4.7) that we can eliminate I_2:

$$\begin{aligned} I_2' &= \sin^{-1}[n(\lambda)\sin I_2] \\ &= \sin^{-1}[n(\lambda)\sin(I_1' - A)]. \end{aligned} \tag{4.11}$$

Recalling the trigonometric identity

$$\begin{aligned} &\sin(I_1' - A) = \sin I_1' \cos A - \cos I_1' \sin A; \\ &I_2' = \sin^{-1}(n \sin I_1' \cos A - n \cos I_1' \sin A), \\ &I_2' = \sin^{-1}\left[n \sin I_1' \cos A - (n \sin A)\left(1 - \sin^2 I_1'\right)^{1/2}\right], \\ &I_2' = \sin^{-1}\left[\sin I_1 \cos A - (\sin A)\left(n^2 - \sin^2 I_1\right)^{1/2}\right]. \end{aligned} \tag{4.12}$$

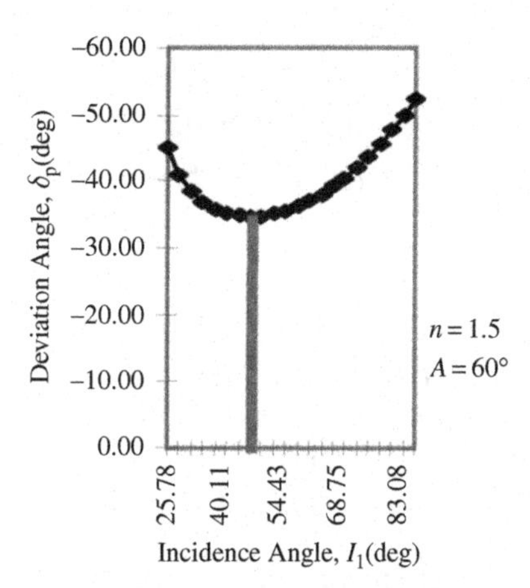

Figure 4.22 Graph of deviation angle versus incident angle.

Substituting Equation (4.10) into Equation (4.8):

$$\delta_p = A - I_1 + \sin^{-1}\left[\sin I_1 \cos A - (\sin A)\left([n(\lambda)]^2 - \sin^2 I_1\right)^{1/2}\right]. \quad (4.13)$$

Since the index of refraction is a function of wavelength ($n(\lambda)$), the deviation is different for different wavelengths or colors of light. Equation (4.13) is the general expression for ray deviation through a prism of apex angle A and incident angle I_1.

If the deviation angle is plotted against incident angles, there is a range of deviation angles that are impossible to obtain. A plot of deviation angle (δ_p) versus incident angle is shown in Figure 4.22 for a prism with a fixed index of refraction. There is a minimum deviation for each index of refraction for a fixed apex angle. Therefore, if one does an experiment with collimated light on a prism while changing the incident angle (I), the deviation angle will increase to a maximum value, and then it will start to become smaller. This provides a means of measuring the index of refraction for an unknown glass.

4.3.4 *Minimum deviation for a triangular prism*

The minimum deviation with respect to the angle of incidence can be found by differentiating Equation (4.13) and setting the result to zero. However, it is much more convenient to differentiate Equations (4.9) and (4.10) to find this minimum:

$$\begin{aligned} \delta_p &= A - I_1 + I_2', \\ \frac{d\delta_p}{dI_1} &= \frac{dA}{dI_1} - 1 + \frac{dI_2'}{dI_1} = 0, \end{aligned} \quad (4.14)$$

since

$$\begin{aligned} dA/dI_1 &= 0, \\ dI_2'/dI_1 &= 1. \end{aligned} \tag{4.15}$$

From Equation (4.9):

$$\frac{dA}{dI_1} = \frac{dI_1'}{dI_1} - \frac{dI_2}{dI_1} = 0, \tag{4.16}$$

$$\frac{dI_2}{dI_1'} = 1. \tag{4.17}$$

Differentiating Snell's law, via Equations (4.6) and (4.7), at each surface of the prism shown in Figure 4.21:

$$\cos I_1 \, dI_1 = n(\lambda) \cos I_1' \, dI_1', \tag{4.18}$$

$$n(\lambda) \cos I_2 \, dI_2 = \cos I_2' \, dI_2'. \tag{4.19}$$

Now divide Equations (4.18) and (4.19) with the appropriate transpose, and substitute Equation (4.15) and (4.17) to yield

$$\frac{\cos I_1}{\cos I_2'} = \frac{\cos I_1'}{\cos I_2}. \tag{4.20}$$

Using the Pythagorean identity $\sin^2 u + \cos^2 u = 1$ and Snell's law directly:

$$\frac{\sqrt{1 - \sin^2 I_1}}{\sqrt{1 - \sin^2 I_2'}} = \frac{\sqrt{1 - \sin^2 I_1'}}{\sqrt{1 - \sin^2 I_2}} \tag{4.21}$$

$$\frac{\sqrt{1 - \sin^2 I_1}}{\sqrt{1 - \sin^2 I_2'}} = \frac{\sqrt{n(\lambda)^2 - \sin^2 I_1}}{\sqrt{n(\lambda)^2 - \sin^2 I_2'}} \tag{4.22}$$

$$\frac{1 - \sin^2 I_1}{1 - \sin^2 I_2'} = \frac{n(\lambda)^2 - \sin^2 I_1}{n(\lambda)^2 - \sin^2 I_2'}. \tag{4.23}$$

The left-hand side of Equation (4.23) cannot be equal to the right-hand side except when:

(1) $n(\lambda) = 1$, which is trivial and non-existent; or
(2) $I_1 = -I_2'$ for a prism,

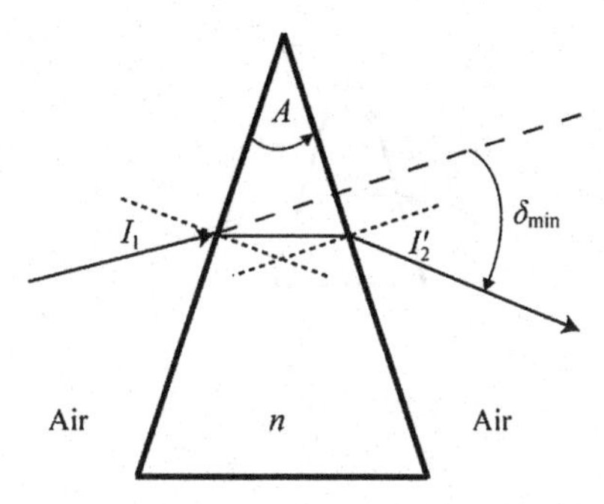

Figure 4.23 Thin prism.

so

$$I_1 = -I_2' \tag{4.24}$$

and therefore

$$I_1' = -I_2. \tag{4.25}$$

Using Equation (4.13) with these constraints:

$$\delta_{\min} = A - 2\sin^{-1}\left[n\sin\left(\frac{A}{2}\right)\right]. \tag{4.26}$$

For minimum deviation, the angle of incidence on the first surface is equal to the angle of incidence on the second surface. This also forces the ray inside the prism to be parallel to the base of the prism (from geometry).

A useful equation for determining the index of refraction of an unknown material by measuring the apex angle (A) and ray direction can be derived by rewriting Equation (4.26) as follows:

$$n = \frac{\sin\left[\frac{1}{2}(A - \delta_{\min})\right]}{\sin(A/2)}. \tag{4.27}$$

A thin prism, such as the one shown in Figure 4.23, is defined as having a small apex angle ($\sin A \approx A$). Using the small angle approximation, the deviation angle is

$$\delta_{\min} = -A(n-1). \tag{4.28}$$

4.3.4.1 Minimum deviation of a thin prism for different wavelengths

The spread in deviation angle versus wavelength can now be determined as the difference in ray deviation, as shown in Figure 4.24. The expression for the spectral spread for a thin prism is:

$$\mathrm{d}\delta_{\min} = -A\mathrm{d}n \tag{4.29}$$

$$= \frac{\delta_{\min}}{n-1}\mathrm{d}n, \tag{4.30}$$

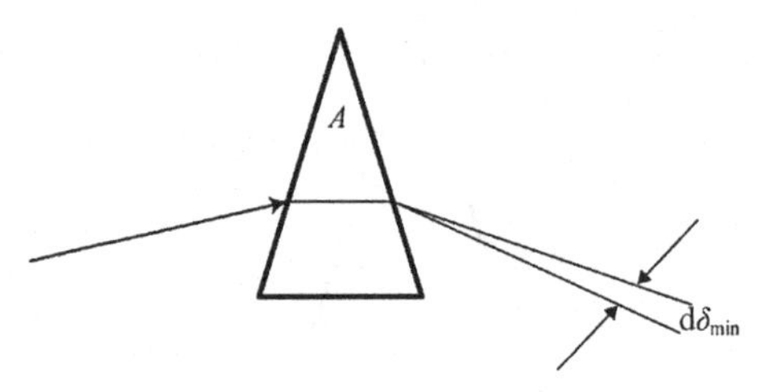

Figure 4.24 Difference in ray deviation.

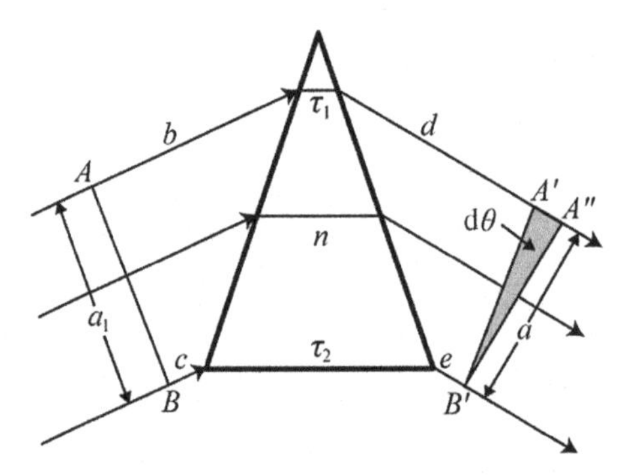

Figure 4.25 Plane wavefront emerges as a plane wave for each wavelength.

where we define the minimum deviation for d light as δ_{min}, and the differential index, dn, is just the refractive index variation from F to C light ($n_F - n_C$):

$$\text{d}\delta_{\text{min}} = \delta_{\text{min}} \frac{n_F - n_C}{n_d - 1}. \tag{4.31}$$

Using Equation (4.5) for the Abbe number:

$$\text{d}\delta_{\text{min}} = \delta_{\text{min}} / V^{\#}. \tag{4.32}$$

This is the deflection, or spread in angle, of F and C light around the minimum deviation of d light (another way of describing the angular spread of the rainbow of colors).

4.3.4.2 Prism base size effects

Consider a plane wavefront AB passing through a prism symmetrically, as shown in Figure 4.25. It passes through the prism and emerges as a plane wave for each wavelength.

Setting up the optical path lengths (OPL) from A to A' and B to B', which must be equal, and A to A'' for the different wavelengths:

$$b + n\tau_1 + d = c + n\tau_2 + e, \tag{4.33}$$

$$b + (n + \delta n)\tau_1 + d + a\,d\theta = c + (n + \delta n)\tau_2 + e, \tag{4.34}$$

and subtracting the two *OPL*s:

$$\delta n \tau_1 + a\,\mathrm{d}\theta = \delta n \tau_2$$
$$a\,\mathrm{d}\theta = \delta n(\tau_1 - \tau_2). \tag{4.35}$$

Then if δn is interpreted as the differential index, dn,

$$\frac{\mathrm{d}\theta}{\mathrm{d}n} = \frac{\tau_1 - \tau_2}{a}. \tag{4.36}$$

If $\tau_1 = 0$, then the prism just fills the plane height a_1, the plane wave out has a cross-sectional height of a, and τ_2 equals the base dimension of the prism. The variation of angle with wavelength can be written, using the above equation as

$$\frac{\mathrm{d}\theta}{\mathrm{d}\lambda} = \frac{\mathrm{d}\theta}{\mathrm{d}n}\frac{\mathrm{d}n}{\mathrm{d}\lambda}, \tag{4.37}$$

$$\frac{\mathrm{d}\theta}{\mathrm{d}\lambda} = \frac{\tau_2}{a}\frac{\mathrm{d}n}{\mathrm{d}\lambda}. \tag{4.38}$$

If we approximate the variation of the index of refraction versus wavelength shown in Figure 4.20, using the formula

$$n = k_1 + \frac{k_2}{\lambda^2}, \tag{4.39}$$

where k_1 and k_2 are constant for a given glass, the differential of the refractive index with respect to wavelength is then

$$\frac{\mathrm{d}n}{\mathrm{d}\lambda} = -\frac{2k_2}{\lambda^3}. \tag{4.40}$$

Now, the angular dispersion in Equation (4.36) can be written as

$$\frac{\mathrm{d}\theta}{\mathrm{d}\lambda} = -\frac{\tau_2}{a}\frac{2k_2}{\lambda^3} = -\frac{\tau_2}{\lambda^3}\frac{2k_2}{a}. \tag{4.41}$$

Therefore, angular dispersion is proportional to the base size τ_2 of the prism and inversely proportional to the wavelength cubed $(1/\lambda^3)$. Therefore, blue light is more angularly dispersed than red light, as was shown in Figure 4.17.

4.3.5 Prism pairs

The use of two cascaded prisms can provide a means to obtain two important optical situations: (1) deviation without dispersion, and (2) dispersion without deviation. The pair consists of a crown prism and a flint glass prism. For the

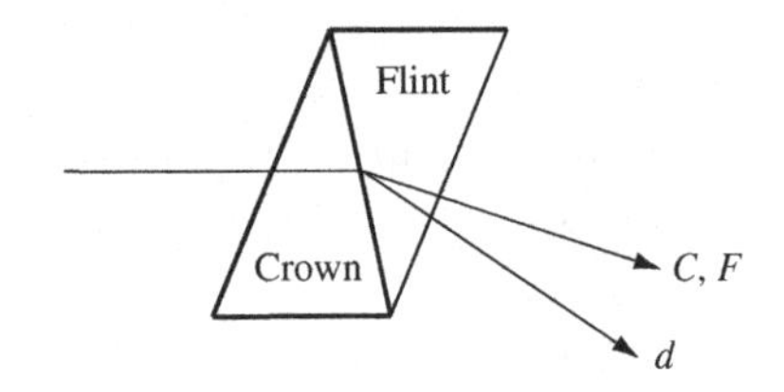

Figure 4.26 Deviation without dispersion (chromatic correction for C and F light).

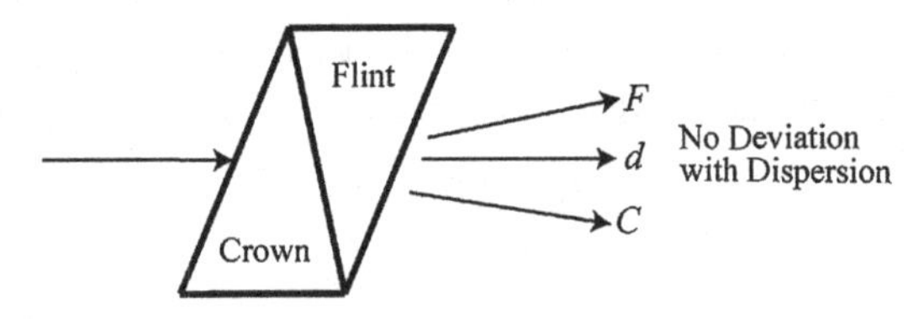

Figure 4.27 Dispersion with no deviation of d light.

case of no dispersion of C and F wavelengths, as shown in Figure 4.26, this suggests:

$$\delta_C^{\text{Crown}} + \delta_C^{\text{Flint}} = \delta_F^{\text{Crown}} + \delta_F^{\text{Flint}}, \tag{4.42}$$

$$\delta_d = \delta_d^{\text{Crown}} + \delta_d^{\text{Flint}}. \tag{4.43}$$

For the case of no deviation of d light, as shown in Figure 4.27, the deviation of the crown prism is compensated for by an equal but opposite deviation by the flint prism. Often it is necessary to linearize the deviation as a function of wavelength. Appendix A gives an example of a linearized pair of prisms.

4.4 Glass

Glass is an amorphous, solid, silicon-based material used for its transparent optical properties as well as to make containers and for art décor. Glass has a transparent hard surface which is very smooth and, in most cases, impervious to chemical and biological material. It is composed mostly of silicon dioxide (SiO_2). Its salient property in optics is its transparency, which is due to the absence of homogeneity and lack of electronic energy states in its atomic structure. Window glass and most optical glasses transmit across the visible wavelengths (400–700 nm) with greater than 95% transmission. In the ultraviolet (300–400 nm), there is some absorption, and in the vacuum ultraviolet (wavelengths shorter than 300 nm), glass basically blocks all radiation. Fused quartz, which is pure silicon dioxide (SiO_2), transmits in the ultraviolet. The additives or other compounds mixed with SiO_2 are the cause of the lack of ultraviolet transmission. Amorphous silicon dioxide (also referred to as "silicon native oxide") is used in the manufacture of solid state imagers, such as charge coupled devices (CCDs) and complementary metal oxide silicon (CMOS) imagers. It acts as an insulator for integrated circuits in silicon substrates because it is electrically neutral and is readily fabricated onto the silicon.

Glass has been around for over 5000 years. It was developed by the Phoenicians about 3000BC. Glassmakers learned to color the glass using different metallic additives. The Roman Empire spread glass fabrication techniques throughout Europe. From the Renaissance to the present day, the island of Murano, near Venice, Italy has been the home of glass blowers that fabricate very beautiful and expensive Venetian glass.

4.4.1 Chemical composition

Typically, other substances are added to the silicon dioxide (SiO_2) to lower the melting point of the glass to around 1100 °C. Quartz (pure SiO_2) has a melting point close to 2000 °C. Only a limited number of inorganic oxides are available for glass making that will lower the temperature without affecting other properties (see Table 4.2).

The various constituents are added to SiO_2 in different combinations; however, the amount of SiO_2 is not less than approximately 60–75% of the total composition. The proportion of the constituents added to SiO_2 varies the optical properties of interest. Hundreds of different glasses can be produced by various combinations, each with different indices of refraction and Abbe numbers.

Table 4.2. *Constituents added to SiO_2 when making glass*

Na_2O
K_2O
CaO
MgO
BaO
PbO
B_2O_3
Al_2O_3

Probably the most common glass is soda-lime glass made with Na_2O or K_2O and CaO. The soda lowers the melting point of SiO_2, and the lime makes the glass more durable to chemical reduction. This glass is used in windows, mason jars, and eyeglasses.

Lead-alkali glasses are made by replacing the lime (CaO) with lead oxide (PbO). This lowers the melting point below that of soda-lime glass. This technique provides glass blowers with a viable artistic glass. Lead glasses, called flint glasses, are more colorful because of their higher refractive indices and greater dispersion.

Borosilicate glass was developed for its lower coefficient of thermal expansion. In general, it has a higher melting temperature than soda-lime or lead-alkali silicate glasses. These are the main types of glasses that are in use today.

Color in glass is produced by adding a metal to the constituents. The quantity of metal added is about 3–4%. The additive produces color either by scatter from these particles or by absorption of certain wavelengths of light. For example, iron absorbs wavelengths in the infrared spectrum, so such ferrous glasses are used for heat rejection in intense light projectors. Table 4.3 lists some metal dopants and the corresponding color of glass produced.

4.4.1.1 Crown glasses

Crown glasses are produced from soda-lime silicates with about 10% CaO or 10% K_2O. They have a low index of refraction (<1.6) and low dispersion (Abbe number, $V^{\#} > 55$). A common Schott glass, N-BK7, is a crown glass used in precision lenses ($n_d = 1.517$, $V^{\#} = 64.7$). Borosilicates, which are crown glasses, have good optical and mechanical properties, as well as resistance to

Table 4.3. *Glass dopants producing color*

Metal	Color of glass
Nickel	Purple
Cobalt	Blue
Chromium	Green
Uranium	Greenish yellow
Ferrous iron	Green (infrared absorber)
Gold	Red
Selenium	Red (most common)
Sulfur	Yellowish
Manganese	Purple
Copper	Turquoise
Cadmium	Yellow
Silver	Orange

chemical damage. Additives that make crown glasses are zinc oxide, phosphorous pentoxide, barium oxide, and fluorites.

4.4.1.2 Flint glasses

Lead-alkali glasses are called "flint glasses." Flint glasses have a relatively high index of refraction ($n_d > 1.6$) and a high dispersion (low Abbe number, $V^{\#} < 50$). Historically, flint glasses contained lead oxide (PbO); however, due to lead's pollution effects, titanium oxide or zirconium oxide are used in modern flints to obtain the same optical properties. Flint glasses are often used in rhinestone jewelry because they glitter brilliantly and mimic diamonds.

Optical glasses are categorized by using six or nine digit glass numbers or by the letter–number code, initiated by Schott, a major German optical glass company, in the latter part of the nineteenth century. The glass number nomenclature uses the digits after the decimal point of the index of refraction of *d* light for the first three digits, and the Abbe number to the first decimal place as the second three digits. If the nine-digit nomenclature is used, the last three digits represent the density in grams per cubic meter. For example, if a glass has a *d* light index of 1.523 ($n_d = 1.523$) and an Abbe number of 58.8, with a density of 3.23 g cm^{-3}, you will find it described either as 523588 (a six-digit glass number) or 523588.323 (a nine-digit glass number).

Alternatively, a letter–number code that indicates its composition and whether it is crown or flint can be used. For example, a crown borosilicate would be called N-BK7; where the B and K are used to represent the words boron and crown in German (i.e. bor and krone). The code SF4 would indicate a high dispersion flint, where S and F represent silicon (silizium) and flint (feuerstein). The glasses are typically categorized by composition, index of refraction, Abbe number, and as crowns or flints.

4.4.2 Glass charts and plots

Three plots are very important for an elementary understanding of optical glasses used for lenses. They are:

(1) the dispersion curve, or index of refraction vs. wavelength (we have already highlighted this in Section 4.3);
(2) the glass chart, in which the refractive indices for *d* light of all glasses are plotted against the Abbe number; and
(3) the secondary color plot of partial dispersion versus Abbe number.

These three plots contain all the information needed to develop first-order optical designs. Each glass type (e.g. N-BK7, PSK56) has a unique dispersion curve of n versus λ (Figure 4.20). As stated earlier, depending on the composition, there are an infinite number of different glasses. However, glass manufacturers have limited their production to only a few types, and try to reproduce the characteristics for a particular type in each melt of glass they fabricate. There is some error ($< \pm 0.002$) in these characteristics within a given glass from one melt to another. The homogeneity within a melt is better than $1(10^{-4})$ for the index of refraction. Due to environmental and toxicity concerns, the number of available glasses has been declining. Presently about 200 glass types are available.

There are several dispersion formulae that fit the glass dispersion curve over a limited spectral range, which includes the visible region (0.4–0.7 μm) (Born and Wolfe, 1959). The formula most frequently used is the Sellmeier formula (Fischer and Tadic-Galeb, 2000), given below, which is valid from 0.36 μm to 1.5 μm, where the wavelength (λ) is in microns and the six coefficients are given by the glass manufacturer for that glass type:

$$n^2 - 1 = \sum_{j=1}^{3} \frac{a_j \lambda^2}{\lambda^2 - b_j}. \tag{4.44}$$

This equation is accurate to $1(10^{-5})$ for the indexes of refraction in the wavelength range 0.36–1.5 μm for a given melt. Table 4.4 lists constants for the Sellmeier formula for some common Schott glasses, along with other parameters.

Figure 4.28 shows a glass chart produced by the Schott Company, showing all of their glasses. Other glass manufacturers produce similar charts. Each glass represents a point on this glass chart, and the chart is a plot of increasing refractive index versus increasing dispersion. Note that the Abbe number is plotted in descending values. This is counterintuitive with the Abbe values starting at 90 at the origin and decreasing to 20. The reason for this is the lower the dispersion (spread in colors or wavelengths), the higher the Abbe number. So a very high dispersive glass (one that breaks the rainbow into a large angular spread) would have a low Abbe number ($V^{\#}$). Many of the glasses in Figure 4.28 have been discontinued, so they may not be available from Schott in the future.

In Figure 4.28, the region bounded by the two lines forming the shape of a boomerang is called the glass line. The glasses in this region are typically called the standard glasses. They follow the general trend of low refractive index and low dispersion or high refractive index and high dispersion. The glasses in the visible region vary in refractive index from about 1.45 to 2.0, while the Abbe numbers run from 20 to 95, as shown in Figure 4.28. Typically, when using

Table 4.4. *Physical constants of five Schott glasses*

Parameter	N-BK7	N-SF11	N-LASF9	N-BAKI	F2
n_d	1.516 80	1.784 72	1.850 25	1.572 50	1.620 04
Melt to melt index tolerance	$\pm 10^{-3}$	$\pm 10^{-3}$	$\pm 10^{-4}$	$\pm 10^{-3}$	$\pm 10^{-3}$
Abbe number ($V^{\#}$)	64.17	25.76	32.17	57.55	36.37
Constants of formal dispersion					
a_1	1.039 612 12	1.738 484 03	2.000 295 47	1.123 656 62	1.345 353 59
a_2	0.231 792 344	0.311 168 97	0.298 926 89	0.309 276 85	0.209 073 18
a_3	1.010 469 45	1.174 908 71	1.806 918 43	0.881 511 96	0.937 357 16
b_1	0.006 000 699	0.013 606 86	0.012 142 6	0.006 447 43	0.009 977 44
b_2	0.020 017 914	0.061 596 05	0.053 873 62	0.022 228 44	0.047 045 08
b_3	103.560 653	121.922 711	156.530 829	107.297 751	111.886 764
Density g cm^{-3}	2.51	4.74	4.41	3.19	3.60
Glass number	517642.251	785258.474	850322.441	572576.319	620364.360

(Schott Glass website, http://www.us.schott.com)

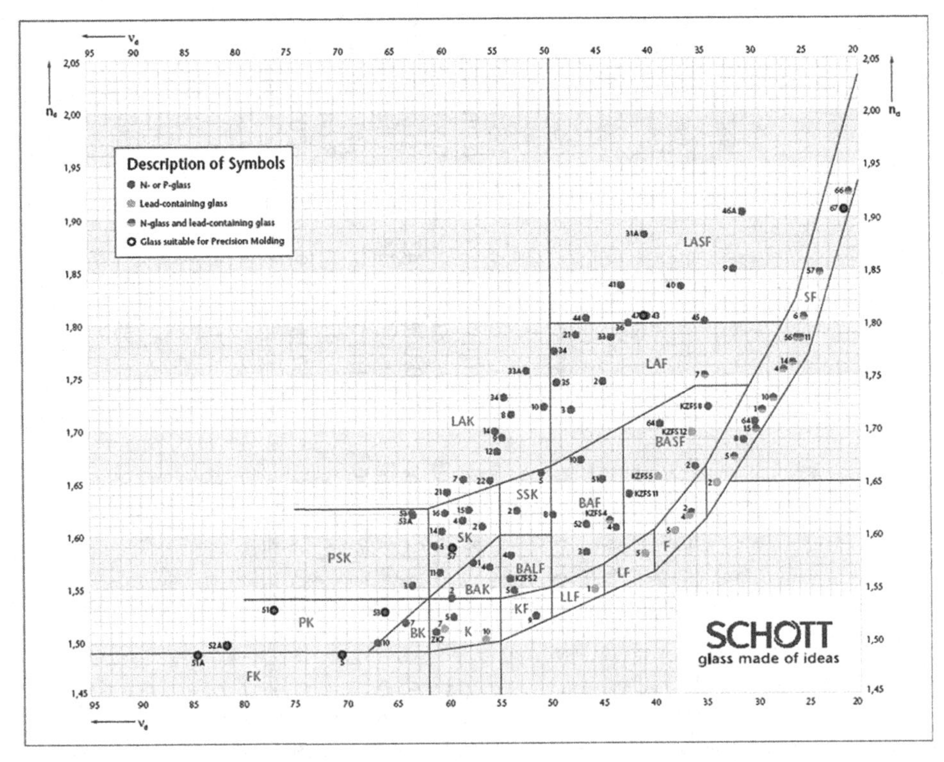

Figure 4.28 Schott Abbe diagram (Schott Glass website, http://www.us.schott.com/).

multiple lenses of different glasses, a large difference in Abbe numbers, such as PK, FK, PSK, SF, F, is preferred to correct color aberrations.

It is usually cost effective to use standard glasses in a design since special glasses such as LaSFN30 or KZFSN4 may be 10–50 times more expensive than N-BK7 glasses. However, there are cases where non-standard glasses are required for secondary color aberration correction. It is always best to use the preferred glasses. Standard glasses are generally held in stock, while non-standard glasses usually require a special order.

Another important plot is the partial dispersion chart for glasses. The partial dispersion, as given in Section 4.3, is

$$p = \frac{n_d - n_C}{n_F - n_C}. \tag{4.45}$$

A plot of partial dispersion versus Abbe number is shown in Figure 4.29 for selected Schott and Ohara common glasses. The standard glasses fit a straight line, as shown in Figure 4.29, with a slope of about 1/2400.

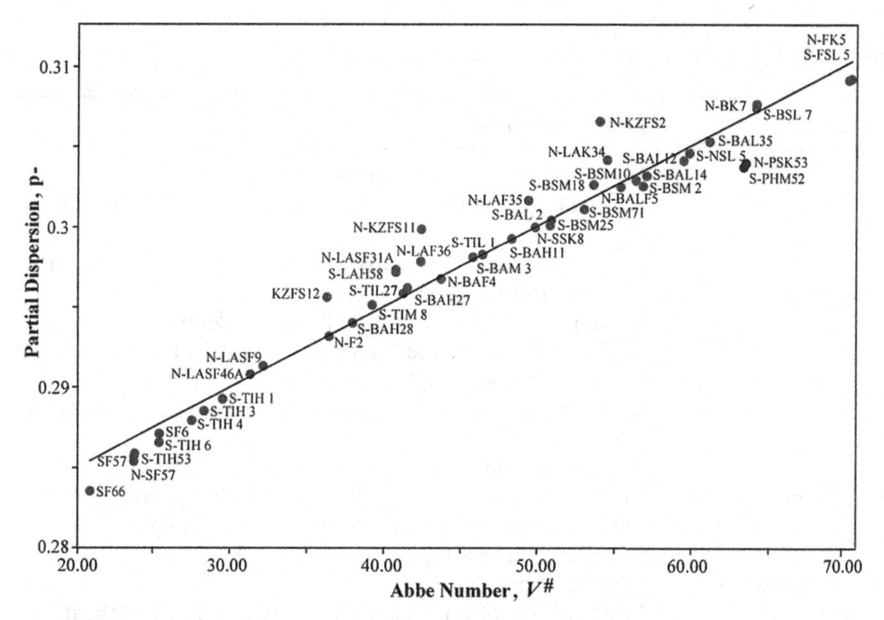

Figure 4.29 Abbe number versus partial dispersion for selected Schott and Ohara common glasses.

4.5 Plastic optical materials

Plastic lenses have become very popular in the last few years, especially for eyeglasses, where weight is a major concern. Other advantages of plastic are its low cost of materials and fabrication, impact resistance, and flexible molding. The mechanical mounts can be integrated into the lens assembly.

The disadvantages of plastic optical materials are:

- low heat resistance,
- surfaces are less durable,
- limited number of materials (see Table 4.5),
- high temperature coefficient of expansion.

The refractive index for plastic varies with temperature about 50 times more than that of glass. Moreover, the refractive index decreases with temperature in plastics, while it increases with temperature in standard glasses. Table 4.5 lists some commonly used plastics with their characteristics (Fischer and Tadic-Galeb, 2000).

Table 4.6 shows some common optical molding resins for plastic optics (Beich, 2002).

Table 4.5. *Commonly used plastic materials*

Plastic type	Characteristics
Acrylic	Most common and important, low cost Good clarity and very good transmission in the visible range High Abbe number (55.3) Easy to machine and polish; good for injection molding
Polystyrene	Cheaper than acrylic Higher absorption than acrylic in the deep blue region Higher refractive index than acrylic (1.59) but has a lower Abbe number (30.9) Lower resistance to UV than acrylic. Scratches more easily than acrylic Acrylic and polystyrene make an achromatic pair
Polycarbonate	More expensive than acrylic, but very high impact strength Performs well over a broad temperature range Poor scratch resistance
COC (cycloolefin copolymer)	Similar to acrylic Water absorption is much lower, and it has a high heat distortion temperature Brittle

Table 4.6. *Common optical molding resins*

Resin	Refractive index
Acrylic (PMMA or polymethyl methacrylate)	1.49
Styrene	1.59
Polycarbonate	1.58
Topas (cycloolefin copolymer)	1.53
Zeonex (cyclo-olefin polymer)	1.52
NAS (methyl methacrylate styrene)	1.533–1.567
SAN (styrene acryonitrile)	1.567–1.571

Problems

4.1 In order for a 6 ft tall man to see himself from head to toe, how tall does a mirror need to be when placed at the following distances from the man:

(a) 6 ft;

(b) 10 ft;

(c) 20 ft?

4.2 What are the glass numbers for:
   (a) N-BK7;
   (b) SF11;
   (c) LaSFN9?
4.3 Which glass has a higher dispersion in the visible spectrum, crown or flint?
4.4 For a penta prism:
   (a) Draw the layout of the tunnel diagram to get a PPP.
   (b) If the input surface of the penta prism is 1 cm wide, what is the total *OPL*, if the glass index is 1.5?
4.5 The new "Back-Saving Horizontal Reading Glasses" for watching television while lying down are being sold. As a student of geometrical optics, who is studying prisms, you should be able to answer the following questions.
   (a) What does the optical diagram for using a right angle prism to view the television look like?
   (b) How do these eyeglasses work, taking into consideration the parity and handedness of the image?
4.6 What is the parity for an image which has traversed:
   (a) a penta prism;
   (b) a dove prism;
   (c) a corner cube?
4.7 Calculate the angle of deviation for *F*, *d* and *C* light for a prism made of SF4 (755276) glass with a 22° apex angle and 30° angle of incidence.
4.8 Derive the expression for the angle of deviation for a thin prism.
4.9 What is the handedness (parity) and orientation of an image seen through the following prisms? How will the image be changed when looking though the prism at the letter R?
   (a) Abbe prism.
   (b) Penta prism.
   (c) Right angle prism:
      (1) Light reflected by the hypotenuse.
      (2) Light reflected at the two smaller faces (Porro).
4.10 For the setup below, what is the correct orientation and description of the R in the final space?

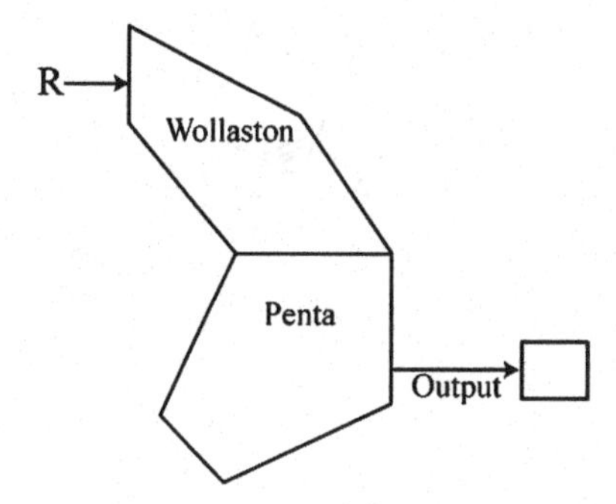

4.11 A student has two glass rods. Rod *A* is # 300800 glass and rod *B* is # 900400 glass.
   (a) What is the ratio of the velocities of light in glass *A* and glass *B*?
   (b) In which rod is the velocity of light slower, *A* or *B*?
   (c) Which is the crown glass?

4.12 There are two pieces of glass, each 10 cm in length. Glass *A* has a glass number of 200700 and glass *B* has a glass number of 600350.
   (a) What is the ratio of the velocities of light in glass *A* and glass *B*?
   (b) In which glass is the speed of light higher?
   (c) Which is the flint glass?

4.13 Plot the deviation angle (δ) for a prism made of F2 glass with an apex angle of 25° versus the incident angle (I) using *d* wavelength light.
   (a) What is the minimum deviation for *d* light?
   (b) What is the minimum deviation for *d* light if we assume a thin lens?

4.14 For SF11 glass, what is the index of refraction calculated to five decimal places for the following wavelengths:
   (a) 400 nm;
   (b) 500 nm;
   (c) 600 nm;
   (d) 700 nm?

4.15 For the following optical systems, what is the orientation of the letter at the observation plane?

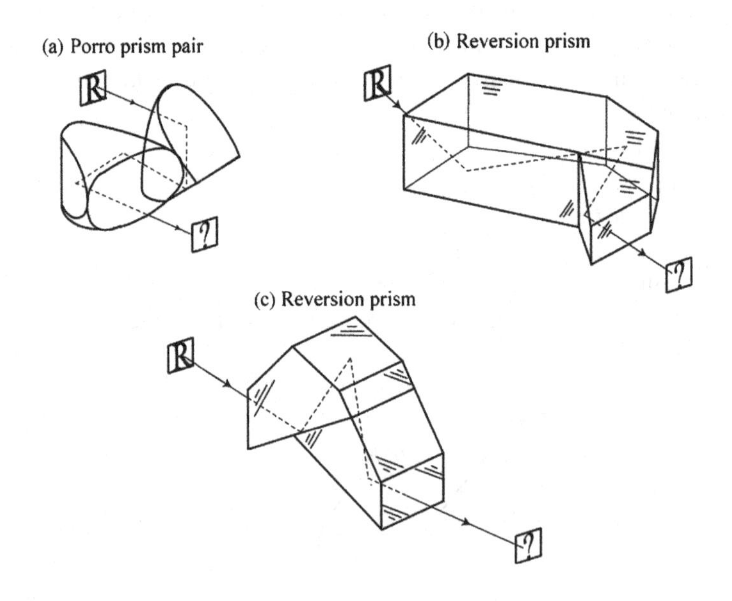

4.16 What is the minimum refractive index that a right-angle prism must have to reflect light by 90° with TIR?

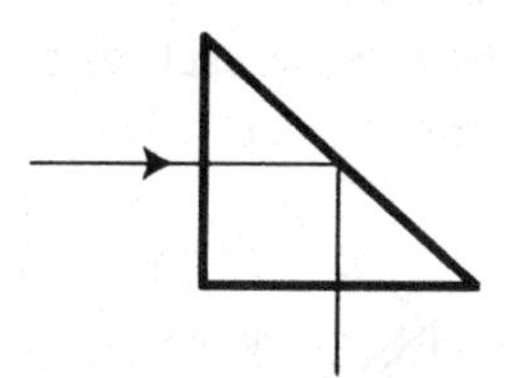

4.17 For the drawing below using an isosceles glass prism ($n = 1.5$), the apex angle (2.5°) is such that the ray is parallel to the axis inside the prism. What is the optical path length for the ray? What is it along the axis?

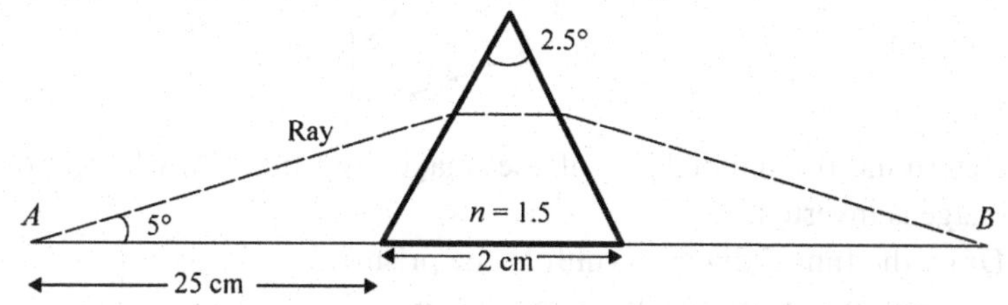

4.18 What are the Abbe numbers and glass numbers for:
(a) SF59;
(b) F14;
(c) BK-1;
(d) BaK4;
(e) FK54?

4.19 A crown (BaK4) thin prism with an apex angle of 15° is to be combined with a flint prism (SF12) so as to produce no net deviation for d light.
(a) Find the apex angle for the contact flint prism.
(b) Find the angular deviation for C light for this prism combination.

4.20 A hollow (and empty) 60° apex angle is immersed in a liquid of refractive index 1.74. What is the angle of minimum deviation?

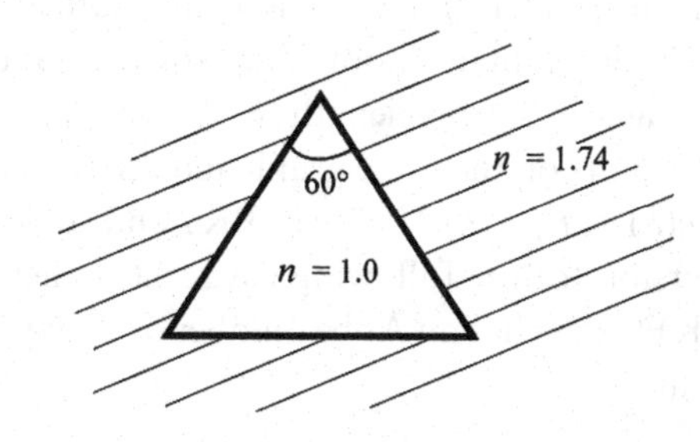

4.21 A glass prism with $n = 4.62$ has an angle of minimum deviation of 48.2°. What is the apex angle?

4.22 A 5 ft tall lady would like to buy a mirror to use while dressing. She requires that it be large enough so that she can see her shoes and hair simultaneously. If she sets the mirror at distances of: (a) 3 ft, (b) 5 ft, and (c) 10 ft away, what is the smallest mirror she should obtain for each case?

4.23 A Pechan prism has the following side view:

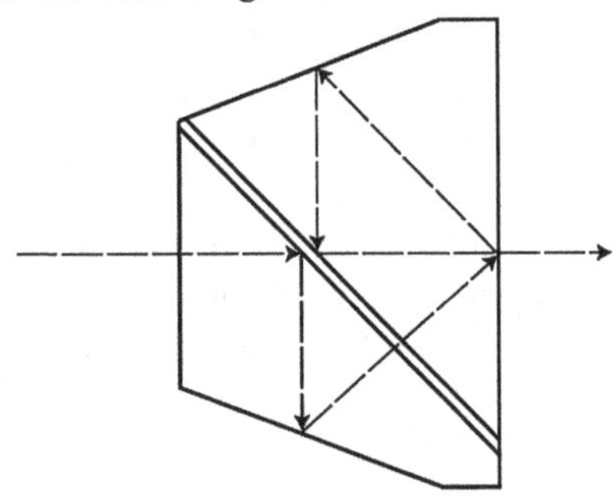

(a) Determine the handedness of the image, and indicate whether or not the image is inverted.
(b) Draw the tunnel diagram (unfold the prism).
(c) Can you see any applications for this prism?

4.24 Plot the deviation angle (δ) versus incident angle (I_1) for a prism with an apex angle of 50°, made of N-BK7 glass, using *d* light. At what angle does the minimum deviation occur?

4.25 A prism with an apex angle of 90° is shown in the figure below. A ray enters this prism at 30° and exits at 50°. What is the index of refraction of the prism?

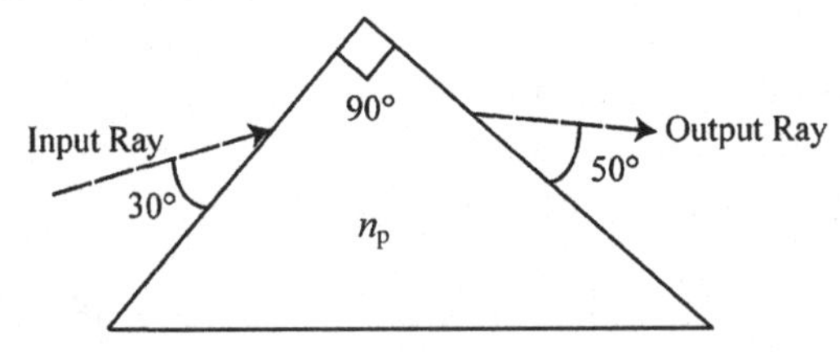

4.26 Given a prism with an apex angle $A = 6.3°$ and an Ohara glass number of 497816, what is the angle of minimum deviation, δ_{min}, for *C*, *d*, and *F* light?

4.27 Plot the index of refraction vs. wavelength (*F*, *d*, *C* only) for N-BK7, N-FK5, and F5 glasses. Plot all points on the same graph and use wavelength in nanometers. Use the equation $n(\lambda) = K_1 + K_2/\lambda^2$ for N-BK7 and solve for K_1 and K_2.

4.28 Describe the dispersion for the following glasses as either higher or lower than N-KF9 glass. (N-KF9 glass has an Abbe number of about 52). Classify each type as a crown or a flint:
(a) N-FK5;
(b) N-BK7;
(c) F5;
(d) N-LaSF45.

4.29 For a prism made of F2 glass with an apex angle of 25°, plot the deviation angle (δ) versus incident angle (I) using d wavelength light, C light, and F light.
(a) What is the angular spread of C to F light rays for an incident angle of 35°?
(b) What is the angular spread of d to F light rays for an incident angle of 35°?
(c) What is the minimum deviation for d light?
(d) What is the minimum deviation for d light if we assume a thin lens?

4.30 For a 3 in prism made of N-BK7 glass, with a 3 in base and height of 4 in, compare the dispersion at F, C and d light. (Hint: Use K_2 from problem 4.27.)

4.31 For the following deviating prisms, draw the correct orientation of the object, R.

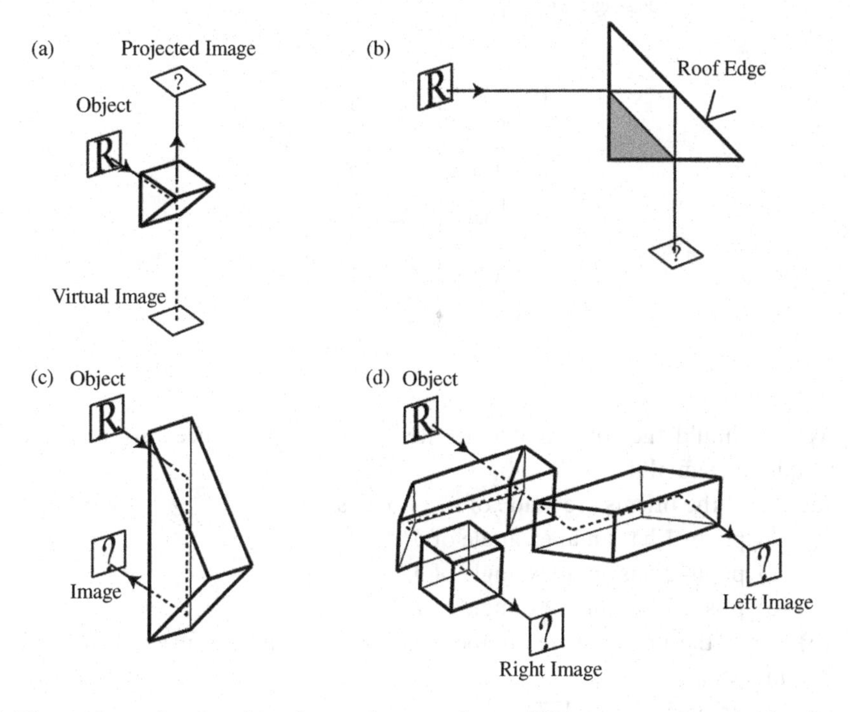

4.32 The index of refraction for a glass at three different wavelengths is given in Table 4.7.
(a) What is the Abbe number?
(b) What is the glass number?

Table 4.7.

λ (nm)	n
486.1	1.525
587.6	1.517
656.3	1.514

4.33 Given a right-angle prism with an apex angle of 7°, what is the angle of deviation for a prism made from:
(a) N-BK7;
(b) F4;
(c) NSF57?

4.34 The prism stack below is used to focus optical radiation onto a solar cell, which is 2 mm in diameter (assume rotational symmetry).

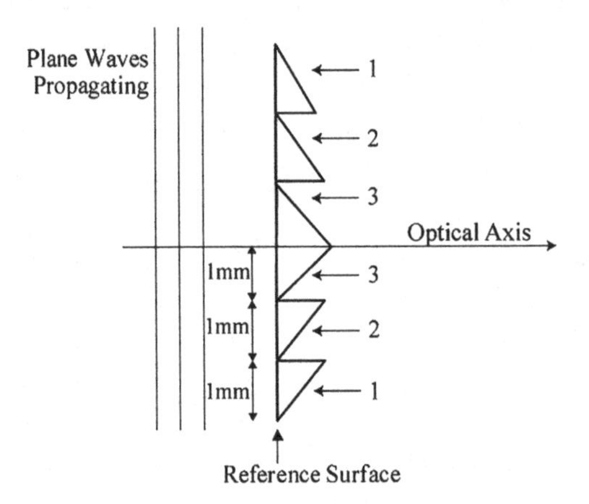

Where should the solar cell be located from the reference surface to achieve maximum signal if:
(a) All of the prisms are made of $n = 1.5$ glass, and:
 (1) prism 1 has an apex angle of 5°,
 (2) prism 2 has an apex angle of 3°,
 (3) prism 3 has an apex angle of 1°?
(b) All of the prisms have an apex angle of 5°, but each prism has a different index:
 (1) prism 1 ($n_d = 1.477$),
 (2) prism 2 ($n_d = 1.286$),
 (3) prism 3 ($n_d = 1.0955$)?

4.35 For the following prism types, what is the parity (odd or even)? Make an unfolded (tunnel) diagram:
(a) Penta prism,
(b) a pair of Porro prisms,
(c) Abbe prism?

4.36 The indices of refraction for sapphire (Al_2O_3) for the ordinary and extraordinary rays are (λ in micrometers):

$$n_o^2 - 1 = \frac{1.431\,349\,3\lambda^2}{\lambda^2 - 0.005\,28} + \frac{0.650\,547\,13\lambda^2}{\lambda^2 - 0.014\,238^2} + \frac{5.341\,402\,1\lambda^2}{\lambda^2 - 325.0178},$$

$$n_e^2 - 1 = \frac{1.503\,975\,9\lambda^2}{\lambda^2 - 0.005\,480\,263} + \frac{0.550\,691\,41\lambda^2}{\lambda^2 - 0.014\,999} + \frac{6.592\,737\,9\lambda^2}{\lambda^2 - 402.8951}.$$

(a) Plot the refractive index versus wavelength for wavelengths from 400 nm to 5 μm.

(b) Over this range, which wavelength has the greatest refractive index difference?

4.37 A pulse of white light 1ps wide is incident on a glass fiber optic that has the following values for blue (*F*), green (*d*), and red (*c*) light: $n_F = 1.6$, $n_d = 1.55$, $n_C = 1.5$. What length of fiber would be needed to separate the red (*c*) light from the blue light (*F*) in the output pulse by a time period of 1 ns?

Bibliography

Biberman. L. (1973). *Perception of Displayed Information*. New York: Plenum Press.

Beich, W. S. (2002). Specifying injection-molded plastic optics. *Photonics Spectra*, **36**, 127–133.

Born, M. and Wolf, E. (1959). *Principles of Optics*, sixth edn. Cambridge: Cambridge University Press.

Ditteon, R. (1997). *Modern Geometrical Optics*. New York: Wiley.

Fischer, R. E., Tadic-Galeb, B. (2000). *Optical System Design*. New York: McGraw-Hill.

Hopkins, R., Hanau, R., Osenberg, H., *et al.* (1962). *Military Standardized Handbook 141 (MIL HDBK-141)*. US Government Printing Office.

Katz, M. (2002). *Introduction to Geometrical Optics*. River Edge, NJ: World Scientific.

McCain, W. M. (1973). How to mount a Pellin–Broca prism for laser work. *Applied Optics*, **12**, 153.

Pedrottii, L. S., Pedrotti, L. M. and Pedrotti, F. L. (2006). *Introduction to Optics*, third edn. Harlow: Prentice Hall.

Rose, A. (1973). *Vision, Human and Electronic*. New York: Plenum Press.

Schott North America: http://www.us.schott.com.

Smith, W. J. (2000). *Modern Optical Engineering*, third edn. New York: McGraw-Hill.

US Precision Lens, Inc. (1983). *The Handbook of Plastic Optics*. Cincinnati, OH: US Precision Lens, Inc.

5

Curved optical surfaces

Thus far, we have discussed images and objects qualitatively. Other than in the case of the pinhole camera, we have not examined the actual production of the image of an object, but only what we thought the image orientation and its handedness might be. In order to produce an optical image, the rays or optical radiation must converge or diverge upon refraction or reflection. Only surfaces of optical power, or curved optical surfaces, can have this effect on rays. This book focuses on geometrical optics; diffractive optical elements of power will not be discussed. Before discussing the effects of curved optical surfaces on rays, we will digress and cover two very important subjects: optical spaces and the sign convention used in this text.

5.1 Optical spaces

The concept of "optical spaces" plays a very important role in the understanding of optical systems as well as in the layout of their design. There are multiple optical spaces in most systems; however, the minimum is three: object space, image space, and lens space, as shown in Figure 5.1. One example of this is the idea of object space, which is defined as the space domain where the object lies in a homogeneous medium which has a given index of refraction. For many systems, this medium is typically air. One can visualize a region from negative infinity to the first glass surface of a lens as being the physical space where an object resides. The concept of object space itself is fairly straightforward, but less intuitive is the idea that object space, in fact, lies throughout and beyond the air–glass boundary, and goes to positive infinity with an index of refraction equal to that of object physical space, n.

This paradoxical concept that object space extends from positive to negative infinity is an important one. For instance, consider entrance pupils which, by definition, lie in object space but may be to the right of the physically placed

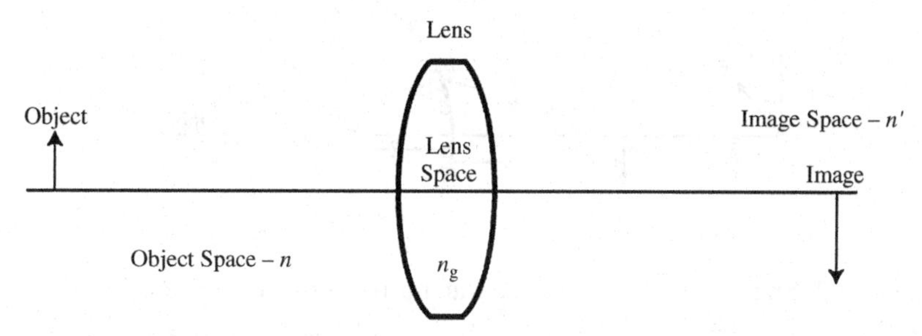

Figure 5.1 Locations of object space, lens space, and image space for a single lens system.

lens. This conflicts with the straightforward illustration of Figure 5.1, which defines everything to the right of a lens as residing in image space. Image space is defined as the collection of points that lie in a location of the optical system associated with the image. This space extends from negative infinity to positive infinity (just like object space), and has an index of refraction equal to that of the image physical space (n').

The region of image space to the right of the last glass surface of the lens in Figure 5.1 has an index of n'; however it mathematically extends to negative infinity. More than just a helpful conceptual model, this image space is a real physically bounded space. One can think of a ray actually traveling in this space. Real images can be formed and radiant power transferred. When we relate something to a quantity in object or image space, we mean it is related to that particular space's index of refraction and spatial location (x, y, z). That is, a ray would not experience refraction, reflection, or bending of any sort in that homogenous medium from negative infinity to positive infinity.

In the simple optical system shown in Figure 5.1, there is also another space: lens space (glass refractive index n_g). Lens space also conceptually stretches over the entire z direction space ($-\infty < z < \infty$), but physically comprises the thickness of the glass making up the lens. The glass material ends, but the glass space (n_g) can be thought of as going on forever. This may seem counter-intuitive, but this conceptual requirement in optical systems is mathematically necessary in order to formally handle image forming systems and subsequently move from one space to the adjacent space.

5.2 Sign convention

In order to be consistent in tracing a ray throughout an optical system, a sign convention for distances and angles must be both clearly stated and followed. There are many sign conventions used for this end, all of which will work as

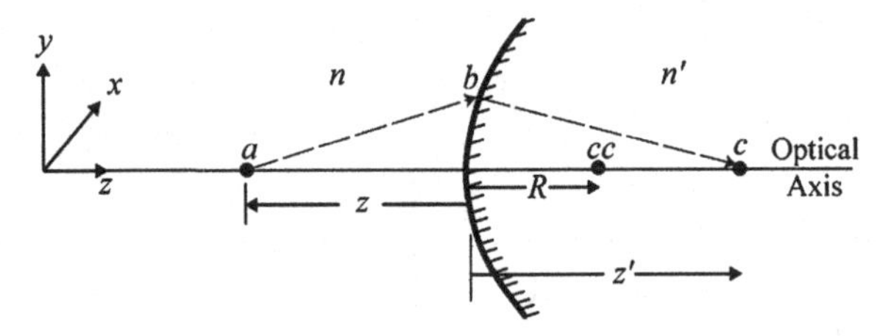

Figure 5.2 Sign convention for ray tracing from object space n to adjacent space n'.

long as they are consistently applied. The sign convention used in this text was chosen because it lends itself to multiple surfaces of power. Consider a single refracting surface, as shown in Figure 5.2, for a ray originating from point a going to point b on the surface where it is refracted (Snell's law: $n \sin I = n' \sin I'$) to point c on the optical axis.

A coherent system of drawing conventions is also obligatory. In this text, all distances and angles are directed. A single arrowhead is used to indicate whether their direction is positive or negative. This sign convention agrees with the right-hand Cartesian coordinate system (x, y, z coordinates), where the positive direction is shown with the arrows in Figure 5.2. This sign convention is used in order to facilitate cascading several optical elements during a ray trace. Other sign conventions which may be used cannot be extended to include more than a single lens optical power without greatly increasing the complexity of keeping track of object and image locations.

When using equations developed for ray tracing and image formation, one must substitute the algebraic values for distances and angles along with their sign. Figure 5.3 shows various examples of the sign convention for angles and distances. Note that the counterclockwise (CCW) angles are positive and clockwise (CW) angles are negative.

Table 5.1 lists the ten rules this text follows regarding sign conventions.

5.3 Ray tracing across a spherical surface

Figure 5.4 illustrates a bundle of rays coming from infinity, represented as a group of parallel rays, which is equivalent to a plane wave incident on an optical surface. In Figure 5.4(a), the PPP does not affect the ray bundle, and all rays remain parallel with each other after they transfer through the PPP. Therefore, that component has no optical power. The plane wave did not change its convergence or divergence.

In the case of a curved surface, shown in Figure 5.4(b), the rays change direction and converge to a point after refracting at the curved surface. Since

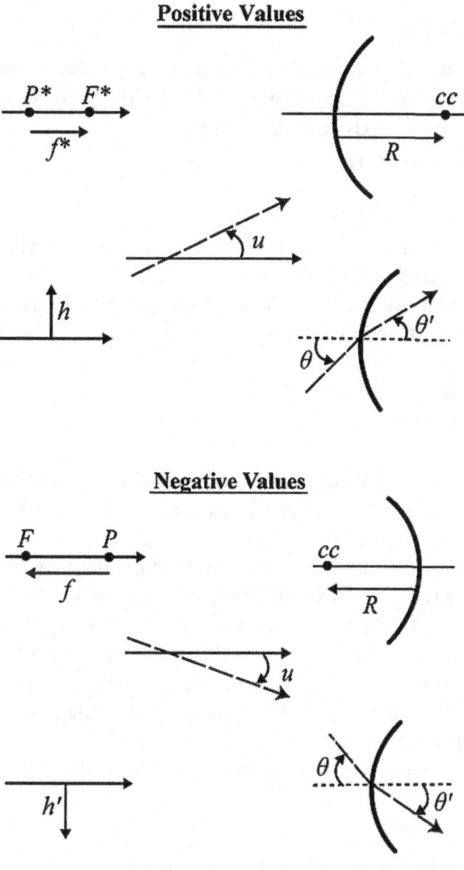

Figure 5.3 All directed distances and angles are identified by arrows with the tail of the arrow at the reference point, line, or plane.

they form a common point after the surface, the surface is said to have positive optical power. In Figure 5.4(c), the parallel rays incident on the negative radius surface (recall the sign convention) diverge after the surface from a point that seems to appear to the left of the surface. Thus, this surface has negative optical power.

We will restrict ourselves in the present discussion to spherical surfaces because they are easier to fabricate. In fact, if two surfaces rub against each other for a long enough time, both a concave and a convex surface with equal radii of curvature will be produced. As we will discuss later, there are also several aspherics and conics (paraboloid, hyperboloid, ellipsoid) that can be produced and are surfaces with optical power.

Optical surfaces can be used to either converge or diverge a bundle of parallel rays, and all surfaces can be approximated as spheroids near the optical (z) axis. A convex spherical surface changes the curvature of a

Table 5.1. *Sign convention*

(1) The optical axis is the z axis, which is positive to the right of the figure. Rotational symmetry exists around the z axis.
(2) Light travels from left to right ($-z$ to $+z$):
  left to right $\rightarrow$ (+) index of refraction,
  right to left $\rightarrow$ (−) index of refraction.
(3) The y axis is in the plane of the drawing. The positive y axis points up; heights are positive in the upward direction.
(4) Distances measured to the left of a reference point are negative, while those measured to the right are positive.
(5) Focal lengths:
  converging lens $\rightarrow$ positive,
  diverging lens $\rightarrow$ negative.
(6) Surface radii:
  positive means that the center of curvature (*cc*) is to the right of the surface;
  negative means that the center of curvature (*cc*) is to the left of the surface.
(7) Angles:
  measured counterclockwise from a reference are positive;
  measured clockwise from a reference are negative.
(8) Signs of all indices of refraction are reversed following a reflection.
(9) Signs of all distances following a reflection are consistent with our sign convention. Recall rule (2) and $n = c/v$, but velocity is negative.
(10) When making drawings, mark all distances and angles as directed distances. Mark angles with a single arrowhead.

wavefront incident on it, and produces a spherical wavefront convergence, as illustrated in Figure 5.4(b). This spherical wavefront is converging (collapsing) to its center. The refracting surface which produces this wavefront change can be approximated as a smooth sphere, which is a good approximation for the region close to the optical axis, which is called the paraxial region.

Consider a ray from point a to point b, as shown in Figure 5.5, which is incident on a spherical surface with radius R (shown in the y–z plane; however, keep rotational symmetry in mind). The ray follows Snell's law of refraction on traversing to point c located on the optical axis. The ray trace process can be accomplished in two ways:

- graphically,
- algebraically.

The graphical approach is worthy of mentioning only for the sake of completeness. A ray can be accurately and exactly traced through an optical system with drafting equipment, a compass and protractor. However, this is not practiced in today's world of powerful computers. The algebraic approach

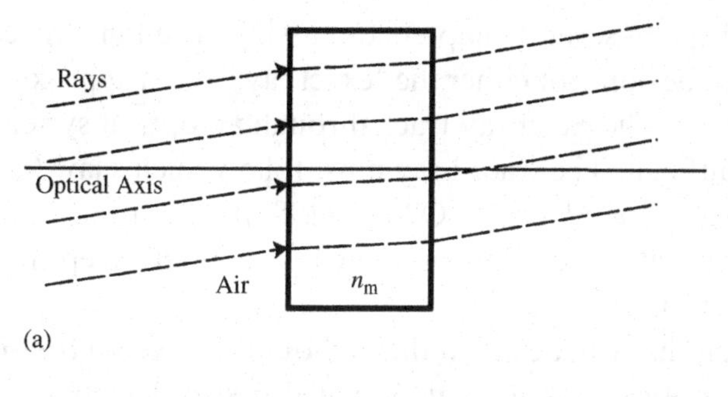

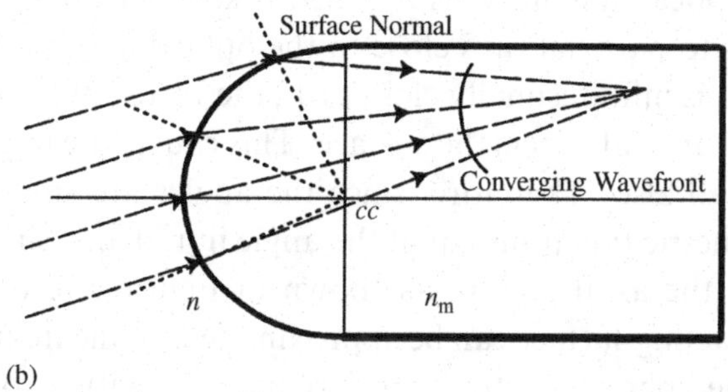

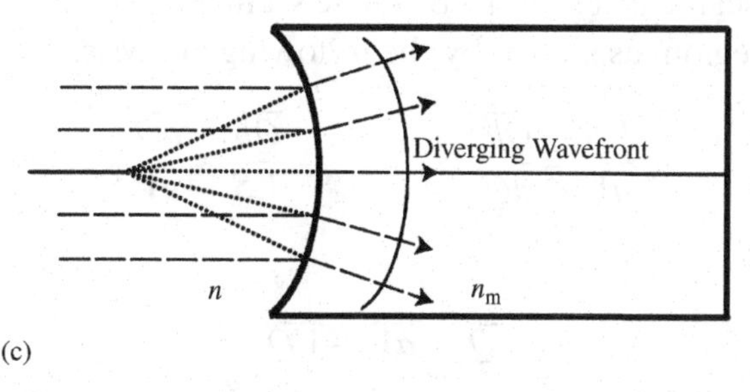

Figure 5.4 Parallel ray bundle incident on optical surfaces: (a) PPP – no optical power; (b) spherical surface of radius R – positive optical power; (c) spherical surface of radius R – negative optical power.

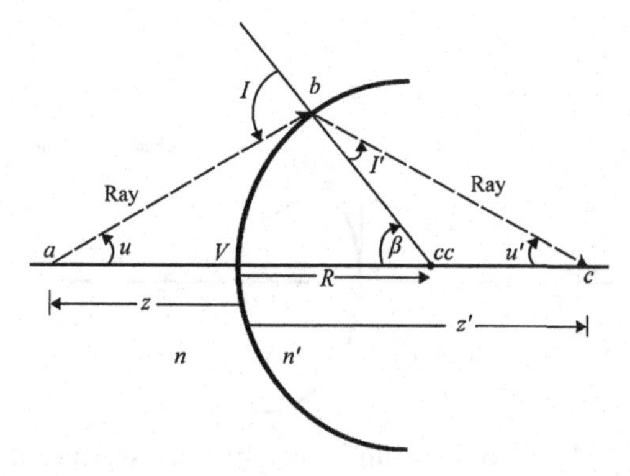

Figure 5.5 Refraction of a ray at a spherical surface.

is what we will spend several chapters discussing. The numeric calculation for the ray trace can be done for either the "exact ray," or an approximate ray called the "paraxial ray." The exact ray trace through an optical system is laborious, iterative, and difficult. The traced ray must follow Snell's law ($n\sin I = n' \sin I'$) exactly across each boundary (see O'Shea (1985)). The "exact" ray trace is what is done with computer programs, which can accurately keep track of 16-place floating-point numbers.

The approach that will be taken throughout this text is a first-order approximation to the optical system. This is called first-order optics. Thus, we will restrict ourselves to paraxial rays close to the optical axis, such as point b in Figure 5.5, which is infinitesimally close to V, the vertex of the optical surface where the optical axis intersects the surface. This causes the angles u, u', I, and I' to be very small also. Therefore, since the angles are small, the sine and tangent trigonometric functions equal the angle in radians: $\sin u = \tan u = u$.

By expanding the axial region, as shown in Figure 5.6, the paraxial ray height at the refracting surface can be approximated. In the figure, the paraxial ray intersects the spherical refracting surface at b, with two heights to be considered: the segments $\overline{Qb}$ and $\overline{VS}$. These segments, however, are equivalent in the paraxial region, as shown by the following analysis:

$$\frac{\overline{VS}}{\overline{aV}} = \frac{\overline{Qb}}{\overline{aQ}} \quad \text{or} \quad \frac{\overline{Qb}}{\overline{VS}} = \frac{\overline{aQ}}{\overline{aV}} \tag{5.1}$$

and

$$\overline{aQ} = \overline{aV} + \overline{VQ}, \tag{5.2}$$

where $\overline{VQ}$, the sag of a spherical surface, will be shown (see Equation (5.15)) to be

$$\overline{VQ} = \overline{Qb}^2/2R, \tag{5.3}$$

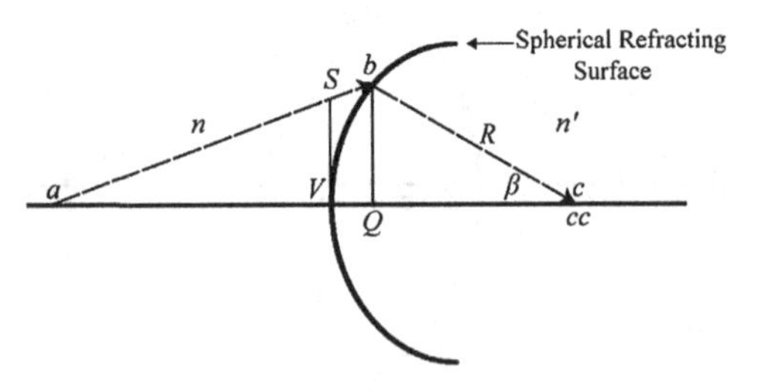

Figure 5.6 Paraxial approximation of height: intersection of a paraxial ray and a refracting surface.

$$\overline{Qb}/R = \sin\beta;\ \overline{Qb} = R\sin\beta. \tag{5.4}$$

Using the small angle approximation, it follows that

$$\overline{VQ} = R^2\beta^2/2R = R\beta^2/2. \tag{5.5}$$

Therefore, $\overline{VQ}$ is small since β is small and is squared; so from Equation (5.2), $\overline{aQ} = \overline{aV}$, and from Equation (5.1),

$$\frac{\overline{Qb}}{\overline{VS}} = 1 \quad \text{and} \quad \overline{Qb} = \overline{VS}. \tag{5.6}$$

Therefore, the paraxial ray height in the vertex plane is equivalent to the ray height at the surface. Figure 5.7(a) shows the paraxial ray refraction and the actual ray refraction at the optical surface. As shown in Figure 5.7(b), the sag is not taken into account for the paraxial ray. The paraxial ray refraction takes place at the plane of the vertex of the refracting surface. Spherical surfaces are now considered flat planes located at the vertex, as modeled

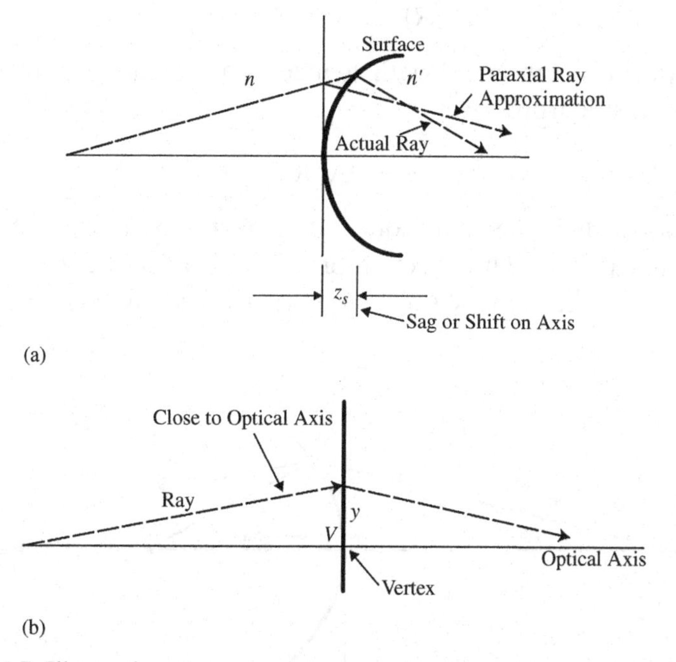

Figure 5.7 Illustrating the paraxial approximation: (a) paraxial ray trace diffraction takes place at the vertex plane of the surface; (b) paraxial surface with optical power representing a spherical surface.

pictorially in Figure 5.7(b). For this paraxial approximation, Snell's law (since the sine of the angle equals the angle) now becomes

$$n'I' = nI. \tag{5.7}$$

5.4 Sag of spherical surfaces

Thus far, we have been discussing paraxial rays and assuming the refraction exists at the vertex of the surface. However, there is some displacement between the vertex and where the ray actually refracts. The amount of displacement may be significant if we move away from the paraxial domain.

For a spherical surface, the displacement can be calculated from the equation of a sphere or circle of radius R. The equation of a sphere is

$$x^2 + y^2 + (z - R)^2 = R^2, \tag{5.8}$$

where a cross section of this sphere is shown in Figure 5.8 in a two-dimensional plot. Since we are only considering rotationally symmetric surfaces, only a two-dimensional cross section is needed.

Rearranging Equation (5.8),

$$(z - R)^2 = R^2 - y^2 - x^2. \tag{5.9}$$

Since we are only concerned with a meridional plane (y–z plane), the x dependence can be dropped:

$$z - R = \pm\sqrt{R^2 - y^2}. \tag{5.10}$$

This equation can be derived from the geometry of Figure 5.8 using the Pythagorean Theorem. The direction of the sag is from the spherical surface to the y axis, and the negative sign is used to conform to the sign convention:

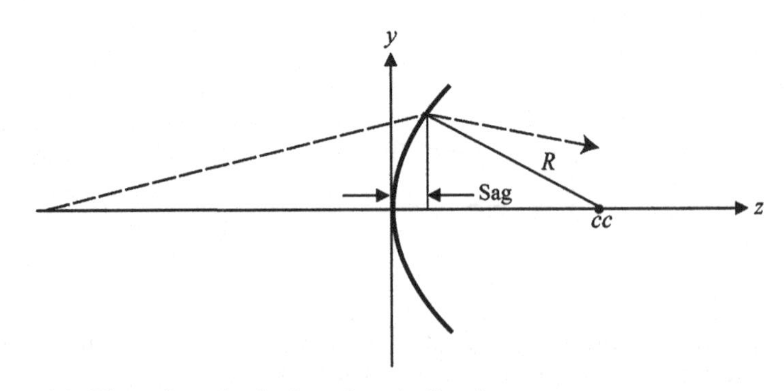

Figure 5.8 Plot of a spherical surface indicating sag.

$$z = R - \sqrt{R^2 - y^2}; \tag{5.11}$$

z takes on the meaning of the sag of a spherical surface. We can rewrite the radical in Equation (5.11) giving

$$z_s = z = R - R\sqrt{1 - \frac{y^2}{R^2}} = \text{sag} \tag{5.12}$$

Since y is small and y/R is squared, the radical can be expressed as $\sqrt{1-\alpha}$ and expanded by the Taylor series. Recall that, for small α, the Maclaurin series (a special case of the Taylor series) is

$$\sqrt{1-\alpha} = 1 - \frac{\alpha}{2 \times 1!} - \frac{\alpha^2}{2^2 \times 2!} - \frac{3\alpha^3}{2^3 \times 3!} - \frac{15\alpha^4}{2^4 \times 4!} - \ldots \frac{\alpha^n f^n(0)}{n!} \tag{5.13}$$

or

$$z_s = R - R\left(1 - \frac{y^2}{2R^2} - \frac{y^4}{8R^4} - \ldots\right). \tag{5.14}$$

Using the first two terms of the series to obtain the expression for sag:

$$z_s = \frac{y^2}{2R} = \frac{y^2 C}{2} = \text{sag}, \tag{5.15}$$

where the curvature, C, is given by $C = 1/R$.

This is the distance lost by paraxial approximation. It is the approximation of the sag for a spherical surface with radius of curvature R for a zone of radius y. The exact sag is given by Equation (5.11).

5.5 Paraxial ray propagation

To determine the direction of rays as they move from one homogeneous medium to another requires a digression into the linear equations of lines. Each ray in any space can be expressed as a linear equation. What we wish to develop is the formal mathematical relationship between a ray in one space and the equation of that same ray in the adjacent space. To this end, we need to know the indices of refraction of the two spaces, the curvature C or radius of curvature R of the surface, and the spatial location at which the ray enters the second medium. This information, together with the direction of an input paraxial ray, can be used to determine that ray's direction in the adjacent space. Figure 5.9 illustrates a paraxial ray trace across an n/n' boundary which has a radius of curvature R.

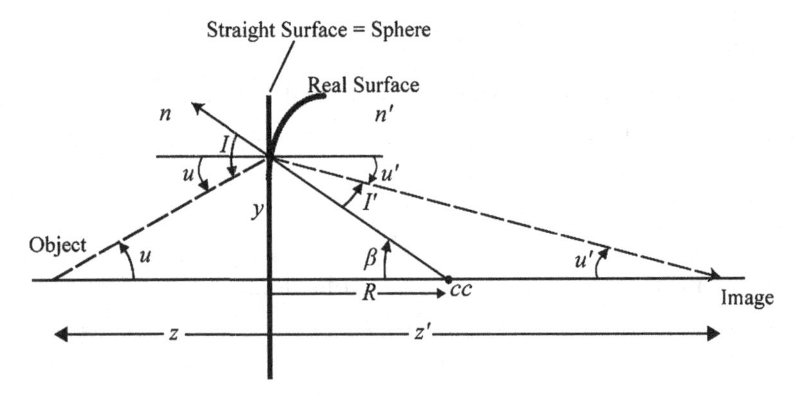

Figure 5.9 Paraxial ray trace.

5.5.1 Refraction equation of paraxial optics

The spherical surface in the paraxial region was shown as a flat surface in Figure 5.7(b), with the incident ray height of y. From the geometry shown in Figure 5.9:

$$u = \frac{y}{-z}; \qquad -u' = \frac{y}{z'}; \qquad -\beta = \frac{y}{R}. \tag{5.16}$$

It can be also observed from Figure 5.9 that

$$I = u - \beta, \quad \beta = u' - I' \quad \text{or} \quad I' = u' - \beta. \tag{5.17}$$

Snell's law ($n' \sin I' = n \sin I$) in the paraxial domain reduces via the small angle approximation to

$$n'I' = nI; \tag{5.18}$$

substituting Equation (5.17) into Equation (5.18) one gets

$$\begin{aligned} n'u' - n'\beta &= nu - n\beta \\ n'u' &= nu + (n' - n)\,\beta, \end{aligned} \tag{5.19}$$

and substituting for β, from Equation (5.16),

$$n'u' = nu - (n' - n)\,\frac{y}{R}. \tag{5.20}$$

This is the "refraction equation" for paraxial optics. It provides a means to define the direction angle (u') of the refracted ray across a boundary. Note this is just Snell's law if the surface is flat ($R = \infty$). This refraction is for paraxial rays only. Also note that we have lost the angles of incidence and angle of refraction on the surface. This equation only uses angles in radians relative to the optical axis.

5.5.2 Optical power

Equation (5.20) holds for any ray traversing a boundary from one homogeneous medium to a second homogeneous medium, even if the surface of that boundary is, in fact, flat (i.e. with no optical power). The refraction equation also applies, for the general case, for any ray as shown in Figure 5.10.

From Figure 5.10,

$$\begin{aligned} u &= I + \beta, \\ u' &= \beta - I'. \end{aligned} \tag{5.21}$$

Rearranging for I and I',

$$\begin{aligned} I &= u + \beta, \\ I' &= u' - \beta. \end{aligned} \tag{5.22}$$

Using Snell's law for small angles,

$$\begin{aligned} n'u' - n'\beta &= nu - n\beta, \\ n'u' &= nu + (n' - n)\beta. \end{aligned} \tag{5.23}$$

Substituting $-\beta = y/R$,

$$n'u' = nu - (n' - n)\,\frac{y}{R}. \tag{5.24}$$

This rederived refraction equation relates the refracted ray direction (u') to the incident ray angle (u). The primed quantities are the values in adjacent space, and the unprimed quantities refer to the corresponding values in object space. For a two-space system like the simple lens system examined thus far, this approach is very conducive to ray tracing through many optical surfaces.

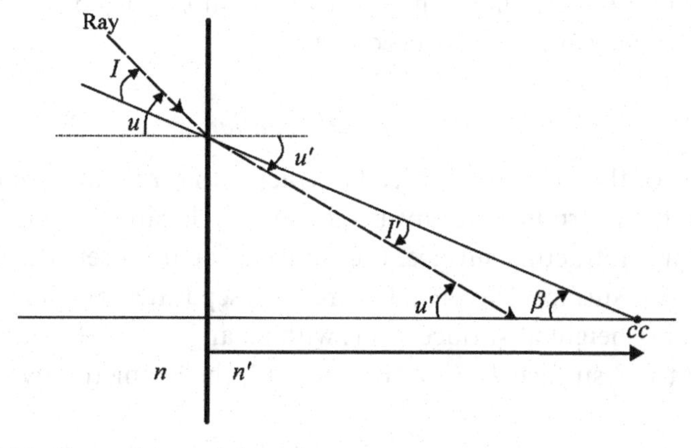

Figure 5.10 General paraxial ray trace case.

The second term in Equation (5.24) (the refraction equation) is related to the optical power (ϕ) of the surface. The optical power of a surface with radius R, separating two indices of refraction, is defined as:

$$\phi = \frac{n' - n}{R}. \tag{5.25}$$

The units of optical power are inverse length. If the radius of curvature of the surface is given in meters, then the power is in diopters (m^{-1}). The optical power of a surface may be positive or negative diopters, depending on the sign of the radius and the index changes.

Example 5.1

What is the optical power of a glass surface, in air, with radius of curvature = +10 cm, and refractive index = 1.5?

$$\phi = \frac{n' - n}{R} = \frac{1.5 - 1}{10} = 0.05\,\text{cm}^{-1} \quad \rightarrow \quad \phi = 5\,\text{diopters}$$

Figure 5.11 illustrates both a positive and a negative optical power surface where, in each case, the ray is going from a less dense to a more dense homogeneous medium. For the case shown in Figure 5.11(a), the radius is positive and the refractive index change is positive, since $n' > n$, so the optical power is positive. Similarly, in the case of Figure 5.11(b), the optical power is negative.

Further analysis of the refraction equation (first stated as Equation (5.20)) reveals that if the height of the ray y on the refracting surface is zero ($y = 0$), the equation is simply Snell's law. The expression can be considered as a case of reduced angles, nu, similar to reduced distances.

5.5.3 Transfer equation

The height, y, of the ray is thus crucial to tracing that ray through the system. So, in order to determine the ray height at any location along a ray, and particularly at a refracting surface, the transfer equation sets the height of y. Consider the two surfaces shown in Figure 5.12, separated by some distance, t'. If the known ray height at surface 1 is y, with an angle u', it is necessary to find the ray height y at surface 2. From Figure 5.12, the geometry gives:

$$y_1 - y_2 = -t' \tan u', \tag{5.26}$$

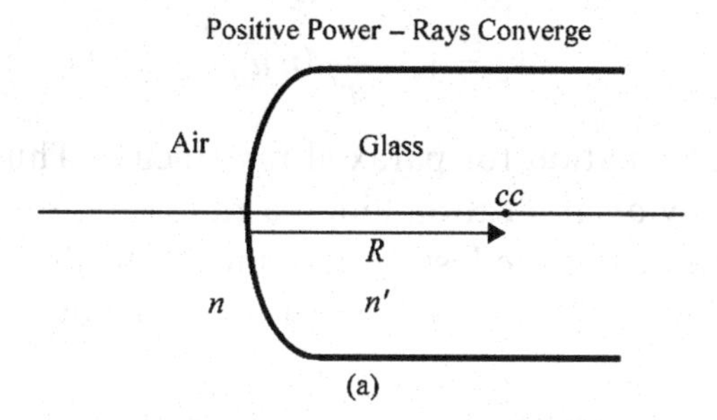

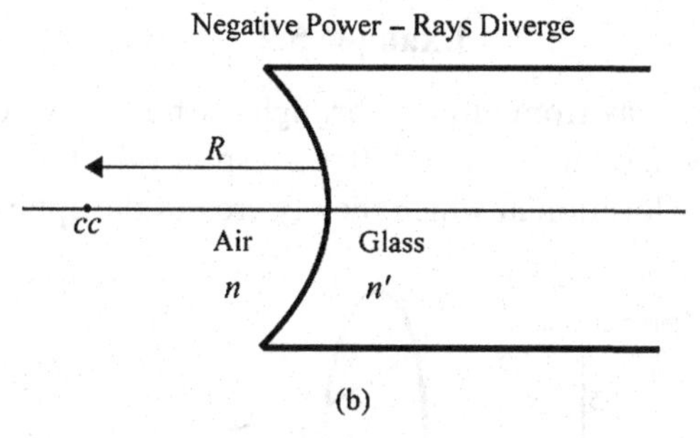

Figure 5.11 Surfaces with: (a) positive optical power for $n' > n$ and (b) negative optical power for n.

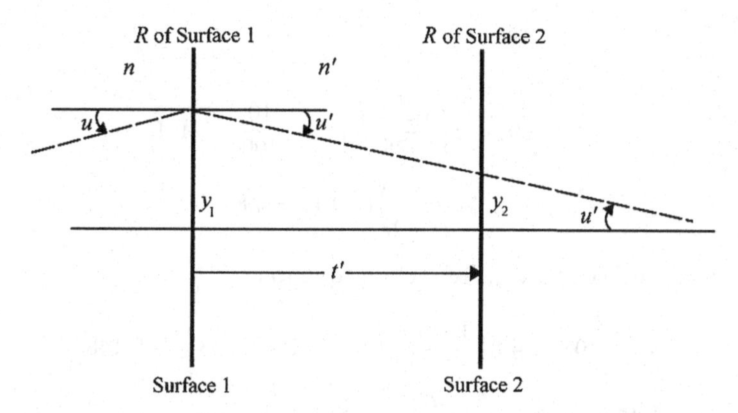

Figure 5.12 Ray height as a function of position along the z axis in a homogeneous medium.

where $\tan u'$ can be approximated by u' for small paraxial angles.

$$y_2 = y_1 + t'u'. \tag{5.27}$$

If we modify the equation by introducing the reduced thickness and the reduced angle, Equation (5.27) becomes

$$y_2 = y_1 + \frac{t'}{n'}(n'u'). \tag{5.28}$$

This is the transfer equation for paraxial ray tracing. Thus, it predicts the ray height (y) at any axial location along a ray. If y_2 is chosen to be zero ($y_2 = 0$), the distance from the last optical surface to the image location is called the back image distance. The distance t' is the distance to the image location.

Example 5.2

Paraxially ray trace a ray from infinity through a series of two optical surfaces separated by 3 cm with radii +25 and −50 cm, respectively. Find the distance from the last vertex to the location at which the ray crosses the optical axis (t) in air.

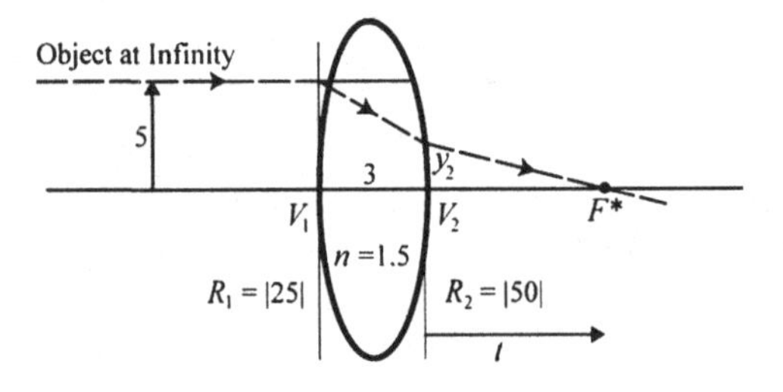

$$n'u' = n(0) - 5\left(\frac{1.5-1}{25}\right) = -\frac{10}{100} = -0.1,$$

$$y_2 = 5 + \left(\frac{3}{1.5}\right)(-0.1) = 4.8.$$

At the second surface apply the refraction equation:

$$n'u' = -0.1 - 4.8\left(\frac{1-1.5}{-50}\right) = -0.1 - 0.048 = -0.148.$$

$$\text{At } F^*; y_F = 0 = y_2 + \frac{t}{n'}(n'u') = 4.8 + \frac{t}{n'}(-0.148)$$

$$t = 32.432.$$

The value (t) of the distance from the last surface to where this ray crosses the optical axis is called the back focal distance (BFD).

5.5.4 Paraxial ray trace equations

The two equations which are used for paraxial ray tracing are the refractive and transfer equations, which are repeated here for convenience. Paraxial ray tracing can use these equations to define ray heights and locations along the z propagation direction:

$$n'u' = nu - y\left(\frac{n' - n}{R}\right); \tag{5.29}$$

$$y_2 = y_1 + \frac{t'}{n'}(n'u'). \tag{5.30}$$

5.6 Gaussian equation of a single surface

The refraction equation (Equation (5.29)) can be rewritten in terms of object distance z and image distance z' using Figure 5.9:

$$\frac{n'y}{-z'} = \frac{ny}{-z} - \frac{(n' - n)}{R}y. \tag{5.31}$$

Note that y (the height) cancels out, and therefore the refraction equation is independent of ray height in the paraxial domain, leaving

$$\frac{n'}{z'} = \frac{n}{z} + \frac{n' - n}{R}. \tag{5.32}$$

Recalling the definition of optical power (ϕ), Equation (5.25):

$$\phi = \frac{n' - n}{R}, \tag{5.33}$$

the refraction equation has been reduced to

$$\frac{n'}{z'} = \frac{n}{z} + \phi. \tag{5.34}$$

This is the Gaussian equation relating object z and image z' distances in media of n and n', respectively, for a single refracting surface of optical power ϕ.

Example 5.3

What radius for a glass rod is necessary to form an image inside the glass at 37.5 cm for an object 50 cm away from the curved surface ($n = 1.5$)?

Using $n'/z' = n/z + \phi$ and substituting in the values $z' = 37.5$ cm, $z = -50$ cm, and $n' = 1.5$:

$$\frac{1.5}{0.375} - \frac{1}{-0.5} = \phi \rightarrow \phi = 4 + 2 = 6\,\text{diopters};$$

$$\phi = \frac{1.5 - 1}{R} = 6 \rightarrow R = \frac{1}{12}\text{m} = 8.3333\,\text{cm}.$$

5.7 Focal lengths and focal points

The focal point (F^*) is the location at which the image of a point infinitely far away is located. Keep in mind that rays from an object at infinity propagate parallel to the z axis with a plane wavefront, as shown in Figure 5.13. Thus, a more formal definition of a focal point would be the point at which a ray from infinity (parallel to the optical axis) in one space passes through a focal point in adjacent space, located on the optical axis.

Using Equation (5.34), with its definition which forces the image distance z' to equal the back focal length, f^*:

$$\frac{n'}{f*} - \frac{n}{(-\infty)} = \phi \tag{5.35}$$

$$\phi = n'/f*. \tag{5.36}$$

Optical power, ϕ, is the index of refraction of the space divided by the focal length in that space. Conversely, if a point source object is placed at the front focal point (F), as shown in Figure 5.14, the image is at infinity. From Equation (5.34) the optical power is

$$\phi = -n/f. \tag{5.37}$$

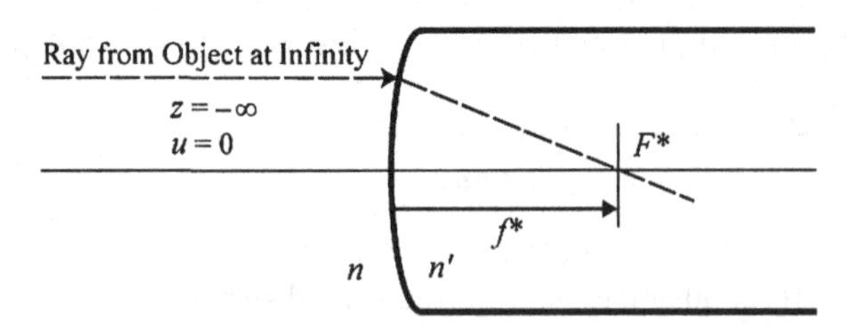

Figure 5.13 Image formed of an object at infinity at the back focal point, F^*.

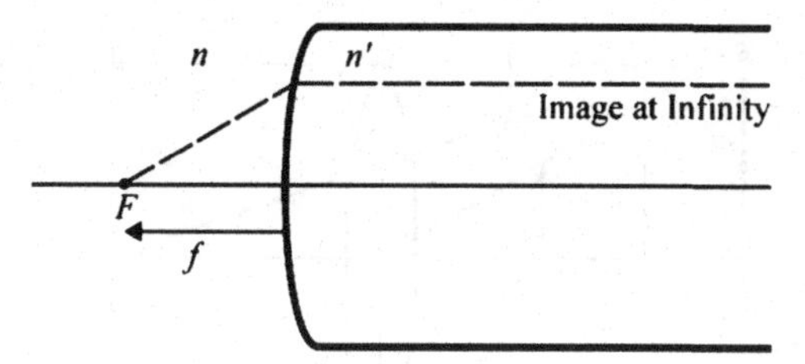

Figure 5.14 Object located at front focal point.

Since the power is the same for both cases:

$$\phi = -\frac{n}{f} = \frac{n'}{f*}$$

$$\frac{f*}{f} = -\frac{n'}{n}. \tag{5.38}$$

The ratio of the back focal length (f^*) to the front focal length (f) is equal to the negative ratio of their indexes of refraction. If both object and image are in air ($n = 1$), then the back focal length equals the front focal length in magnitude, but is opposite in sign.

Example 5.4

For an optical power of 15 diopters and an object at a distance of 20 cm to the left of the vertex of a glass rod, where is the image relative to this vertex ($n = 1$, $n' = 1.5$)?

$$\frac{n'}{z'} = \frac{n}{z} + \phi \rightarrow \frac{1.5}{z'} = \frac{1}{-0.2} + 15 = 10$$

$z' = 1.5/10 = 0.150$ m $= 15$ cm measured from the vertex.

5.8 Transverse magnification

In Section 5.6 we developed the Gaussian equation for the location of an image created by a surface with optical power. We now need to determine the size of the image. The transverse magnification is defined as the ratio of the lateral image height to the lateral object height. There are two important examples of image formation with lenses. The most common, for centuries, has been the use of a lens as a magnifying glass, as shown in Figure 5.15.

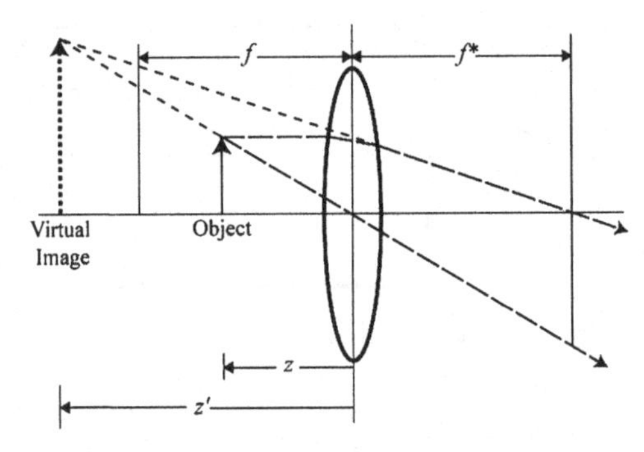

Figure 5.15 Lens used as a magnifier.

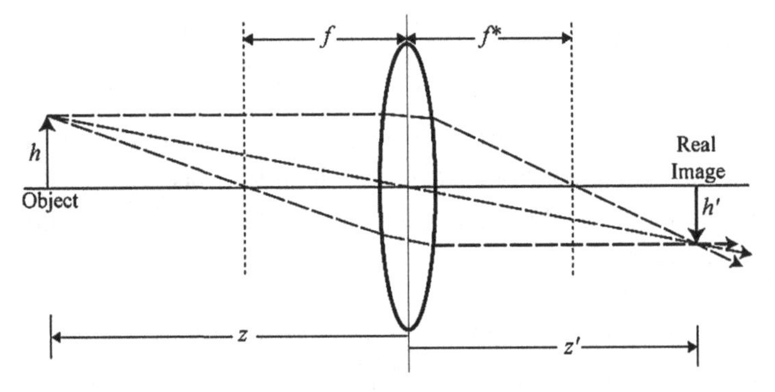

Figure 5.16 Real image magnified.

Note that, if the object is located inside the front focal point ($z < f$), the image distance (z') becomes negative, so the image will be positioned on the same side of the lens as the object. Although this kind of image, known as a virtual image, cannot be projected onto a screen, an observer looking through the lens will see an image larger than the original object (positive magnification). A magnifying glass forms this kind of image, enabling Grandma to read the newspaper.

In the second case, an object at a finite distance is focused to a real image via the Gaussian equation (Equation (5.39)), as shown in Figure 5.16. The plane perpendicular to the lens axis, situated at a distance f from the lens, is called the focal plane, while the image is formed at the image plane.

Distances from the object to the lens are negative to the left, and represented by z, while the image is positive to the right, and represented by z':

$$\frac{1}{z'} = \frac{1}{z} + \frac{1}{f^*}. \tag{5.39}$$

Therefore, an object placed at a distance greater than the front focal length ($|z| > -|f|$) along the axis in front of a positive lens will form an image at a distance, z', behind the lens. The image in this case is called a real image, and can be recorded on a detector or screen for viewing. The transverse magnification of the lens is given by the image height divided by the object height:

$$M_t = \frac{h'}{h} = \frac{z'}{z} = \frac{f}{f - z} = \frac{f^* - z'}{f*}. \tag{5.40}$$

M_t is the transverse magnification. The sign convention shows whether M_t is negative or positive. If $|M_t|$ is greater than 1, the image is larger than the object. For real images, the image is upside-down with respect to the object. For virtual images (Figure 5.15), M_t is positive and the image is upright. In the special case that $z = -\infty$, then $z' = f^*$ and $M_t = -f/\infty = 0$. This corresponds to a collimated beam being focused to a single spot at the focal point.

A single spherical glass surface forms an image inside the glass, as demonstrated in Figure 5.17.

From the refraction equation: $n'u' = nu - y\phi = nu$, since $y = 0$ for the ray through the center of the lens. Therefore:

$$n'\frac{h'}{z'} = n\frac{h}{z}. \tag{5.41}$$

The general equation for transverse magnification can be expressed as the image height divided by object height:

$$\frac{h'}{h} = \frac{z'/n'}{z/n} = M_t. \tag{5.42}$$

Transverse magnification is the ratio of reduced image distance to reduced object distance for a thin lens (similar triangles), where z'/n' and z/n may be thought of as reduced thickness! Axial or longitudinal magnification (M_z)

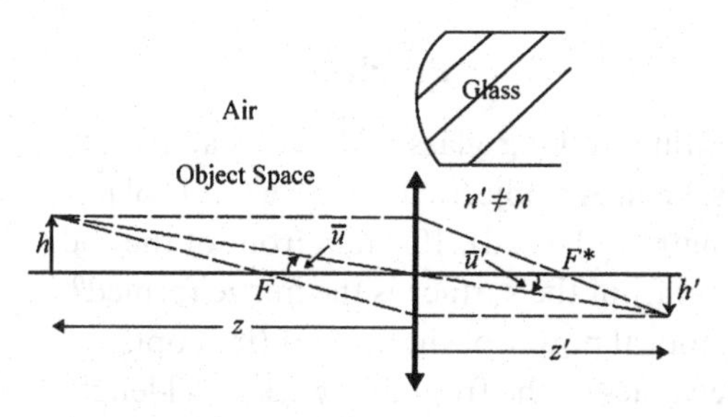

Figure 5.17 Image in a medium other than the object medium.

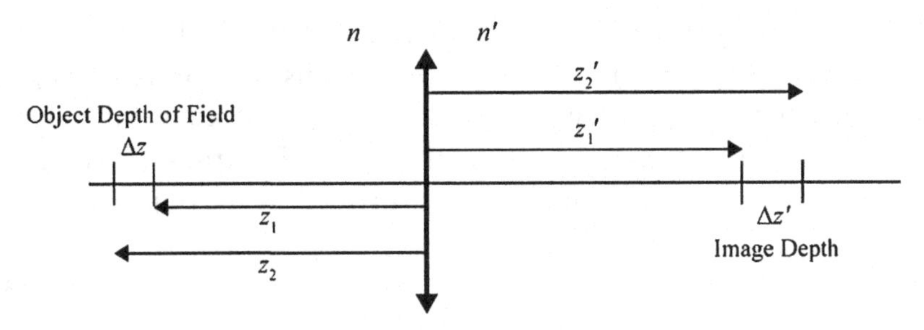

Figure 5.18 Change in size and direction of z

may be thought of as the change in size in the z direction, as shown in Figure 5.18:

$$M_z = \Delta z' / \Delta z. \tag{5.43}$$

From our Gaussian equation,

$$\frac{n'}{z'} = \frac{n}{z} + \phi.$$

To find the change in axial magnification, take the differential axial change of object and image distances:

$$\frac{-n'\mathrm{d}z'}{z'^2} = \frac{-n\mathrm{d}z}{z^2}$$

$$\frac{\mathrm{d}z'}{\mathrm{d}z} = \frac{n}{n'}\frac{z'^2}{z^2} \tag{5.44}$$

$$M_z = \frac{n}{n'}\frac{z'^2}{z^2} = M_\mathrm{t}^2\frac{n'}{n}.$$

Axial or longitudinal magnification (M_z) is transverse magnification (M_t) squared times the refractive index ratio of the image space to the object space.

Problems

5.1 The end of an infinitely long glass rod of refractive index 1.7 has a spherical surface of radius 3 cm. An object 2 mm in height is placed in air on the axis (the axis down the length of the rod), 10 cm in front of the rod.
   (a) How far (+/−) from the surface is the image formed?
   (b) What is the optical power of the surface (in diopters)?
   (c) What are the values of the front and back focal lengths?
   (d) What is the transverse magnification?

5.2 A plastic rod of refractive index 1.45 has a spherical radius of 15 cm. A 2 cm high object is located (in air) 20 cm in front of the rod on the extension of the axis (down the length of the rod).
(a) What are the front and back focal lengths?
(b) What is the transverse magnification?

5.3 Some algae are growing at the bottom of a lake, and a student would like to bring the image of the algae out of the water with a glass rod, such that the image would be formed on the end of the rod. The rod's refractive index is 1.6 with a convex radius on its end. The algae are 1 mm high, located in water about 12 cm away from the rod, and the rod is 30 cm long, extending above the water.

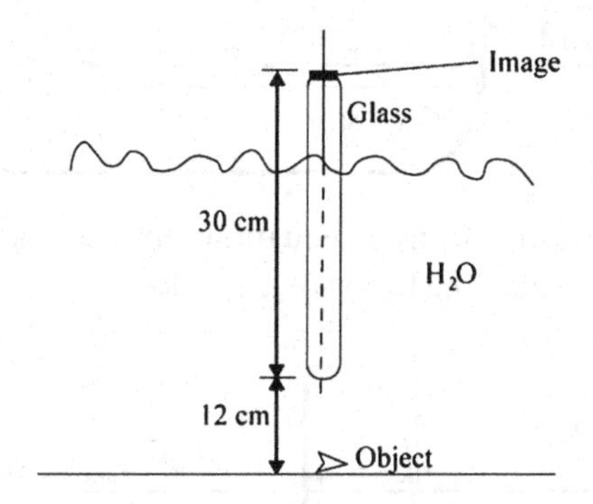

(a) What radius is needed on the end of the rod?
(b) What is the transverse magnification?
(c) What are the front and back focal lengths of the rod's surface of optical power?

5.4 A spherical surface with a radius of 2.75 cm is on a glass rod of refractive index 1.5. Find the optical power in diopters of this rod when placed in:
(a) air;
(b) water ($n = 4/3$);
(c) oil ($n = 1.63$).

5.5 In the figure below, find the axial point (d) where the paraxial ray crosses the optical axis in image space by using refraction and transfer equations. (All dimensions are in centimeters.)

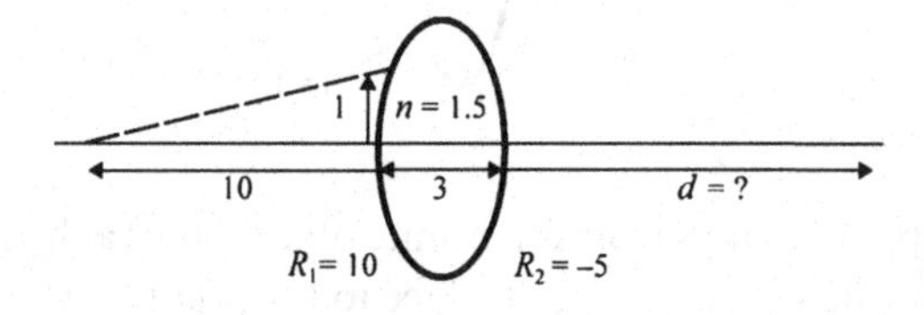

5.6 Calculate the optical power (in diopters) of a surface with $|R| = 2$ in for both positive and negative values of radius (refractive index = 1.5).

5.7 What is the sag of a spherical surface ($R = 20$ cm) for a ray 2 cm from the optical axis:
(a) exact sag;
(b) approximate sag?

5.8 Find the location at which a ray of *d* light (587.6 nm) crosses the optical axis for a glass rod, as shown below
(a) for an exact ray;
(b) for a paraxial ray.

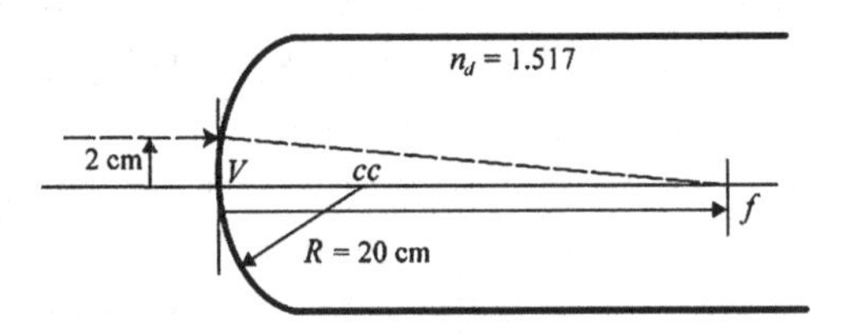

5.9 Using the refraction and transfer equations for paraxial ray tracing, find the following for the ray emerging from the lens below (all dimensions in centimeters).

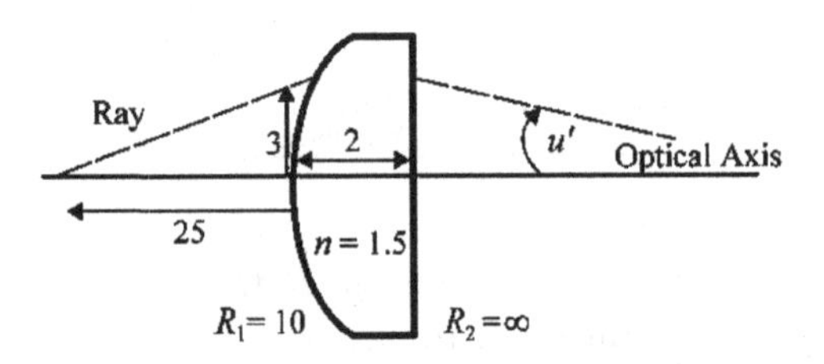

(a) What is the location, relative to the last surface, at which the ray crosses the axis?
(b) What is the value of u' in image space?

5.10 Set up a numerical (algebraic) paraxial ray trace for the following lens (all dimensions are in centimeters).

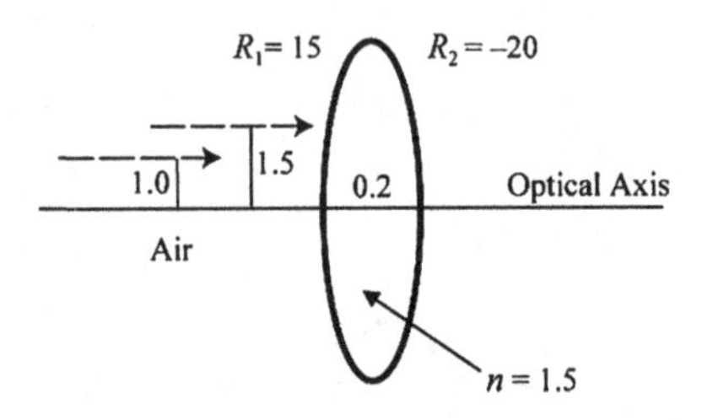

(a) Trace two paraxial rays from an infinite object point at heights of 1.5 and 1.0.
(b) Find the distance from the last surface to the point at which the paraxial ray crosses the axis in each case (BFD).
(c) What is the point called at which these rays cross the optical axis?

5.11 Consider a single refracting surface (radius 10 cm and refractive index 1.5) with a ray from infinity incident at a height of 5 cm.

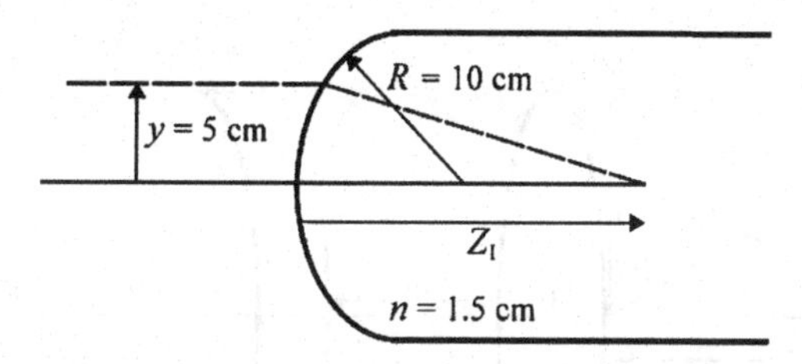

(a) Where does this paraxial ray cross the axis (Z_I)?
(b) What is the sag of this spherical surface at the ray height of 5 cm?
(c) How does this affect Z_I?

5.12 There's a fish in an aquarium (call him Julius) being interviewed after the underwater basket weaving competition for non-mammalian species. The surface of the fish bowl has a radius of 5 in (neglect glass thickness, and consider only the water ($n = 1.3\overline{3}$) to air interface). An image is formed of Julius on a screen outside of the bowl at 25 in.
(a) How far is Julius from the surface?
(b) What is the transverse magnification?
(c) Is the image real or virtual?

5.13 Derive the refraction equation ($n'u' = nu - y\phi$) from the following sketch, using only the angles shown.

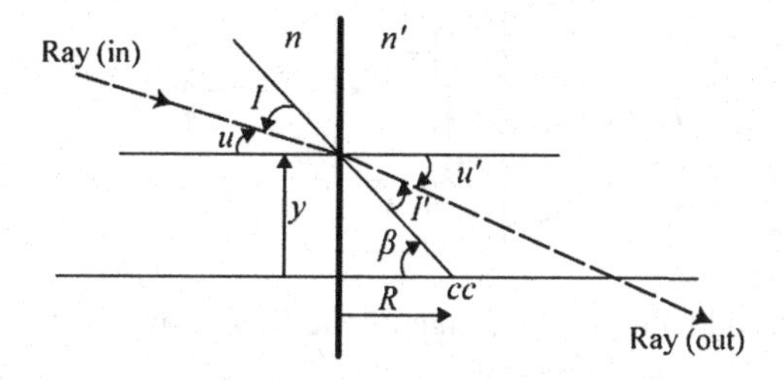

5.14 What is the optical power (in diopters) of the following optical surfaces made of 755276.479 material placed in different environments:
(a) air, radius = 25 cm;
(b) water, radius = 25 cm;
(c) oil ($n = 1.55$), radius = 12.5 cm;
(d) water, radius = 5 in;
(e) air, radius = 0.1 ft?

5.15 A glass with a refractive index of 2 ($n = 2$) forms an image in the glass 25 mm to the right of the vertex, while the object is 25 mm in front of the vertex.
(a) What is the optical power of the surface (in diopters)?
(b) What is its radius of curvature in millimeters?
(c) What is the value of the front focal length (f) in millimeters?
(d) What is the value of the back focal length (f^*) in millimeters?

5.16 Ray trace the rays shown (A and B) in the figure below from object to image space, where it crosses the optical axis (N-BK7 = 517642.251; SF66 = 923209.602).

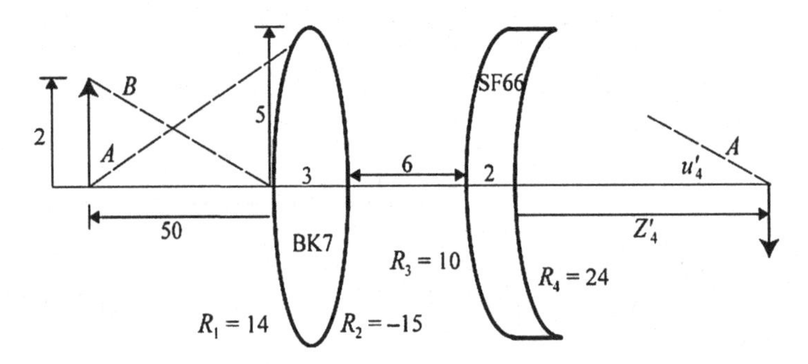

(a) What is the Z'_4 distance?
(b) What is the u'_4 angle?

5.17 A clever student is using a glass mixing rod to view the effect acid has on metal. He mounts his film on the end of a 1 cm diameter rod faceplate. The radius of curvature on the end of the rod is 4 cm, and the metal in the acid is 26 cm away from the end of the rod (see sketch below). How long should the rod be to form an image on the film (faceplate)?

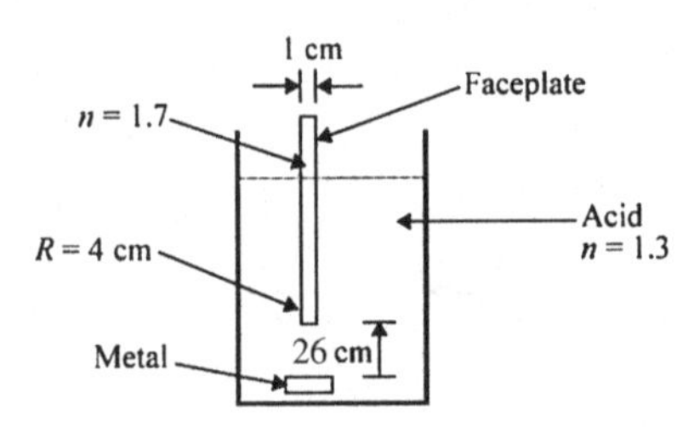

5.18 A glass rod is used to form an image of the Moon inside the glass ($n = 1.5$). The radius of curvature on the end of the glass rod is 20 cm. The Moon is 2160 miles in diameter and 240 000 miles from the Earth. Find the image diameter (cm) and its location inside the glass rod (cm).

5.19 Using the refraction equation and the transfer equation, what are the values of SS' and hh' in the diagram below?

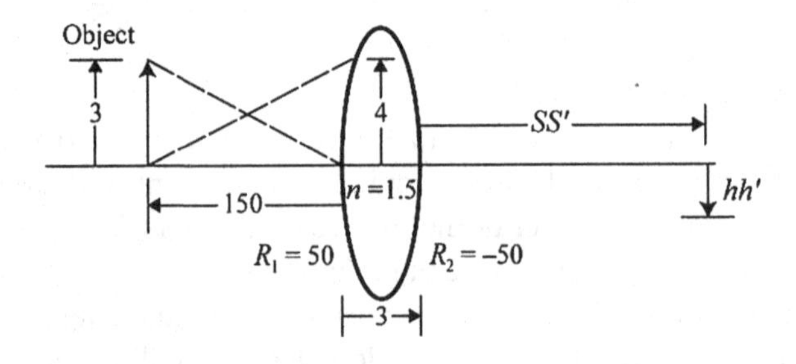

5.20 A fish in a spherically-shaped fish bowl (1 m diameter) is swimming around and periodically goes through the center of the sphere. Neglecting the effect of the wall's refractive index, find for this center position:
(a) where the fish's image is located relative to the aquarium surface;
(b) the lateral magnification of the fish.

5.21 A 1 cm cube has its near surface 16 cm in front of a rod made of 755276.479 glass. The glass rod has a radius of curvature on its end equal to 4 cm.
(a) Where is the image inside the glass?
(b) What is the transverse magnification of the near surface?
(c) What is the transverse magnification of the far surface?
(d) What is the axial magnification?

5.22 For a glass rod ($n = 1.5$) with a radius of curvature on its end of 5 cm find:
(a) the optical power (in diopters);
(b) the front focal length (in centimeters);
(c) the back focal length (in centimeters).

5.23 Using the refraction and transfer equations for paraxial ray tracing, find the ray angle in image space (water) shown below.

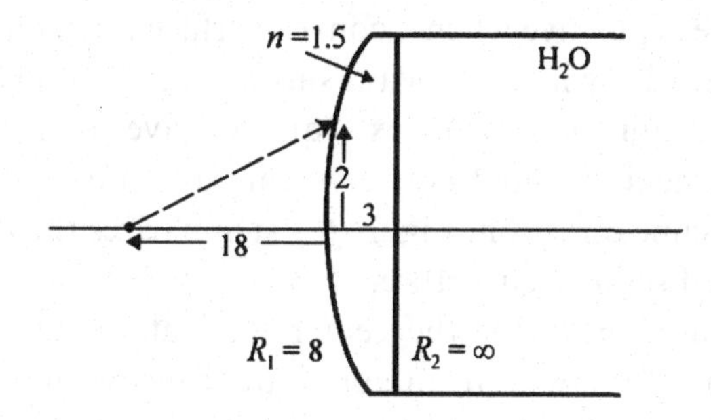

Bibliography

Gradshteĭn, I. S., Ryzhik, I. M., and Jeffrey, A. (1980). *Table of Integrals, Series, and Products*. New York: Academic Press.

O'Shea, D. C. (1985). *Elements of Modern Optical Design*. Atlanta: Wiley-Interscience.

6

Thin lenses

6.1 Lens types and shape factors

Lenses can have many different shapes in cross section. For instance, their surfaces may be convex, concave or flat (plano). The curved surface of a lens has a defined radius of curvature (R). When light travels from left to right, the radius of curvature is positive when a convex surface is encountered first, and the curvature is negative when a concave surface is encountered. There are six basic lens shapes: equi-(or bi-)convex, equi-concave, plano-convex, plano-concave, concave-meniscus, and convex-meniscus. In a meniscus lens, both sides curve in the same direction. Figure 6.1 summarizes the basic lens shapes and their associated radii of curvature.

Convex lenses are thicker at the center than at the circumference, and decrease the radius of the wavefront curvature, causing the rays to converge (Figure 6.2(a)). These lenses are known as converging, or positive lenses. Concave lenses are thinner in the center and increase the radius of the wavefront curvature, causing the rays to diverge (Figure 6.2(b)). These lenses are known as diverging, or negative lenses.

Lenses with many different radii can give the same optical power. This is demonstrated in the lens maker equation, where n_g is the glass refractive index:

$$\phi = (n_g - 1)\left(\frac{1}{R_1} - \frac{1}{R_2}\right). \tag{6.1}$$

There are a range of values for R_1 and R_2 that will produce the same optical power. The lens shapes that yield the same power are described in terms of a shape factor. The shape factor $(\mathcal{S})$ is defined as

$$\mathcal{S} = \frac{R_1 + R_2}{R_2 - R_1} = \frac{C_1 + C_2}{C_1 - C_2}, \tag{6.2}$$

where the curvature $C = 1/R$.

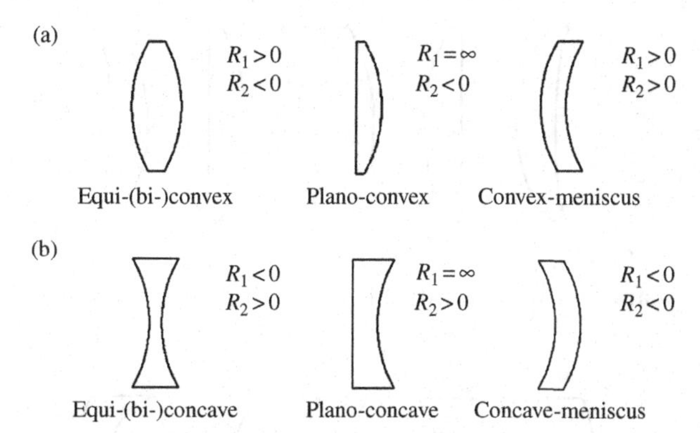

Figure 6.1 Basic lens shapes and their associated radii of curvature: (a) convex lenses; (b) concave lenses.

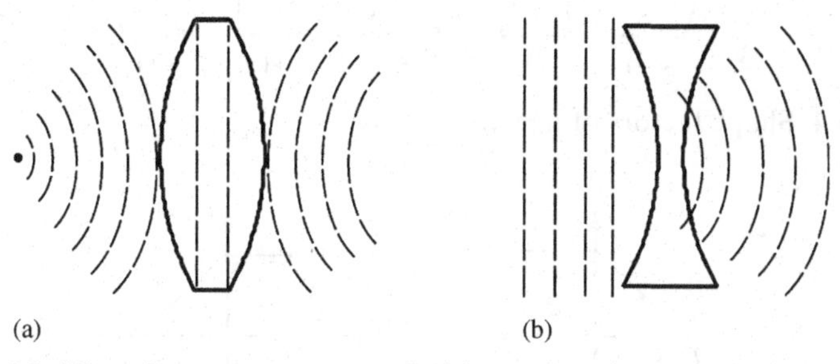

Figure 6.2 Wavefront curvatures of: (a) convex (or converging) and (b) concave (or diverging) lenses.

Figures 6.3 and 6.4 show the shape factors for positive and negative lenses. Each lens in the two series of lenses with different shape factors will give the same focal length, and thus the same optical power, as any other lens in that series.

6.2 Gaussian optics – cardinal points for a thin lens

A lens is defined by six cardinal points. These are the front and back focal points (F and F^*), the principal points (P and P^*), and the nodal points (N and N^*). The six cardinal points are essential in describing a thick lens. Here we will briefly discuss the focal points and principal points of a thin lens, saving in-depth coverage of principal and nodal points for thick lenses for Chapter 7.

A ray from infinity and parallel to the axis in one space will refract in the adjacent space after passing through an optical surface. The focal point is defined as the point at which this refracted ray crosses the optical axis.

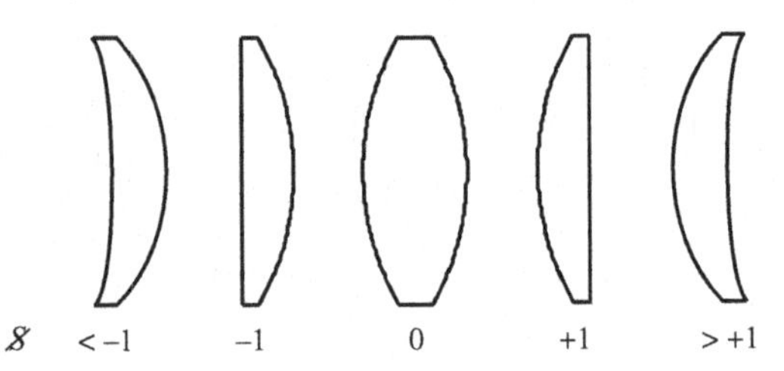

Figure 6.3 Shape factors of positive lenses.

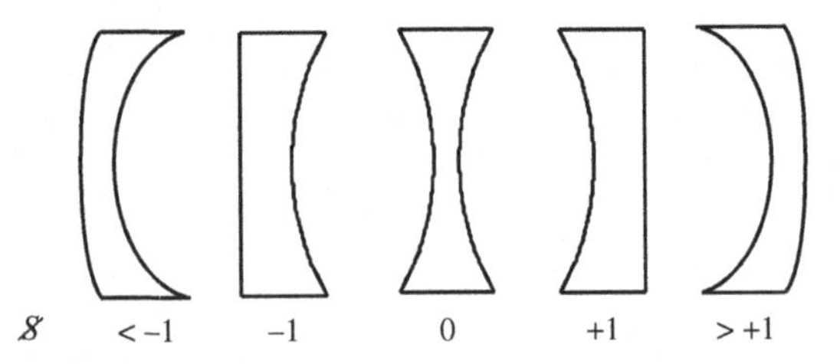

Figure 6.4 Shape factors of negative lenses.

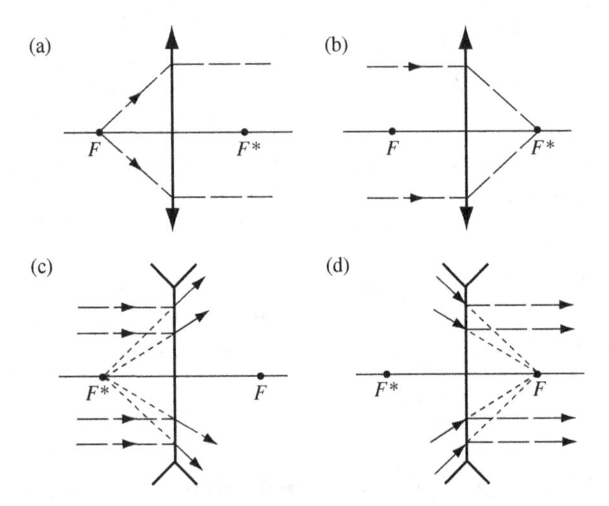

Figure 6.5 Focal points for positive and negative lenses: (a) object at the front focal point of a positive thin lens; (b) object at infinity of a positive thin lens; (c) object at infinity of a negative thin lens; (d) object at the front focal point of a negative thin lens.

In Figure 6.5(a), diverging rays from the front focal point, F, emerge parallel to the optical axis in the adjacent space. In Figure 6.5(b), parallel rays converge and come to focus at the back focal point F^*. The locations of the focal points for a negative lens are found by back tracing the rays from the adjacent space (Figure 6.5(c) and (d)).

The principal planes are defined by the intersection of the rays entering a system with the rays leaving a system. The intersections of the principal planes with the optical axis are called the principal points, P and P^*. In a thin lens, P and P^* are coincident with the lens itself. In Chapter 7, we will see that the location of the principal planes is dependent on the shape factor of the lens as well as its thickness. The principal planes also have unit transverse magnification between them.

6.3 Mapping object space to image space

6.3.1 Non-linear mapping of a positive lens

A positive lens has six mapping possibilities. The mapping regions of a positive lens are summarized in Figure 6.6 and Table 6.1. In the first case, an object at infinity forms an image at F^*. From ray tracing, rays drawn parallel to the optical axis converge and pass through the back focal point. The second case is when the object is between negative infinity and $2F$. The image forms between F^* and $2F^*$. This image has a transverse magnification between 0 and –1. A special case exists when the object is exactly at $2F$. The image then forms at exactly $2F^*$, as shown in Figure 6.6(c), producing a one-to-one transverse magnification. In a laboratory setting, this case is often used to determine the placement of a lens that will focus a light source onto an aperture. An object located between $2F$ and F forms an image beyond $2F^*$. The transverse magnification is between negative infinity and –1. This fourth case and the second case (Figure 6.6(d) and (b)) represent the two lens positions which give conjugate planes whose distances $z+z'$ are equal. In the fifth case, an object exactly at F produces an image at infinity. This mapping has an important application in the laboratory, because this lens placement produces collimated light from a point source. Likewise, case (1) can be used to focus collimated light down to a point source. The last case occurs when the object is inside the front focal point. The object is mapped beyond infinity to a virtual image to the left of the lens. This is the principle behind a simple magnifier.

For a positive thin lens, any finite object-to-image distance yields two locations where an in-focus image can be produced. In order for an image to form, the distance between the object and image L must be greater than or equal to $4f$ ($|L| \geq 4f$). Examination of Figure 6.6(a)–(f) reveals that, as an object at negative infinity is brought closer to the lens, the image moves from the lens until eventually the image is at positive infinity.

Table 6.1. *Mapping regions of a positive lens*

Case	Object location	Image location	Figure
(1)	$-\infty$	F^*	Figure 6.6(a)
(2)	Between $-\infty$ and $2F$	Between F^* and $2F^*$	Figure 6.6(b)
(3)	$2F$	$2F^*$	Figure 6.6(c)
(4)	Between $2F$ and F	Beyond $2F^*$	Figure 6.6(d)
(5)	F	∞	Figure 6.6(e)
(6)	Inside F	Beyond ∞ to virtual image, $\lvert z'\rvert > \lvert z\rvert$	Figure 6.6(f)

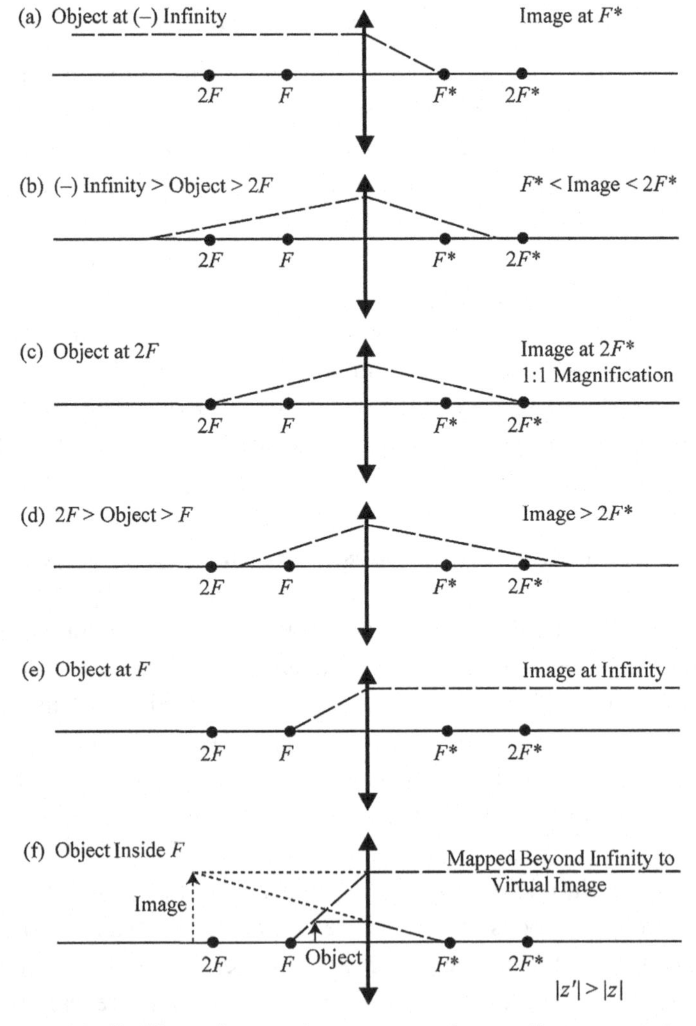

Figure 6.6 (a)–(f) These figures demonstrate the nonlinear mapping of six non-linear possible object locations to conjugate image locations for a positive lens.

6.3.2 Non-linear mapping of a negative lens

Recall that the focal points of a negative lens have spatial locations opposite from those of a positive lens. The back focal point is located to the left of the lens, and the front focal point to the right. The general Gaussian equation, which relates the image distance (z') to the object (z) for a given power, works, provided the correct sign notation is used:

$$\frac{n'}{z'} = \frac{n}{z} + \phi. \tag{6.3}$$

That is, distances to the left are negative and those to the right are positive. The optical power of a negative lens is given by

$$\phi = \frac{n}{-f}$$

As we shall see in the ZZ' diagram discussion (Section 6.6), real objects will always produce virtual images with a negative lens.

A negative lens can have four different object/image mapping regions, which are summarized in Figure 6.7 and Table 6.2. In the first case, consider an object located at negative infinity as in Figure 6.7(a). The image forms at F^*:

$$\frac{1}{z'} = \frac{1}{-\infty} + \frac{1}{-f} \Rightarrow z' = -f = F^*. \tag{6.4}$$

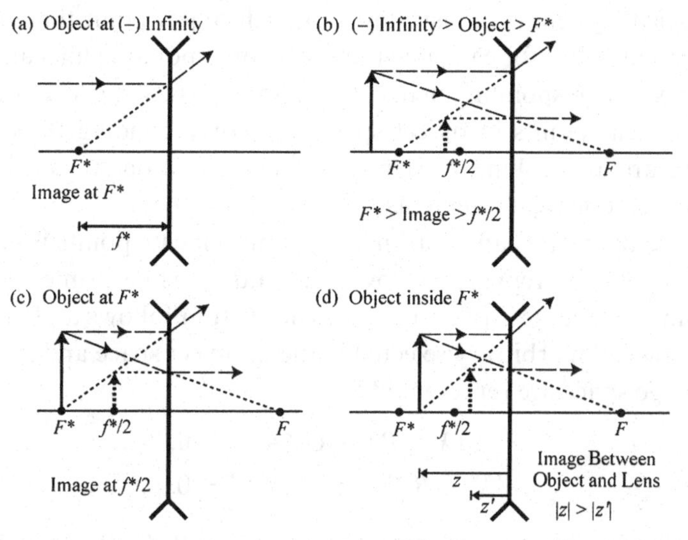

Figure 6.7 (a)–(d) These figures demonstrate the non-linear mapping of object to conjugate image locations for a negative lens.

Table 6.2. *Mapping regions of a negative lens*

Case	Object location	Image location	Figure
(1)	$-\infty$	F^*	Figure 6.7(a)
(2)	Between $-\infty$ and F^*	Between F^* and $f^*/2$	Figure 6.7(b)
(3)	F^*	$f^*/2$	Figure 6.7(c)
(4)	Inside F^*	Between object and lens, $\lvert z\rvert > \lvert z'\rvert$	Figure 6.7(d)

If the object is located between negative infinity and F^*, the image forms between F^* and $f^*/2$ (Figure 6.7(b)). The third case (Figure 6.7(c)) occurs when the object is exactly at F^*. The image forms exactly at $f^*/2$ as given by

$$\frac{1}{z'} = \frac{1}{-f} + \frac{1}{-f} \Rightarrow z' = \frac{-f}{2} = \frac{f^*}{2}. \tag{6.5}$$

Finally, when the object is located inside F^*, the image is always located between the object and the lens such that $\lvert z\rvert > \lvert z'\rvert$.

6.3.3 Collinear transformation in first order optical design

The technique of mathematically mapping geometrical objects from one space to another space, e.g. object space to image space, can be developed purely on the basis of constraints and symmetry requirements. A necessary and sufficient condition for mapping between two spaces is that points in either space must be in a one-to-one correspondence between the two spaces, which is basically what an optical system demonstrates. For each object point there is a corresponding image point, for each line there is a corresponding line, and for each plane there is a corresponding plane. The mapping is collinear if, for every set of three collinear points in object space, the corresponding three points in image space are also collinear. Elements that are in a one-to-one correspondence are called conjugate elements.

We will use the Cartesian coordinate system to locate points in each space. Object space will be represented by x, y, and z, while primed values will represent image space (x', y', and z'). The derivation follows a plane-to-plane correspondence. An arbitrary selected plane in object space and its conjugate plane in image space are represented by

$$\begin{aligned} AX + BY + CZ + D &= 0, \\ A'X' + B'Y' + C'Z' + D' &= 0. \end{aligned} \tag{6.6}$$

For a location in object space, represented by (A, B, C, D), there is a corresponding set of conjugate points (A', B', C', D') in image space.

Consider two planes represented by M and M' which are implicit functions of x, y, z and x', y', and z', respectively:

$$\begin{aligned} AX + BY + CZ + D &= M, \\ A'X' + B'Y' + C'Z' + D' &= M', \end{aligned} \tag{6.7}$$

where M and M' represent families of planes in object and image space respectively. However, only when M and M' are zero do we have a conjugate relation between the planes in the two spaces. For the condition M is zero, M' must also be zero, which can only be true if M' contains M as a factor or:

$$\begin{aligned} M' &= \frac{M(x, y, z)}{P(x, y, z)}, \\ M &= M'/P, \end{aligned} \tag{6.8}$$

where P can be interpreted as the transverse magnification in optical systems, but here it is just a mathematical requirement for collinear transformation. Substituting Equation (6.7) into Equation (6.8):

$$\frac{A'X'}{P} + \frac{B'Y'}{P} + \frac{C'Z'}{P} + \frac{D'}{P} = AX + BY + CZ + D. \tag{6.9}$$

Since each variable is linear on both sides of the equation, individually they must be linear:

$$\begin{aligned} \frac{1}{P} &= a_0x + b_0y + c_0z + d_0, \\ \frac{x'}{P} &= a_1x + b_1y + c_1z + d_1, \\ \frac{y'}{P} &= a_2x + b_2y + c_2z + d_2, \\ \frac{z'}{P} &= a_3x + b_3y + c_3z + d_3. \end{aligned} \tag{6.10}$$

Solving for x', y', and z' provides general equations relating point $P(x, y, z)$ in object space to point $P'(x', y', z')$ in image space for collinear transformations which are (Kreyszig, 1992):

$$\begin{aligned} x' &= \frac{a_1x + b_1y + c_1z + d_1}{a_0x + b_0y + c_0z + d_0}, \\ y' &= \frac{a_2x + b_2y + c_2z + d_2}{a_0x + b_0y + c_0z + d_0}, \\ z' &= \frac{a_3x + b_3y + c_3z + d_3}{a_0x + b_0y + c_0z + d_0}. \end{aligned} \tag{6.11}$$

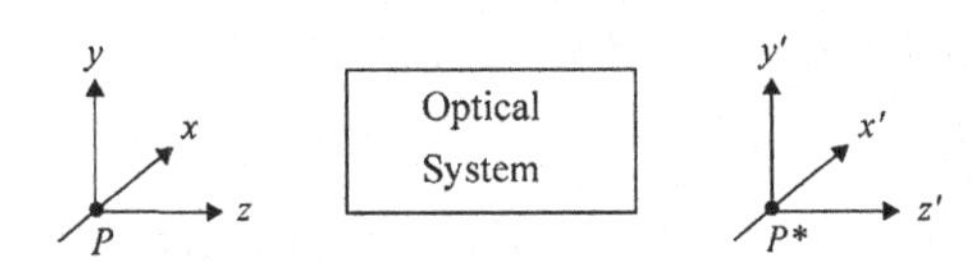

Figure 6.8 Coordinate systems in the object and the image space.

A mapping of one three-dimensional space to another is shown pictorially in Figure 6.8 in which the z axes are collinear.

These equations are the most general for any orientation of the object. We are interested in some subset of these planes being mapped, namely rotationally (axially) symmetric optical systems that are collinear. These constraints eliminate many terms in the equation set, which can then be rewritten as

$$\begin{aligned} x' &= \frac{a_1 x}{c_0 z + d_0}, \\ y' &= \frac{a_1 y}{c_0 z + d_0}, \\ z' &= \frac{c_3 z + d_3}{c_0 z + d_0}. \end{aligned} \tag{6.12}$$

These equations arise due to axial symmetry. x' can only have x terms which have $\sqrt{x^2 + y^2}$ functional dependence and no offset, so $d_1 = 0$. Similarly y' can only have y terms due to $\sqrt{x^2 + y^2}$ functional dependence and no offset, so $d_2 = 0$. This also forces the x' and y' coefficients to be equal, $a_1 = b_2$. In addition, z' must contain only z terms, since both coordinate systems must be collinear in the z terms.

At this point there are two choices for the origin of the coordinate systems: (1) should we make the origin of one space correspond to infinity in the other space; or (2) should we set the origins of the coordinates at unit magnification? One approach leads to the Newtonian equation and the other to the Gaussian equation for first-order optics.

6.3.3.1 Gaussian equation

If we define the origin of each coordinate system ($z = 0$, $z' = 0$), as the plane of unity magnification, similar to the principal planes (P, P^*), this forces $a_1 = d_0$ and $d_3 = 0$. Rewriting Equations (6.12) with a division by d_0, the new equations become

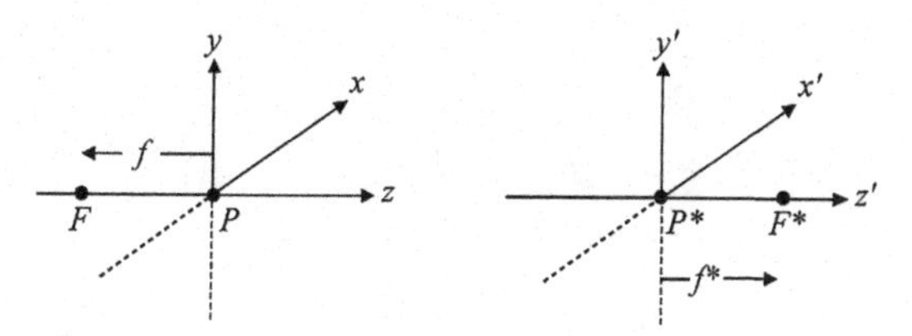

Figure 6.9 Origin of two coordinate systems in the object and the image space.

$$x' = \frac{x}{kz + 1},$$
$$y' = \frac{y}{kz + 1}, \qquad (6.13)$$
$$z' = \frac{k_3 z}{kz + 1}.$$

where k and k_3 are to be determined.

These equations produce a collinear transformation of an axially symmetric system with a coordinate origin for each space at the principal points. This is shown in Figure 6.9.

The transverse magnification is

$$M_t = \frac{y'}{y} = \frac{1}{kz + 1} = \frac{x'}{x} = \frac{1}{kz + 1}. \qquad (6.14)$$

Applying what we know about Gaussian optics, an object at infinity ($z = -\infty$) is imaged at the back focal point, so

$$z' = f^* = k_3/k. \qquad (6.15)$$

This is found by applying L'Hôpital's rule in calculus for ∞/∞. For an object at the front focal point ($z = -f$), the image is at infinity ($z' = \infty$). So

$$z = f; \; z' = k_3/0 = \infty. \qquad (6.16)$$

This can only happen if the denominator is zero. So $kz + 1$ must be zero, or

$$k = -1/z = -1/f \qquad (6.17)$$

since $z = f$. Now from Equation (6.15) we find that

$$k_3 = -f^*/f. \qquad (6.18)$$

By substitution, the collinear transformation equations then become

$$x' = \frac{x}{1 - z/f},$$
$$y' = \frac{y}{1 - z/f}, \tag{6.19}$$
$$z' = \frac{(f^*/f)z}{1 - z/f}.$$

We can now explicitly find the transverse magnification, M_t, from x'/x or y'/y as

$$M_t = \frac{1}{1 + z/f} \tag{6.20}$$

From the third of Equations (6.19), the Gaussian equation can be found:

$$\frac{z'}{z} = -\frac{f^*}{f - z}$$
$$\frac{f - z}{z} = -\frac{f^*}{z'}, \tag{6.21}$$

$$-\frac{f^*}{z'} = \frac{f}{z} - 1. \tag{6.22}$$

Since $f^* = -f$,

$$\frac{1}{z'} = \frac{1}{z} + \frac{1}{f^*}. \tag{6.23}$$

This is the Gaussian object–image relationship for distances measured relative to the principal planes (Equation (5.39)).

6.3.3.2 Newtonian equation

For this derivation we set the origins at the focal points of the two spaces. In this case, we have the origins of object space being conjugate to infinity in image space and vice versa ($z = 0$ when $z' = \infty$, and $z' = 0$ when $z = \infty$). This leads to the equation (its derivation if left as an exercise)

$$zz' = ff^*. \tag{6.24}$$

This is the Newtonian equation for first-order optics.

6.4 Magnification

In optical systems magnification refers to the ratio of the image size to the object size. Three types of magnification are used in describing optical systems: transverse (or lateral) magnification, axial (or longitudinal) magnification, and angular magnification. Transverse magnification is perpendicular to the optical axis, whereas axial magnification is along the optical axis. Angular magnification will be discussed in Chapter 8. These three types of magnification describe the three-dimensionality of the images.

6.4.1 Transverse magnification

Transverse magnification occurs perpendicular to the optical axis, and is denoted M_t. If the magnification is negative, the image is inverted. Similarly, a positive magnification indicates the image is upright. The relationship between object and image height may be determined by examining the geometry of the graphical ray trace shown in Figure 6.10.

First, we will assume a thin lens in which object and image spaces have the same refractive indices, $n = n'$. The undeviated ray forms two similar triangles, ABC and $A'B'C$. This yields the following relationship:

$$\frac{AB}{BC} = \frac{A'B'}{B'C} \quad \text{or alternatively} \quad \frac{h}{z} = \frac{h'}{z'}. \tag{6.25}$$

Rearranging, we get

$$\frac{h'}{h} = \frac{z'}{z}. \tag{6.26}$$

Finally, magnification is defined as the ratio of image size to object size, yielding the equation

$$M_t = \frac{h'}{h} = \frac{z'}{z}. \tag{6.27}$$

Thus transverse magnification may be defined as the ratio of image height to object height or as the ratio of image distance to object distance. However, if the refractive indices of the object and image spaces are different, the formula

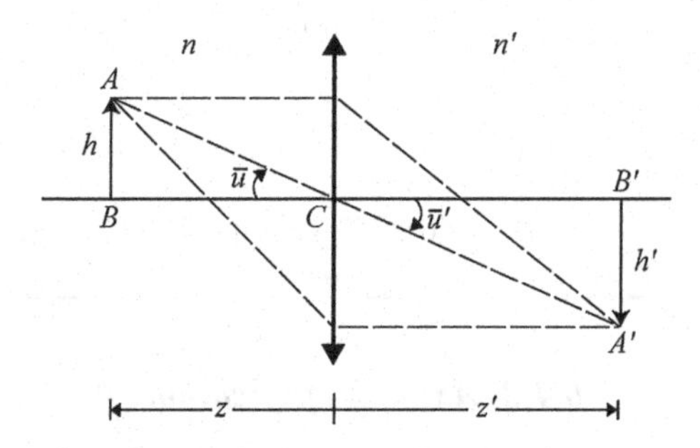

Figure 6.10 Transverse magnification.

must be modified. Again, we use the similar triangles in Figure 6.10. Both $\bar{u}$ and $\bar{u}'$ are clockwise, and therefore are negative angles. Solving for $\bar{u}$ and $\bar{u}'$ using small angle approximations gives

$$\bar{u} = \frac{h}{z} \text{ and } \bar{u}' = \frac{h'}{z'}. \tag{6.28}$$

Notice that, by the sign convention, z and h' are negative, giving negative values for $\bar{u}$ and $\bar{u}'$ as expected. The refraction equation, $n' - u' = n - u - y\phi$, in which $y = 0$ at the image, reduces to

$$n' - u' = n - u. \tag{6.29}$$

Substituting Equation (6.28) into Equation (6.29) we obtain

$$n'\frac{h'}{z'} = n\frac{h}{z}. \tag{6.30}$$

Rearranging, we obtain the general equation for transverse magnification:

$$M_\mathrm{t} = \frac{h'}{h} = \frac{z'/n'}{z/n}. \tag{6.31}$$

Example 6.1

An object 15 mm high is located 200 mm from a negative lens, $f^* = -120$ mm. Calculate the transverse magnification and the image height.
First calculate the optical power of the lens:

$$\phi = \frac{1}{f*} = \frac{1}{-120\,\mathrm{mm}} \cong -0.00833\,\mathrm{mm}^{-1} \cong -8.33 \text{ diopters.}$$

Assuming the lens and object are in air, calculate the image location:

$$z' = \frac{n'z}{n + z\phi} = \frac{-200\,\mathrm{mm}}{1 + (-200\,\mathrm{mm})(-0.00833\,\mathrm{mm}^{-1})} = -75\,\mathrm{mm}.$$

The transverse magnification is then:

$$M_\mathrm{t} = \frac{z'}{z} = \frac{-75\,\mathrm{mm}}{-200\,\mathrm{mm}} = 0.375;$$

and the image height is:

$$h' = \frac{z'h}{z} = M_t h = (0.375)(15\,\mathrm{mm}) = 5.625\,\mathrm{mm}.$$

6.4.2 Axial magnification

Axial magnification, also known as longitudinal magnification, occurs along the optical axis, and is represented by M_z. The relationship between object and

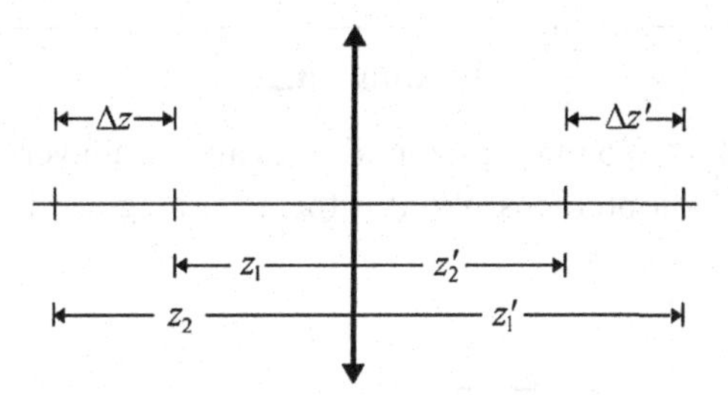

Figure 6.11 Axial magnification.

image width may be determined by examining the relationships in Figure 6.11. The ratio of image to object width is

$$M_z = \Delta z'/\Delta z. \tag{6.32}$$

Taking the partial derivatives of the general Gaussian equation,

$$\frac{n'}{z'} = \frac{n}{z} + \frac{1}{f^*} \tag{6.33}$$

with respect to z' and z yields

$$\frac{\mathrm{d}z' n'}{z'^2} = \frac{\mathrm{d}z\, n}{z^2}, \tag{6.34}$$

which can be rearranged into the form

$$\frac{\mathrm{d}z'}{\mathrm{d}z} = \frac{z'^2/n'}{z^2/n}. \tag{6.35}$$

Substituting Equation (6.35) into Equation (6.32) yields $M_z = z'^2/z^2$. Recalling $M_t = z'/z$, axial magnification may be written as the transverse magnification squared, provided the thickness is small:

$$M_z = \frac{z'^2/n'}{z^2/n} = M_t^2 \frac{n'}{n}. \tag{6.36}$$

If the object and image spaces have different refractive indices, the equation becomes

$$M_z = M_t^2 \frac{n'}{n} = \frac{(z'/n')^2}{(z/n)^2} \times \frac{n'}{n}. \tag{6.37}$$

The expressions for transverse and axial magnification are for objects and images along the optical axis. The image is also not always magnified. Magnification can be less than 1, in which case, the image is smaller than the object.

Example 6.2

A transparent block with 5 mm horizontal sides and 7 mm vertical sides is placed 75 cm in front of a 2 diopter lens. What is the axial magnification?

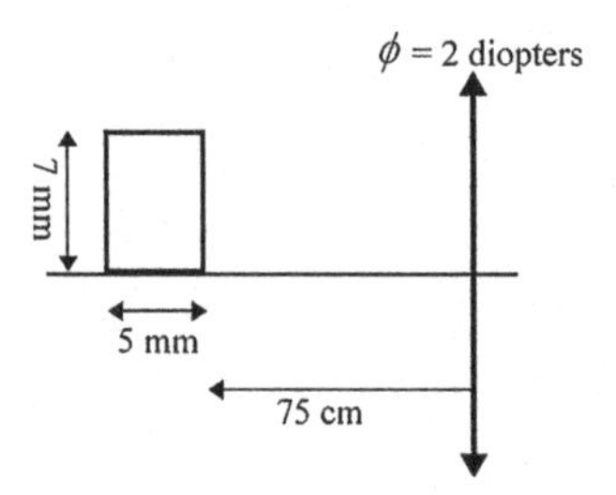

From the information given we can calculate the change in object distance:

$$\Delta z = 0.5\,\mathrm{cm}.$$

We must find the image location for both the right-hand and left-hand surfaces of the block:

$$(\text{right})\ z' = \frac{n'z}{n + z\phi} = \frac{-75\,\mathrm{cm}}{1 + (-75\,\mathrm{cm})(0.02\,\mathrm{cm}^{-1})} = 150.0\,\mathrm{cm},$$

$$(\text{left})\ z' = \frac{n'z}{n + z\phi} = \frac{-75.5\,\mathrm{cm}}{1 + (-75.5\,\mathrm{cm})(0.02\,\mathrm{cm}^{-1})} = 148.0\,\mathrm{cm},$$

$$\Delta z' = 2\,\mathrm{cm},$$

$$M_z = \frac{\Delta z'}{\Delta z} = \frac{2\,\mathrm{cm}}{0.5\,\mathrm{cm}} = 4.0.$$

The axial magnification can also be calculated from the transverse magnification. The M_t from either the left-hand or right-hand surface may be used. Using the right-hand surface:

$$M_\mathrm{t} = \frac{z'}{z} = \frac{150\,\mathrm{cm}}{-75\,\mathrm{cm}} = -2.0,$$

$$M_\mathrm{t}^2 = (-2)^2 = 4 = M_z.$$

6.5 *F*-number

The brightness of an image depends on both the focal length and the lens diameter (more appropriately, the entrance pupil diameter). *F*-numbers describe the light gathering ability of a lens or optical system. There are two types of F-number calculations, F-number (infinity) and F-number (working), represented by $(F/\#)_\infty$ and $(F/\#)_w$ respectively (e.g. $F/2.8$ means the $F/\#$ is 2.8).

6.5.1 *F/# (infinity)*

F-number (infinity) in its simplest form is defined as the ratio of focal length to lens diameter (entrance pupil diameter), yielding the equation

$$(F/\#)_\infty \equiv \frac{f^*}{D_{\text{ent}}} = \frac{\text{focal length}}{\text{lens diameter}} = \frac{\text{focal length}}{\text{entrance pupil diameter}}. \tag{6.38}$$

A 25 mm diameter lens with a 50 mm focal length has an $(F/\#)_\infty = 2$ and is designated as $F/2$. The F-number of a lens refers to an object at infinity. Examining Figure 6.12 and using Equation (6.38) yields the relationship

$$(F/\#)_\infty = \frac{f^*}{d} = \frac{1}{2\tan\theta}. \tag{6.39}$$

However, a spherical wavefront converges to the focal point, obeying the Abbe sine condition. Thus, a more accurate and preferred definition of F-number is

$$(F/\#)_\infty = \frac{1}{2\sin\theta}. \tag{6.40}$$

A small F-number indicates a large lens diameter or aperture and the ability to gather more light. In photography the F-number is often referred to as the "speed" of the lens. Smaller F-numbers are "faster" requiring shorter exposure times. A slow system has a high F-number. A fast speed allows poorly lit or moving objects to be photographed. F-number is also often referred to as the "F-stop" of a lens. When stopping down a lens, the entrance pupil diameter is decreased, and this increases the F-number and results in less light passing through to the image. Although this produces a greater depth of field, it sacrifices image brightness. Camera lenses typically have markings such as 1, 1.4, 2, 2.8, etc. The 1 represents $F/1$, and the F-number is multiplied by $\sqrt{2}$ for

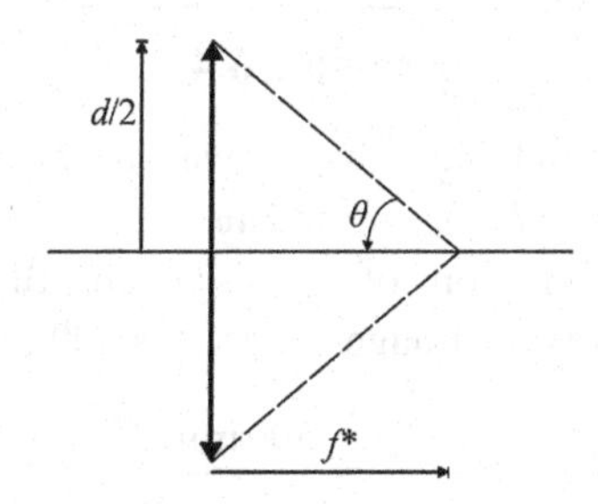

Figure 6.12 F-number (infinity), $(F/\#)_\infty$.

each successive setting. F-number may also be described in terms of the numerical aperture, NA, of a system:

$$NA = n' \sin u', \tag{6.41}$$

where n' is the refractive index of the medium and u' is the half angle in radians. If a medium with a high refractive index is used, the NA may be greater than 1, such as in oil immersion microscope objectives. The numerical aperture is related to F-number by:

$$F/\# = \frac{1}{2NA} = \frac{1}{2n' \sin u'}. \tag{6.42}$$

Example 6.3

A negative lens has a back focal length of –100 mm and a diameter of 20 mm. What is the lens $(F/\#)_\infty$?

$$(F/\#)_\infty = \frac{f^*}{D_{\text{ent}}} = \frac{-100\,\text{mm}}{20\,\text{mm}} = -5 = |5|.$$

6.5.2 F/# (working)

While $(F/\#)_\infty$ is a practical and consistent way to label lenses, not all objects are located at infinity. The working F-number takes into account the real image distance. $(F/\#)_\text{w}$ is defined as the ratio of the image distance to the lens (or entrance pupil) diameter, yielding the equation:

$$(F/\#)_\text{w} \equiv \frac{z'}{D_{\text{ent}}} = \frac{\text{image distance}}{\text{lens (entrance pupil) diameter}}. \tag{6.43}$$

$(F/\#)_\infty$ and $(F/\#)_\text{w}$ have similarities that can be seen by recalling the non-linear mapping relationship for an object at infinity. For an object at infinity, the image forms at the focal point. Therefore, the focal length in this case is identical to the image distance in Equation (6.43).

Example 6.4

An object is placed 400 mm in front of a positive lens ($f = 200$ mm) the diameter of which is 14 mm. What is the $(F/\#)_\text{w}$ of the lens?

Recall, from the mapping regions of a positive lens, that an object placed at $2F$ forms an image at $2F^*$. Thus the image distance is 400 mm.

$$(F/\#)_\text{w} = \frac{z'}{D_{\text{ent}}} = \frac{400\,\text{mm}}{14\,\text{mm}} = 28.57.$$

6.6 ZZ' diagram

6.6.1 Derivation and construction

The relationship between object and image distances and focal length can be graphically represented using the Cartesian coordinate system. The derivation of the ZZ' diagram begins with the recollection of the Gaussian equation for a thin lens in air:

$$\frac{1}{Z'} = \frac{1}{Z} + \phi. \tag{6.44}$$

Rearranging,

$$\frac{1}{Z'} - \frac{1}{Z} = \phi$$

$$\frac{1}{Z'\phi} - \frac{1}{Z\phi} = 1. \tag{6.45}$$

Recall,

$$\phi = \frac{-1}{f} = \frac{1}{f^*}. \tag{6.46}$$

Substituting Equation (6.46) into Equation (6.45) yields

$$\frac{f^*}{Z'} + \frac{f}{Z} = 1. \tag{6.47}$$

Equation (6.47) can be interpreted as the intercept–intercept form of a line. In this form, f^* and f are represented by x and y, respectively, in the Cartesian coordinate system:

$$\frac{x}{Z'} + \frac{y}{Z} = 1, \tag{6.48}$$

where the x and y coordinate intercepts are Z and Z'. If we substitute for a point on this line of (f^*, f), Equation (6.47) is obtained. The foundation of the ZZ' diagram is where the (x,y) coordinates can now be defined as the pivot point (f^*, f).

The Cartesian axes are then constructed from a generic object and image space diagram for a single positive lens (Figure 6.13). Object space is physically $-\infty$ to 0, and image space is 0 to $+\infty$, but, conceptually, both object and image spaces are infinite in length, ranging from $-\infty$ to $+\infty$. This allows the separation of the object and image space axes. As indicated in Figure 6.14, objects and images that lie in physical space are termed real objects ($-\infty$ to 0)

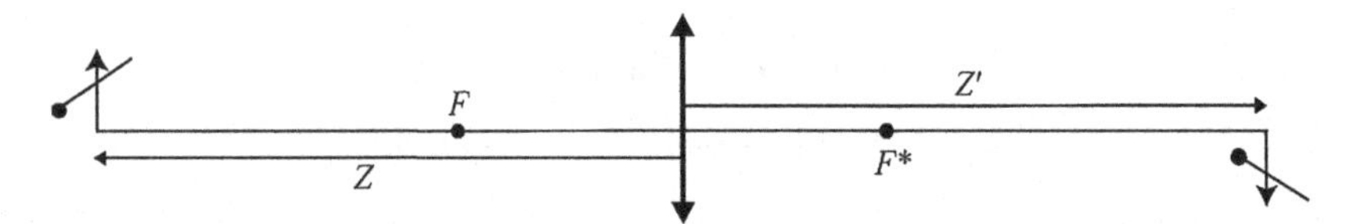

Figure 6.13 Classical object and image space relationship.

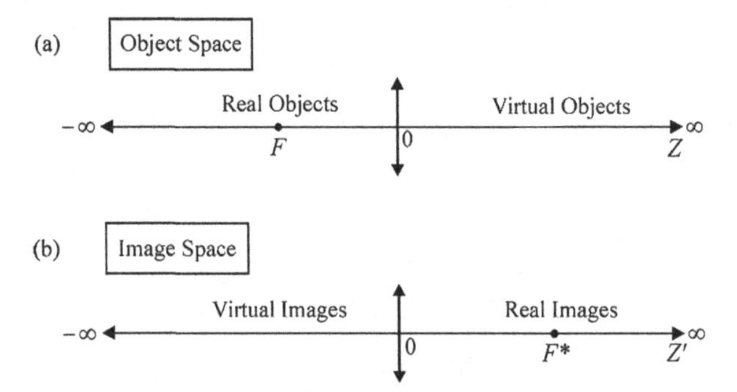

Figure 6.14 Separation of (a) object and (b) image space axes.

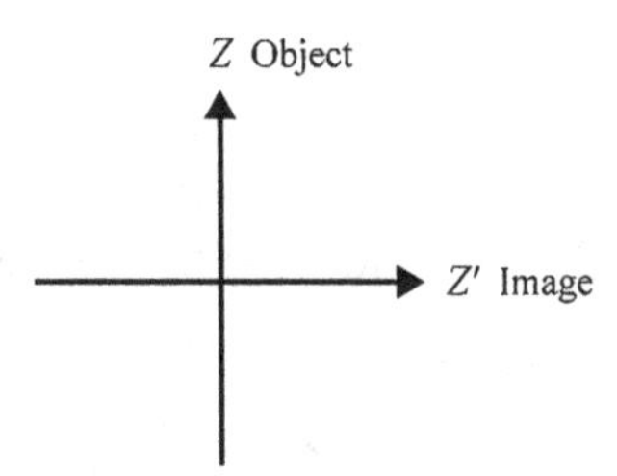

Figure 6.15 ZZ' diagram.

and real images (0 to $+\infty$) in image space. Virtual objects (0 to $+\infty$) and virtual images ($-\infty$ to 0) are shown in Figure 6.14(a) and (b).

The Cartesian axes are constructed by rotating the object space axis counterclockwise through 90° and placing it over the image space axis so that the intersection point or thin lens is at the origin. The y axis represents object distances, and the x axis image distances. This is the ZZ' diagram (Figure 6.15). The ZZ' diagram is valid for both positive and negative lenses, provided a proper sign convention is used.

A positive thin lens on the ZZ' diagram is shown in Figure 6.16. If both object and image distances are known, a straight line connecting the positions will always pass through the point (f^*, f). Likewise, by knowing the values of f^* and f, multiple lines may be drawn through the pivot point (f^*, f) to indicate possible object and image locations and determine whether the object or image

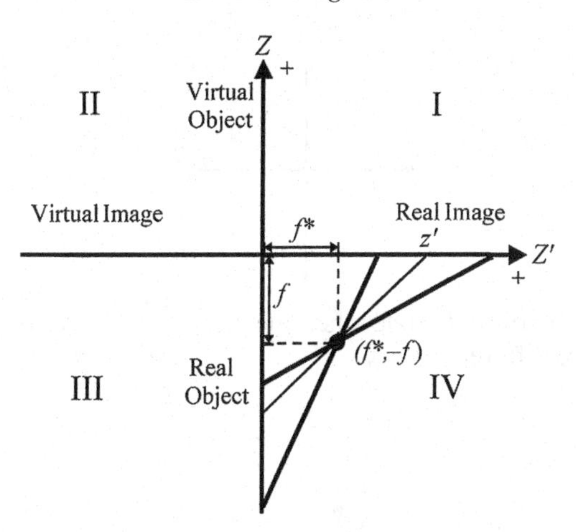

Figure 6.16 ZZ' diagram with a positive thin lens for three object–image conjugates.

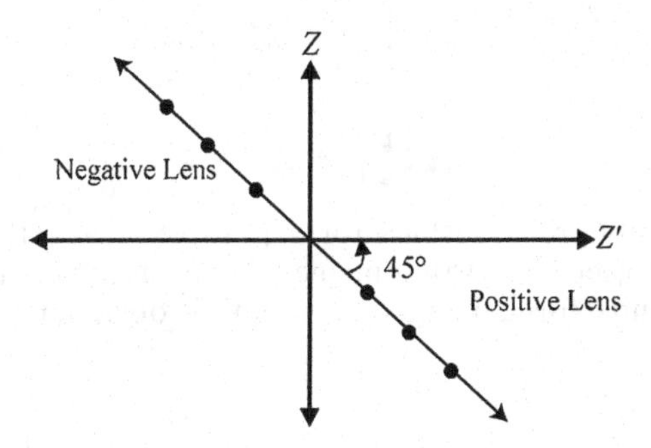

Figure 6.17 The point (f^*, f) lies at 45° to the y axis when the object and image space refractive indices are equal.

is real or virtual. Notice that it is impossible to draw a line through the origin which represents object and image conjugate locations for the positive lens in Figure 6.16.

Quadrant II corresponds to negative lenses, and quadrant IV to positive lenses. Any point in quadrants II or IV corresponds to a lens with the given front and back focal lengths. If the object and image spaces have the same index of refraction, the points (f^*, f) fall on a 45° diagonal, as illustrated in Figure 6.17. If the refractive indices in object and image space are not equal, the slope of the line is not 45°, but is equal to the negative ratio of the indices. If the locus of the position of the thin lens is as shown in Figure 6.18, the angle formed with the z axis is less than 45°; therefore, object space refractive index is greater than image space index ($n'' < n$).

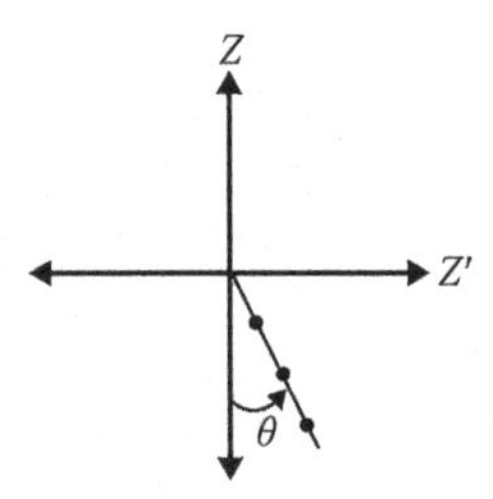

Figure 6.18 Reduction of angle when the refractive indices of the object and image spaces are different.

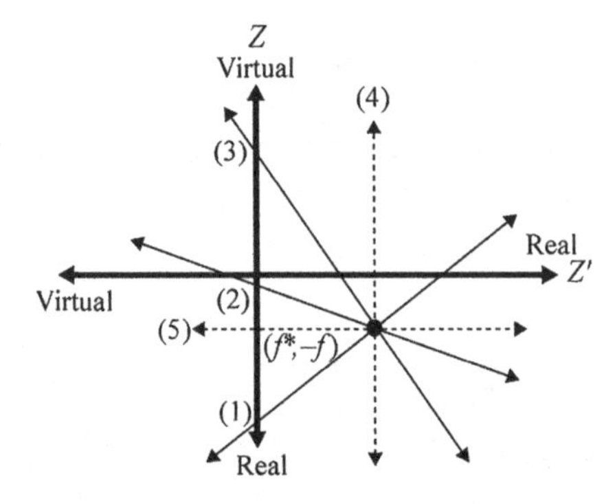

Figure 6.19 Possible relationships for a positive lens: (1) real object and image, (2) real object and virtual image, (3) virtual object and real image, (4) object at infinity forms image at f^*, and (5) object at f forms image at infinity.

6.6.2 Positive lenses

For a positive lens, the pivot point is in quadrant IV of the ZZ' diagram. There are three choices of arrangement of object and image for a positive lens (Figure 6.19).

The various object–image lines in Figure 6.19 all pass through the pivot point $(f^*,-f)$. Line (1), which has a real object, produces a real image. The location (2) object produces a virtual image. Location (3), a virtual object, produces a real image. This diagram readily provides not only the type of image, but also its axial location. The transverse magnification is simply the negative inverse of the slope.

The possible choices for the object–image relationship for a positive lens are shown in Table 6.3. Note it is impossible to have a virtual object and virtual image with a positive lens. Recalling the mapping of object and image spaces, an object at $-\infty$ has an image at f^*, and an object at $-f$ has an image at ∞, which is indicated with dotted lines in Figure 6.19 as (4) and (5).

Table 6.3. *The only realizable object–image relationships for a positive thin lens*

(1) Real object	→	Real image
(2) Real object	→	Virtual image
(3) Virtual object	→	Real image

Example 6.5

An object is 30 cm in front of a positive lens with a focal length of 10 cm. Where is the image located? What is the transverse magnification?

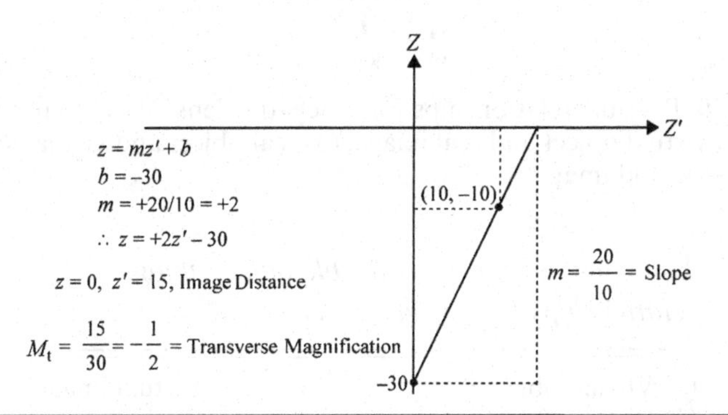

Z
Z′
z = mz′ + b
b = –30
m = +20/10 = +2
∴ z = +2z′ – 30
z = 0, z′ = 15, Image Distance
M_t = 15/30 = – 1/2 = Transverse Magnification
(10, –10)
m = 20/10 = Slope
–30

Example 6.6

An object is 6 cm in front of a positive lens of 25 diopters power. Where is the image located? What is the transverse magnification?

$z = mz' + b$

$f^* = \frac{1}{\phi} = \frac{M}{25} = \frac{100\text{ cm}}{25} = 4\text{ cm}$

$m = \frac{2}{4}, \; b = -6$

$z = \frac{1}{2}z' - 6$

Intercept on z', $z = 0$

$0 = \frac{1}{2}z' - 6$

$z' = 12$

$M_t = \frac{12}{-6} = -2$

Z, Z′, (4,–4), (0,–6)

6.6.3 Negative lenses

For a negative lens, the pivot point is in quadrant II, as shown in Figure 6.20.

There are three possible relationships between object and image, as shown in Table 6.4.

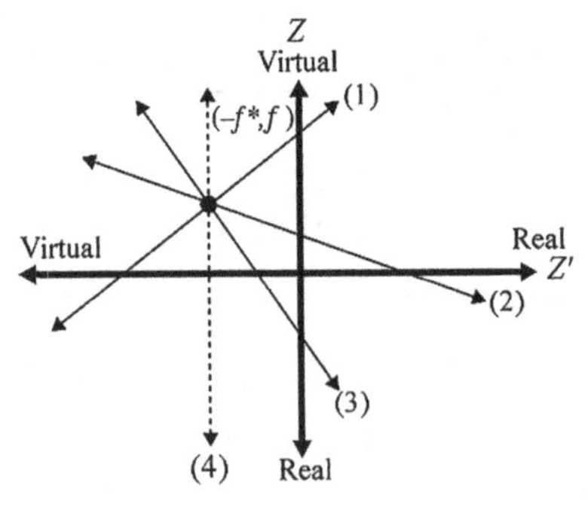

Figure 6.20 Possible relationships for a negative lens: (1) virtual object and image, (2) virtual object and real image, (3) real object and virtual image, (4) object at $-\infty$ and image at $-f^*$.

Table 6.4. *The only realizable object–image relationships for a negative lens*

(1) Virtual object	→	Virtual image
(2) Virtual object	→	Real image
(3) Real object	→	Virtual image

Note: It is impossible for a negative lens to have a real object and real image. The slope of the line can again be used to indicate the transverse magnification.

Example 6.7

An object is 12 cm in front of a negative diverging lens of $-16\frac{2}{3}$ diopters. Where is the image located? Is it real or virtual? What is the transverse magnification?

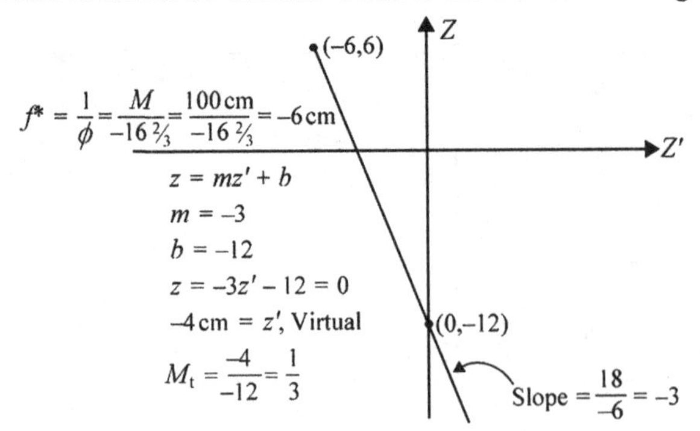

6.7 Thick lens equivalent of thin lens

Optical systems typically consist of more than one lens. The equations for a single optical surface can be applied to multiple lens systems. Consider two thin lenses, L_1 and L_2, separated by a distance or thickness t (Figure 6.21(a)). To determine the final location of an image, one can use the lens maker equation to find the image location for L_1. The location of this image is then calculated relative to L_2. This image is now used as the object for L_2 and its image location is calculated. While this method works, it is very cumbersome when used for multiple lens systems. It is much easier to have equations for finding the thin lens equivalent of a multiple lens element system. In Chapter 10 you will learn how to set up a paraxial ray trace table to simplify the process.

To find the equivalent optical power of lens combinations, consider two thin lenses of optical powers ϕ_1 and ϕ_2 separated by a thickness t, in air (Figure 6.21(a)). These lenses can be reduced to a single thin lens equivalent of total optical power ϕ_{12} (Figure 6.21(b)). Tracing a ray from infinity at height y_1 and letting $n = n' = n'' = 1$, the refraction equation, $n'u' = nu - y\phi$, may be used to derive the total optical power for a single lens equivalent. Recall that a ray from infinity focuses at F^*.

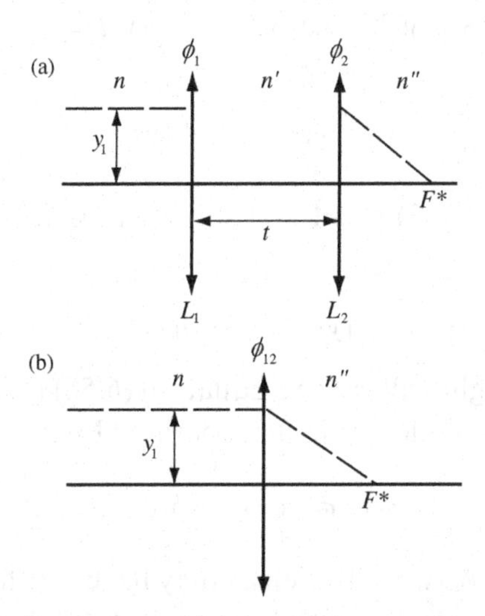

Figure 6.21 (a) Two thin lenses separated (in air) by a distance t; (b) a single thin lens with the equivalent optical power of the two lenses shown in (a).

From Figure 6.21(a),

$$n'u' = nu - \phi_1 y_1 \text{ where } n = n' = 1 \text{ and}$$
$$u = 0 \text{ at } L_1 \text{ reduces to } u' = -\phi_1 y_1, \tag{6.49}$$
$$n''u'' = n'u' - \phi_2 y_2 \text{ where } n' = n'' = 1 \text{ at } L_2 \text{ reduces to}$$
$$u'' = u' - \phi_2 y_2. \tag{6.50}$$

Substituting Equation (6.49) into Equation (6.50) yield:

$$u'' = -\phi_1 y_1 - \phi_2 y_2. \tag{6.51}$$

From Figure 6.21(b), $n''u'' = nu - \phi y_1$ where $n = n' = 1$ and

$$u = 0 \text{ reduces to } u'' = -\phi y_1. \tag{6.52}$$

Substituting Equation (6.52) into Equation (6.51) yields

$$-\phi y_1 = -\phi_1 y_1 - \phi_2 y_2. \tag{6.53}$$

The transfer equation,

$$y_2 = y_1 + n'u'\left(\frac{t}{n'}\right), \tag{6.54}$$

reduces to $y_2 = y_1 + tu'$, and therefore, $u' = (y_2 - y_1)/t$, but from Equation (6.49), $u' = -\phi_1 y_1 = (y_2 - y_1)/t$ becomes $-\phi_1 y_1 t = y_2 - y_1$, and rearranging gives

$$-\phi_1 y_1 t + y_1 = y_2. \tag{6.55}$$

Multiplying Equation (6.53) by –1 and substituting in Equation (6.55) for y_2 yields

$$\phi y_1 = \phi_1 y_1 + \phi_2(-\phi_1 y_1 t + y_1). \tag{6.56}$$

Noting that any y_1 height will cancel, Equation (6.56) reduces to the combination equation for two thin lenses in air separated by t:

$$\phi = \phi_1 + \phi_2 - \phi_1 \phi_2 t. \tag{6.57}$$

However, the medium between the lenses may have a different refractive index where $n = n'' \neq n'$. In this case, Equation (6.49) becomes $n'u' = -\phi_1 y_1$. Substituting for $n'u'$ in Equation (6.54) yields

$$y_2 = y_1 - \phi_1 y_1 t/n'. \tag{6.58}$$

Substituting Equation (6.58) into Equation (6.53) and multiplying by –1 yields

$$\phi y_1 = \phi_1 y_1 + \phi_2 y_1 - \frac{\phi_1 \phi_2 y_1 t}{n'}, \tag{6.59}$$

which can be rearranged into Gullstrand's equation:

$$\phi = \phi_1 + \phi_2 - \phi_1 \phi_2 \frac{t}{n'}. \tag{6.60}$$

Gullstrand's equation, which finds the total optical power of a single thin lens equivalent, works for both thin and thick lenses. For lenses separated by air, $n' = 1$ and drops out of the equation. This equation also works for finding the total optical power of a thin lens in which the two refractive surfaces have a $t = 0$ separation. In that case, the equation reduces to $\phi = \phi_1 + \phi_2$. Thus, for thin lenses in contact, the resultant total power is the sum of the powers of the individual refractive surfaces, recalling $\phi_1 = (n' - n)/R_1$ and $\phi_2 = (n'' - n')/R_2$. Note that the location of this equivalent lens has not been defined. Hence, the exact location of the final image cannot be determined by this equation alone.

Example 6.8

Two thin lenses with the following prescriptions are placed against each other (in contact) at their matching –25 cm radii.

Lens 1: $R_1 = 20$ cm, $R_2 = -25$ cm, $n_1 = 1.4$;
Lens 2: $R_1 = -25$ cm, $R_2 = 100$ cm, $n_2 = 1.8$.

Calculate: (a) the individual optical powers and focal lengths of the two lenses, and (b) the combined optical power and focal length of the doublet.

(a) Lens 1:

$$\phi = \frac{1.4 - 1}{20} + \frac{1 - 1.4}{-25} = 0.036 \text{ cm}^{-1} = 3.6 \text{ diopters},$$

$$f^* = \frac{n'}{\phi} = \frac{1}{0.036 \text{ cm}^{-1}} = 27.8 \text{ cm} = 0.278 \text{ m}.$$

Lens 2:

$$\phi = \frac{1.8 - 1}{-25} + \frac{1 - 1.8}{100} = -0.04 \text{ cm}^{-1} = -4 \text{ diopters},$$

$$f^* = \frac{n'}{\phi} = \frac{1}{-0.04 \text{ cm}^{-1}} = -25 \text{ cm} = -0.25 \text{ m}.$$

(b) The combined optical power is the sum of the individual optical powers calculated above! There are three surface optical powers. The thickness between the surfaces is zero, so the combined power is the sum of the three surface optical powers: $\phi = \phi_1 + \phi_2 + \phi_3$.

$$\phi_1 = \frac{n_{L1} - 1}{R_1} = \frac{1.4 - 1}{20} = 0.02\ \text{cm}^{-1},$$

$$\phi_2 = \frac{n_{L2} - n_{L1}}{R_2} = \frac{1.8 - 1.4}{-25} = -0.016\ \text{cm}^{-1},$$

$$\phi_3 = \frac{1 - n_{L2}}{R_3} = \frac{1 - 1.8}{100} = -0.008\ \text{cm}^{-1},$$

$$\phi = \phi_1 + \phi_2 + \phi_3 = 0.02 - 0.016 - 0.008 = -0.004\ \text{cm}^{-1},$$

$$f^* = \frac{1}{\phi} = \frac{1}{-0.004\,\text{cm}^{-1}} = -250\ \text{cm}.$$

6.8 Newtonian optics

In Gaussian optics the object and image distances are measured from the front and rear principal planes. In a single thin lens, the principal planes are coincident, and the object and image distances are the same as the distance from the lens itself. However, as we saw in Section 6.7 (and as we shall see in Chapter 7), when two or more thin lenses are combined, the use of a thick lens requires additional calculations to determine the location of the principal planes from the lens surfaces. In Newtonian optics, the object and image distances are measured from the foci, eliminating the need to worry about lens thickness and the location of the principal planes. By examining the two sets of similar triangles in Figure 6.22, Newton's formula can be derived.

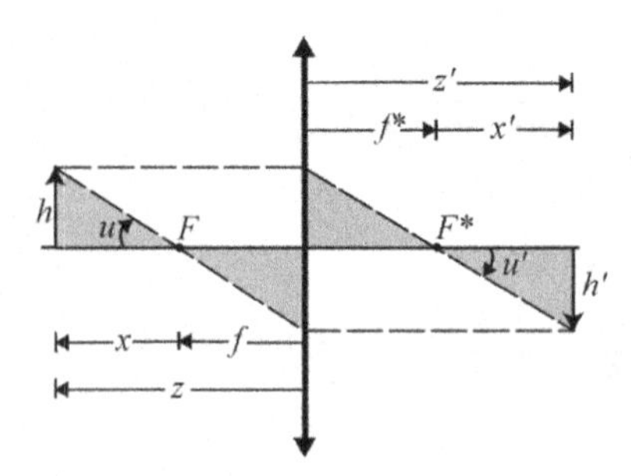

Figure 6.22 Newtonian conjugates.

To the left of the lens, the shaded triangles are related by the equation

$$-u = \frac{h}{-x} = \frac{-h'}{-f}. \qquad (6.61)$$

Rearranging, we obtain

$$\frac{h'}{h} = \frac{-f}{x}. \qquad (6.62)$$

To the right of the lens a similar equation is found:

$$-u' = \frac{-h'}{x'} = \frac{h}{f^*}. \qquad (6.63)$$

Rearranging, we obtain

$$\frac{h'}{h} = \frac{-x'}{f^*}. \qquad (6.64)$$

Combining Equations (6.62) and (6.64) yields

$$\frac{f}{x} = \frac{x'}{f^*}, \qquad (6.65)$$

which can then be rearranged into Newton's formula:

$$xx' = ff^*. \qquad (6.66)$$

The Newtonian formula can be used for both thick and thin lenses since the object and image distances are measured from the focal points and not from a physical lens reference point. The Newtonian distances can be related to the Gaussian distances by

$$z = x + f \quad \text{and} \quad z' = x' + f^*. \qquad (6.67)$$

The transverse magnification equation can also be written in Newtonian form:

$$M_t = \frac{-f}{x} = \frac{-x'}{f^*} = \frac{h'}{h}. \qquad (6.68)$$

6.9 Cardinal points of a thin lens

So far we have discussed the focal points of a lens; however, there are two other sets of points that are also important in order to fully characterize a lens. These

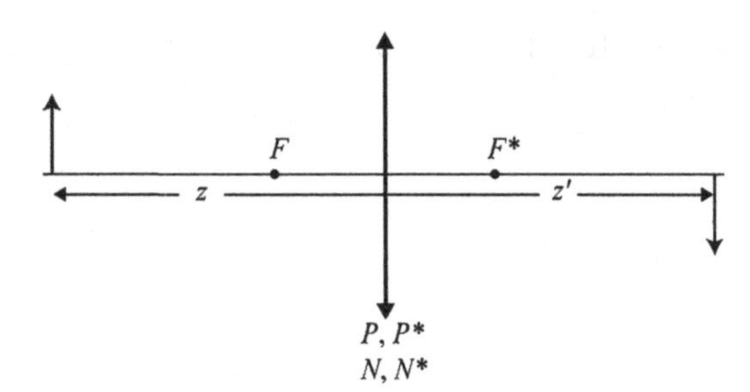

Figure 6.23 Thin lens cardinal points.

two sets of points are the principal points (P, P^*) and the nodal points (N, N^*). Each of these points lies on the optical axis and has a corresponding plane in the x–y direction located perpendicular to the optical axis at these four points on the optical axis. Unlike for focal points, the planes of these principal and nodal points are conjugate planes. Figure 6.23 shows these points for a single thin lens. The principal points and nodal points lie at the thin lens, and all four planes are collocated for this special case.

The principal planes are defined as a pair of planes where unit transverse magnification takes place in an optical system, and they are the reference from which the object and image distances are measured. In other words, any ray arriving at P will exit at P^* with the same height ($M_t = 1$). The nodal points are where the angular magnification is 1. That is, the ray arriving at N with a certain angle, u, will exit the optical system at N^* with the same angle. In more complex optical systems, the principal points do not coincide, and neither do the nodal points.

6.10 Thin lens combinations

Putting together several lenses to obtain a different optical power is a way to avoid the expensive and time-consuming process of fabricating a new lens with different radii of curvature. In addition, we will learn later that using several lenses to obtain a given optical power provides a means of producing better image quality also.

Consider two thin lenses (ϕ_1 and ϕ_2) separated in air by a thickness t. What is the equivalent optical power of the combination? In order to answer that question, let us first consider a ray from infinity, which will be focused by the two-lens combination, and compare that with an equivalent single lens which would produce the same resulting focus point. Shown in Figure 6.24(a) are two thin lenses separated by a thickness, t, with the corresponding equivalent lens in Figure 6.24(b).

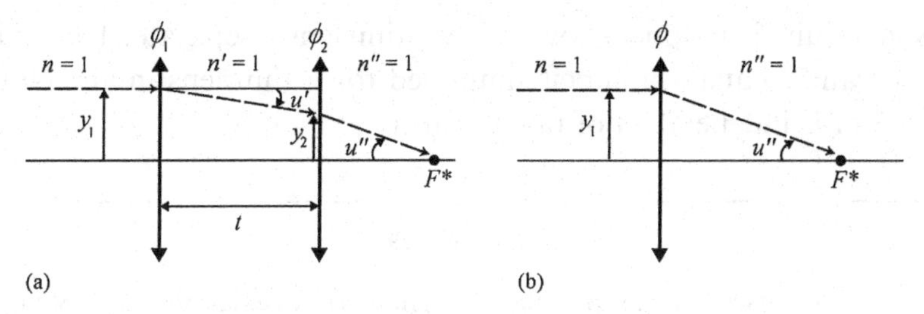

Figure 6.24 (a) Two thin lenses in series separated by air and (b) the equivalent single lenses.

If we choose a ray from infinity ($nu = 0$), for a single refraction in Figure 6.24(b), the ray propagates to the back focal point F^* at some angle u'', which follows from the refraction equation:

$$n''u'' = 0 - y_1\phi,$$
$$\phi = -u''/y_1, \tag{6.69}$$

where ϕ represents the optical power of the equivalent single lens.

Applying the refraction equation to the thin lens combination of Figure 6.24(a):

$$n'u' = 0 - y_1\phi_1,$$
$$u' = -y_1\phi_1. \tag{6.70}$$

From the transfer equation,

$$y_2 = y_1 + tu' \tag{6.71}$$

and applying the refraction equation at the second lens:

$$u'' = u' - y_2\phi_2 \tag{6.72}$$

or from Equations (6.70) and (6.71):

$$u'' = -y_1\phi_1 - \phi_2(y_1 + tu')$$
$$= -y_1\phi_1 - y_1\phi_2 + ty_1\phi_1\phi_2. \tag{6.73}$$

Moving $-y_1$ from the right-hand side of the equation to the left and noting that this is just equal to Equation (6.69), we obtain

$$-\frac{u''}{y_1} = \phi_1 + \phi_2 - t\phi_1\phi_2,$$
$$\phi = \phi_1 + \phi_2 - t\phi_1\phi_2, \tag{6.74}$$

which is the equivalent optical power of two thin lenses separated by air. This is also Gullstrand's equation, albeit simplified for a thin lens in air. Note this optical power is independent of ray height, y.

Example 6.9

Using two thin lenses in air, separated by 4 cm, with focal lengths $f_1^* = 8$ cm, and $f_2^* = 3$ cm, what is the equivalent power?

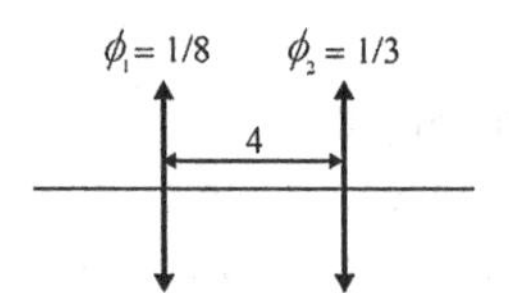

$$\phi = \tfrac{1}{8} + \tfrac{1}{3} - 4\left(\tfrac{1}{8}\right)\left(\tfrac{1}{3}\right) = \tfrac{7}{24}\,(\text{cm}^{-1}),$$
$f = 3\tfrac{3}{7}$ cm, the effective focal length.

6.10.1 *Unique separations of two thin lenses*

There are three special distances between two lenses that are worth discussing when combining two thin lenses:

(a) $t = 0$: no separation;
(b) $t = f_1^*$ or f_2^*: the separation is equal to one of the focal lengths;
(c) $t = f_1^* + f_2^*$.

In the first case ($t = 0$), the equivalent optical power is just the sum of the powers ($\phi = \phi_1 + \phi_2$). This is achieved when two thin lenses are in direct contact. For the second case ($t = f_1^*$ or f_2^*), if the separation is one of the focal lengths, the equivalent optical power is just the value of the reciprocal of the separation. By making the substitution $t = 1/\phi_1$,

$$\begin{aligned} \phi &= \phi_1 + \phi_2 - \left(\frac{1}{\phi_1}\right)\phi_1\phi_2 \\ &= \phi_1 \end{aligned} \tag{6.75}$$

so the equivalent optical power is just ϕ_1, but what good is this? The answer lies in the next section, where we will see that the location of the principal plane changes. This is the case for the eyeglasses worn by many people. The separation distance from the eye to the eyeglasses is the front focal length of the eye.

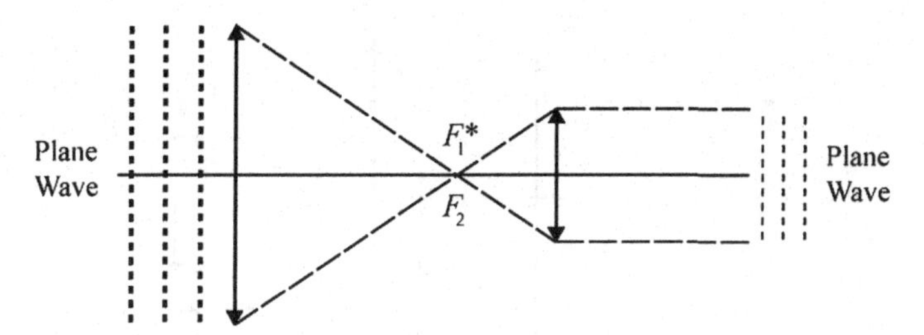

Figure 6.25 Afocal system.

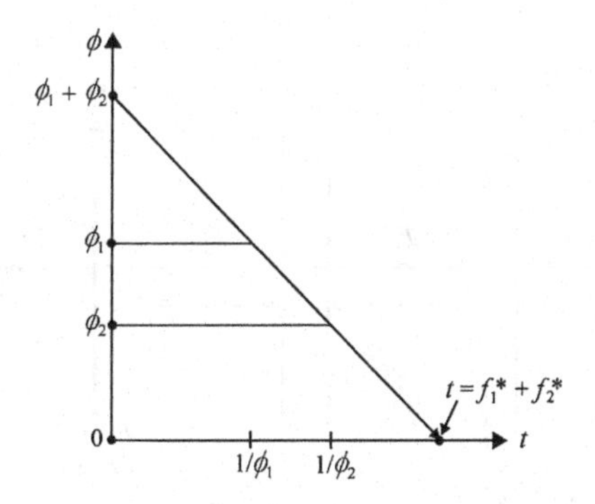

Figure 6.26 Plot of the equivalent optical power of two positive thin lens versus the separation distance.

For the third case ($t = f_1^* + f_2^*$), the sum of the focal lengths is the separation distance. Substituting in Gullstrand's equation (Equation (6.74)):

$$\begin{aligned} \phi &= \phi_1 + \phi_2 - (f_1^* + f_2^*)\frac{1}{f_1^* f_2^*} \\ &= 0. \end{aligned} \tag{6.76}$$

For this case, we have an afocal system (no optical power) typically used in a telescope (e.g., a Keplerian). An afocal system is shown in Figure 6.25.

Although these are three unique thin lens separations, one can plot the equivalent optical power of two positive lenses as a function of separation, as in Figure 6.26. Note that this curve predicts that two positive lenses can be arranged to have a negative optical power.

Now we can produce many powers without having to fabricate a given lens for the required power.

6.10.2 Cardinal points of two thin lenses

In addition to adjusting the total power, changing the separation distance also causes the cardinal points to shift. A ray trace is shown in Figure 6.27 in which

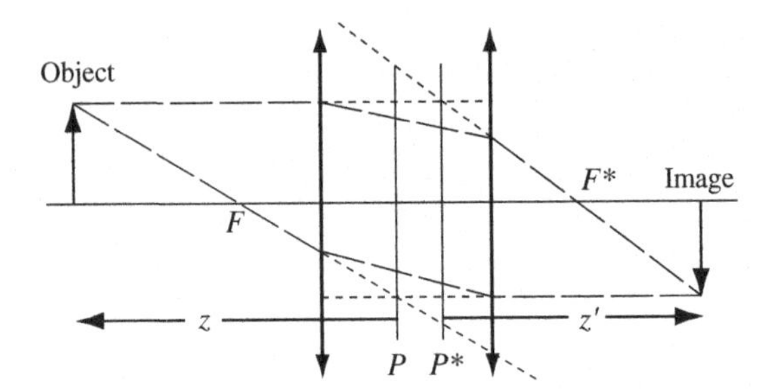

Figure 6.27 Ray trace to find the principal planes.

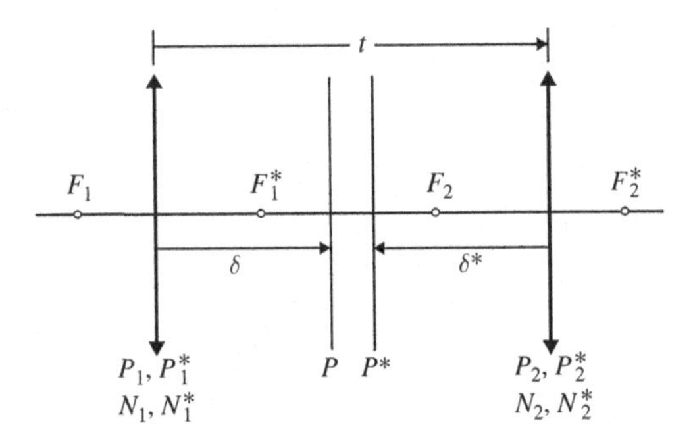

Figure 6.28 Principal planes for two thin lenses.

extending the rays from object space and back-projecting the rays in image space provides a means of locating the equivalent lens back principal plane (P^*). Similarly, if one propagates a ray from $+\infty$ in image space and back-propagates the ray into object space, one locates the front principal plane (P).

The location of the principal planes for two thin lenses is shown in Figure 6.28, which also includes the cardinal points of each lens and the final principal planes for the equivalent lens. This location of the front principal point (P) is determined by:

$$\delta = \frac{\phi_2}{\phi} t, \tag{6.77}$$

which is measured from the front principal point of the first lens (P_1). The location of the rear principal plane (P^*) of the combination is

$$\delta^* = -\frac{\phi_1}{\phi} t. \tag{6.78}$$

δ^* is measured from the rear principal plane of the second lens (P_2^*). The distances f and z are measured from the front principal plane P, while f^* and z' are measured from the back principal plane P^* of the equivalent lens.

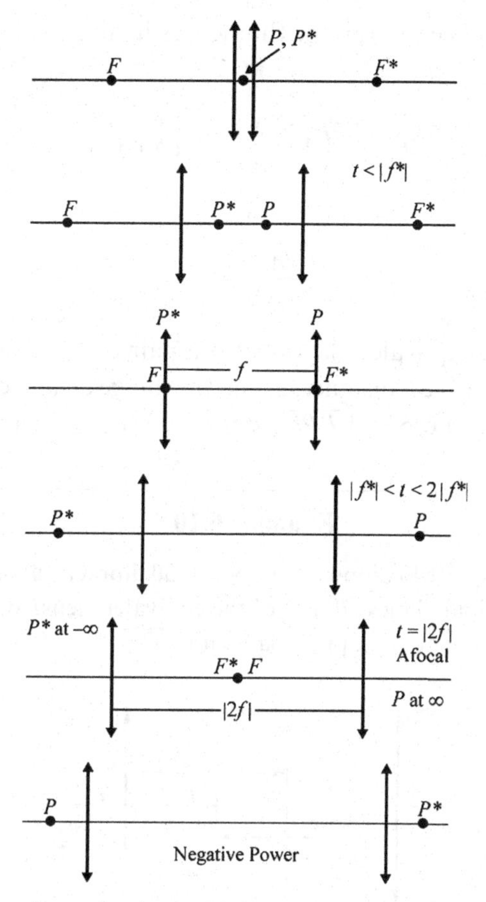

Figure 6.29 Location of principal planes for various separations of two equivalent thin lenses.

Consider two equal thin lenses ($\phi_1 = \phi_2$) whose separation is gradually increased. We plotted the equivalent optical power ϕ as a function of separation t in Figure 6.26. In Figure 6.29, the locations of the principal points are shown for a few selected separations for the case of two equal optical power lenses.

The separation between the front and back principal planes ($\overline{PP^*}$) is given by

$$\overline{PP^*} = t + \delta^* - \delta \tag{6.79}$$

or substituting for δ and δ^*

$$\overline{PP^*} = t - \frac{\phi_1}{\phi} t - \frac{\phi_2}{\phi} t = t\left(1 - \frac{1}{\phi}(\phi_1 + \phi_2)\right) \tag{6.80}$$

and using Gullstrand's equation for thin lenses in air to substitute for $\phi_1 + \phi_2$ in Equation 6.80,

$$\overline{PP^*} = t\left(1 - \frac{1}{\phi}(\phi + t\phi_1\phi_2)\right)$$

or

$$= -\frac{t^2\phi_1\phi_2}{\phi}. \tag{6.81}$$

Figure 6.29 shows equivalent layouts for a pair of thin lenses. Recall that P and P^* are where the object distance, z, and the image distance, z', are measured. It is also important to note that $\overline{PP^*}$ is negative if the powers are all positive.

Example 6.10

If two thin lenses, $\phi_1 = 0.40$ diopters and $\phi_2 = 0.30$ diopters, are separated by 1 m, where are the principal planes (P, P^*) of the equivalent lens? What is the separation between the front and back principal points ($\overline{PP^*}$)? What is the focal length of the equivalent lens?

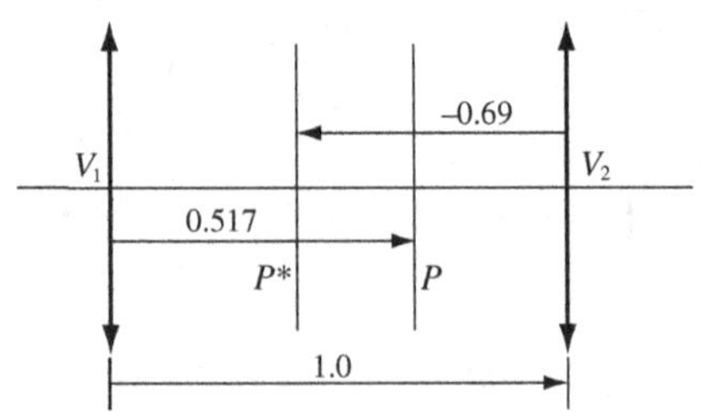

$$\phi = 0.4 + 0.3 - (0.4)(0.3)\frac{1}{1} = 0.58\,\text{m},$$

$$f^* = 1.72\,\text{m},$$

$$\delta = \frac{0.3}{0.58} = 0.517\,\text{m},$$

$$\delta^* = -\frac{0.4}{0.58} = -0.690\,\text{m},$$

$$\overline{PP^*} = -(1)^2\frac{(0.4)(0.3)}{0.58} = -0.207\,\text{m}.$$

6.10.3 Three thin lenses combined

Figure 6.30 shows three thin lenses which are to be combined into one set of principal planes (P and P^*). The approach is to combine two adjacent lenses, find the corresponding location of the principal planes, and then combine the last lens with the first equivalent lens (the combination of the first two). We can combine lenses 1 and 2 or lenses 2 and 3, but not lens 1 and lens 3.

For this problem, we will combine lenses 1 and 2 and find the corresponding principal planes for these: P_{12} and P^*_{12}. Then we will combine P_{12} and P^*_{12} with P_3 and P^*_3 to get the final resulting equivalent lens, as shown in Figure 6.31.

The power of the first two lenses is

$$\phi_{12} = \phi_1 + \phi_2 - t\phi_1\phi_2, \tag{6.82}$$

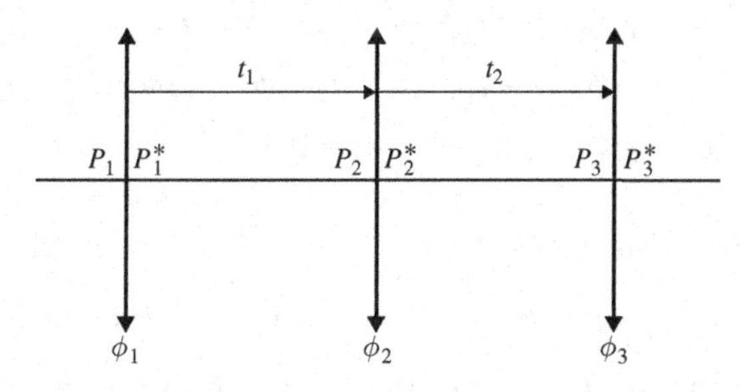

Figure 6.30 Thin lenses separated by air.

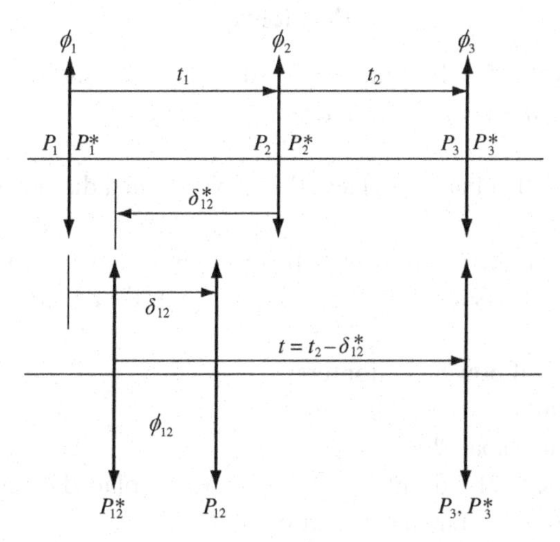

Figure 6.31 Gaussian reduction of lenses 1 and 2 for a three-thin-lens combination.

where the location of the principal plane relative to lenses 1 and 2 is

$$\begin{aligned} \delta^* &= -t\phi_1/\phi \\ \delta &= t\phi_2/\phi. \end{aligned} \tag{6.83}$$

Therefore, the separation between P^*_{12} and P_3 is

$$t = t_2 - \delta^*_{12}. \tag{6.84}$$

The total power resulting from combining the remaining lens (lens 3) follows:

$$\begin{aligned} \phi_{123} &= \phi_{12} + \phi_3 - t\phi_{12}\phi_3 = \phi_{12} + \phi_3 - (t_2 - \delta^*_{12})\phi_{12}\phi_3 \\ &= \phi_1 + \phi_2 + \phi_3 - t_1\phi_1\phi_2 - (t_2 - \delta^*_{12})\phi_{12}\phi_3 \\ &= \phi_1 + \phi_2 + \phi_3 - t_1\phi_1\phi_2 - t_2\phi_{12}\phi_3 + \delta^*_{12}\phi_{12}\phi_3 \\ &= \phi_1 + \phi_2 + \phi_3 - t_1\phi_1\phi_2 - t_2(\phi_1 + \phi_2 - t_1\phi_1\phi_2)\phi_3 - t_1(\phi_1/\phi_{12})\phi_{12}\phi_3 \\ &= \phi_1 + \phi_2 + \phi_3 - t_1\phi_1\phi_2 - t_2\phi_1\phi_3 - t_2\phi_2\phi_3 + t_1t_2\phi_1\phi_2\phi_3 - t_1\phi_1\phi_3 \\ &= \phi_1 + \phi_2 + \phi_3 - t_1\phi_1\phi_2 - t_2\phi_1\phi_3 - t_2\phi_2\phi_3 - t_1\phi_1\phi_3 + t_1t_2\phi_1\phi_2\phi_3 \\ &= \phi_1 + \phi_2 + \phi_3 - t_1\phi_1\phi_2 - t_1\phi_1\phi_3 - t_2\phi_1\phi_3 - t_2\phi_2\phi_3 + t_1t_2\phi_1\phi_2\phi_3 \\ &= \phi_1 + \phi_2 + \phi_3 - t_1\phi_1(\phi_2 + \phi_3) - t_2\phi_3(\phi_1 + \phi_2) + t_1t_2\phi_1\phi_2\phi_3. \end{aligned} \tag{6.85}$$

The three thin lenses added are:

$$\phi_{123} = \phi_1 + \phi_2 + \phi_3 - \phi_1\phi_2t_1 - \phi_2\phi_3t_2 - \phi_1\phi_3(t_1 + t_2) + \phi_1\phi_2\phi_3t_1t_2. \tag{6.86}$$

This is the total equivalent power of three thin lenses separated by air spaces.

Problems

6.1 For a thin lens with $f^* = 10$ mm, sketch and label the cardinal points.

6.2 What is the optical power of a thin lens with $R_1 = R_2 = +25$ mm, with a refractive index of 1.5?

6.3 For a thin lens, with a back focal length of 25 cm and a diameter of 10 cm, what is the $F/\#$?

6.4 Consider an object that is 50 cm away from a plano-convex thin lens that is 5 cm in diameter. The lens has a refractive index of 1.5 and one surface radius of 5 cm. What are:
   (a) the lens' optical power in diopters;
   (b) $(F/\#)$ at infinity;
   (c) $(F/\#)_w$ in image space?

6.5 A negative thin lens of 5 diopters (–) has an object placed 40 cm in front of it.
   (a) What is the transverse magnification?
   (b) Is the image real or virtual?

6.6 A thin lens has radii of $R_1 = 10$ mm and $R_2 = -25$ mm. The lens diameter is 14.3 mm and the lens is made of 517642.251 glass. Calculate the following:
(a) back focal length for *d* light;
(b) $(F/\#)_\infty$;
(c) the optical power (in diopters).

6.7 An equi-convex thin lens is made of 861302.444 glass. Calculate the radius of curvature for the lens surfaces necessary to give an optical power of 4.5 diopters for *d* light.

6.8 An object 15 mm high is located 200 mm from a negative thin lens, $f^* = -120$ mm. Calculate the following:
(a) the optical power of the lens (in diopters);
(b) the image location;
(c) the transverse magnification.

6.9 The following three thin lenses have a refractive index of 1.5:
(1) $R_1 = 10$ cm, $R_2 = \infty$,
(2) $R_1 = 20$ cm, $R_2 = -20$ cm,
(3) $R_1 = 5$ cm, $R_2 = 10$ cm.
(a) Calculate the optical power for each lens (in diopters).
(b) Sketch each lens layout and give each an appropriate name.
(c) What is the shape factor for each lens?

6.10 A negative thin lens in air has a back focal length of –100 mm and a diameter of 20 mm.
(a) Where is the image of a 5 mm object placed in front of the lens at 100 mm?
(b) What is its magnification?
(c) What is the $(F/\#)_\infty$ for this lens?

6.11 A thin meniscus lens made of 517642.251 glass has radii of curvature of +100 cm (R_1) and +50 cm (R_2). This lens is located between air and oil ($n = 1.8$).
(a) What is the optical power of this lens if the oil is at the +100 surface?
(b) What are the values (plus sign) of the front focal length and the back focal length for this configuration?
(c) What is the optical power of this lens if the oil is at the +50 surface?
(d) What is the back focal length for this configuration?

6.12 An object is placed 25 cm from a positive thin lens having a back focal length of 15 cm. Using the following approaches, determine the image distance:
(a) Gaussian optics;
(b) Newtonian optics;
(c) the ZZ' diagram.

6.13 An object to image distance is 100 cm for a setup with a transverse magnification of –4 using a single thin lens in air. What is the back focal length of the thin lens?

6.14 An object 10 mm high is placed 15 cm from a thin plano-convex lens ($n = 1.5$), which has a radius of curvature on the convex side of 15 mm. Calculate the following:
(a) the image distance (z');
(b) the transverse magnification (M_t);
(c) the type of image.

6.15 A negative thin lens-meniscus of refractive index 1.5, with radii of curvature of 100 mm and 50 mm, is held horizontally so the concave side can be filled with water.
(a) What is the optical power without water?
(b) What is the optical power with water?

6.16 Using the ZZ' diagram, find the location of an image for an object 30 cm in front of a positive lens with a back focal length (f^*) of 10 cm.

6.17 An object is 10 cm in front of a negative thin lens that has a back focal length (f^*) of -5 cm.
(a) What is the object distance?
(b) What is the transverse magnification?
(c) Is the image real or virtual?

6.18 A parallel beam of light 16 mm in diameter is incident on a thin lens. Passing through the lens, the beam diverges into a cone with a (total) apex angle of 30°. What is the optical power of the lens?

6.19 Two positive thin lenses are placed 20 cm apart. If the second lens has twice the power of the first lens, and the system is afocal, what are the powers of the two lenses?

6.20 Trace a paraxial ray through the two lenses shown below. Determine the distance to the back focal point relative to the negative lens.

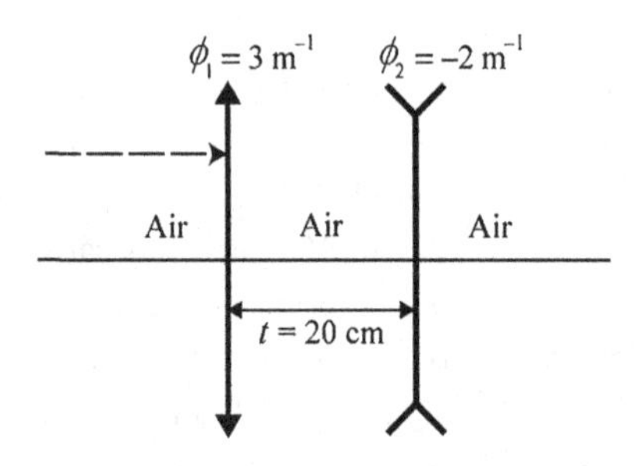

6.21 What is the power of a thin lens combination of two lenses ($efl_1 = 8$ cm, $efl_2 = 20$ cm (efl is the effective focal length)) separated by 10 cm in air?

6.22 A 2 cm object in air is in front of a thin lens of optical power 10 diopters, 40 cm from the front focal point (F).
(a) Where is the image located?
(b) What is its transverse magnification?

6.23 An equi-convex thin lens is made of 755276.255 glass. Calculate the radius of curvature for the lens surfaces necessary to give a power of 2.5 diopters for d light.

6.24 When an object is placed 25 cm in front of a thin lens, a virtual image is formed 5 cm in front of the lens.
(a) What is the back focal length of the lens?
(b) Show this on the ZZ' diagram.

6.25 A +5.00 diopter thin lens forms a real image on a screen placed 100 cm away from the object.
(a) Find the two lens positions at which it is possible to form an image.
(b) What is the transverse magnification for each position?
(c) Show this with two lines on the ZZ' diagram.

6.26 You have two positive thin lenses of effective focal lengths $efl_1 = 5$ cm and $efl_2 = 15$ cm.
(a) For a separation in air of 5 cm, what is the resulting optical power and the location of the principal points (P, P^*)?
(b) For a separation of 15 cm, what is the total resulting optical power and the location of the principal points (P, P^*)?
(c) For a separation of 20 cm, what is the total resulting optical power and the location of the principal points (P,P^*)?

6.27 Two thin lenses in air with *efl* of 20 cm and –5 cm are separated by 3 cm.
(a) What is the effective focal length (*efl*) of the system?
(b) What is the total optical power?
(c) What are the distances from the lenses to the principal points of the equivalent system?
(d) Where are the focal points of the system relative to each lens?

6.28 You have three identical thin lenses, each with a 50 mm back focal length (f^*). The lenses are spaced 50 mm apart.
(a) Calculate the total equivalent optical power.
(b) What is the equivalent focal length for the three lens combination?

6.29 Design a thin glass lens (refractive index = 1.5) with 20 diopters of optical power, for the following shape factors:
(a) –1;
(b) 0;
(c) +3.

6.30 For two thin lenses, $\phi_1 = 10$ diopters and $\phi_2 = 5$ diopters, separated by 10 cm, find the equivalent lens' cardinal points and make a sketch of the lens layout.

6.31 An object is 100 cm in front of a two-thin-lens combination. The two lenses are separated by 2 cm and have optical powers of 20 diopters and 10 diopters, respectively.
(a) Where is the image formed relative to the second 10 diopter thin lens?
(b) What is the transverse magnification?
(c) What is the equivalent focal length of the thin lens combination?

6.32 Two plano-convex thin lenses, each of +4.00 diopters, are placed coaxially and 6 cm apart, their plane sides facing each other. Determine:
(a) the equivalent power;
(b) the location of the principal planes relative to each lens.

6.33 Three thin lenses are each separated by 10 mm of air. The lens' focal lengths are 15, 20, and 25 mm respectively.
(a) What is the equivalent focal length of the assembly?
(b) Where are the principal planes relative to the front vertex?

6.34 Two rulers are spaced 15 cm and 10 cm away from a positive thin lens. You look at the images and discover that 3 mm on one ruler is equivalent to 9 mm on the other. What is the optical power of the lens?

6.35 A glass marble is used to image a star. The star is at an infinite distance from the marble, which has a 1 cm radius. The refractive index of the marble is 2 ($n = 2$).
(a) Where is the image of the star located?
(b) If the radius of the marble is 10 cm, where is the image?

6.36 The two surfaces of an equi-concave thin lens ($t = 0$) in air (see Figure 6.1) have radii of curvature that are of equal magnitude (11 cm) but opposite in sign.

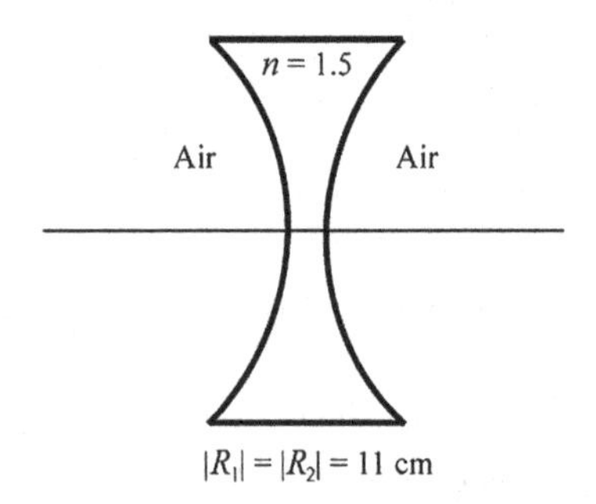

(a) What is its optical power?
(b) What are the values of the front and back focal lengths (f and f^*)?
(c) Make a sketch and locate f, f^*.

6.37 Two positive thin lenses are placed 20 cm apart. If the second lens has twice the optical power of the first lens, and the system is afocal, what are the optical powers of the two lenses?

6.38 Consider a negative thin lens with a back focal length (f^*) of -5 cm. An object is 10 cm in front of the back focal point (F^*).
(a) What is the object distance?
(b) What is the transverse magnification?
(c) Is the image real or virtual?

6.39 For collinear transformation:
(a) Why is x' not a function of y?
(b) Which point in object space is conjugate to the rear focal point?
(c) Why does z' contain only z terms and not x and y?
(d) If an object point is not on the axis, is the conjugate image point necessarily off-axis as well? Why?
(e) An off-axis object point and the axis define an object space meridional plane. What can you say about the conjugate image meridional plane?

6.40 A collinear transformation requires rotational symmetry. Why does this rotational symmetry eliminate some coefficients of variables?

Bibliography

Ditteon, R. (1997). *Modern Geometrical Optics*. New York: Wiley.

Fischer, R. E. and Tadic-Galeb, B. (2000). *Optical System Design*. New York: McGraw-Hill.

Hecht, E. (1998). *Optics*, third edn. Reading, MA: Addison-Wesley.

Jenkins, F. A. and White, H. E. (1976). *Fundamentals of Optics*, fourth edn. New York: McGraw-Hill.

Katz, M. (2002). *Introduction to Geometrical Optics*. River Edge, NJ: World Scientific.

Kreyszig, E. (1992). *Advanced Engineering Mathematics*, seventh edn. New York: Wiley.

Meyer-Arendt, J. R. (1989). *Introduction to Classical and Modern Optics*, third edn. London: Prentice-Hall.

Mouroulis, P. and MacDonald, J. (1997). *Geometrical Optics and Optical Design*, New York: Oxford.

Sears, F. W. (1958). *Optics*, third edn. Cambridge, MA: Addison-Wesley.

7
Thick lenses

A real lens has axial thickness, two radii of curvature, one for each surface (front and back), as well as some non-zero edge thickness. The line connecting the two centers of curvature is the optical axis. Thus far, we have disregarded the axial thickness of a lens by making it zero ($t = 0$). This produced a set of equations for a thin lens that relate the conjugate planes of the object and image. The $t = 0$ assumption gave the thin lens an optical power approximately equal to that of the thick lens. These equations were developed in order to solve paraxial optical relationships with analytical functions instead of ray tracing, resulting in a body of knowledge referred to as Gaussian optics. The real refractive lens has to have some axial thickness in the most common cases, except for the cases of spherical mirrors and single refracting surfaces (SRS), which mimic thin lenses.

Therefore, for real refractive lenses, the question becomes: from what surface or location does one measure the focal lengths, object distances, and image distances for a given setup? Chapter 6 showed that one can determine the optical power of a thick lens; however, the fiducial points from which to measure these distances were not determined. The cardinal points of a thick lens system will be explored in this chapter. Recall there are six cardinal points: two principal points, two focal points, and two nodal points, as was discussed for the thin lens.

7.1 Principal points

The focal length is measured from the principal points. In the case of a thin lens, the focal length (f^*) is measured from the lens or principal points, as shown in Figure 7.1(a) for a thin lens and Figure 7.1(b) for an SRS where the principal points are at the refracting surface. Unfortunately, for a thick lens the focal lengths (f and f^*) are measured to a reference point associated with the lens principal points. Thus, the value of the focal length is not readily obvious upon inspection of the physical lens. As was stated earlier, rays that

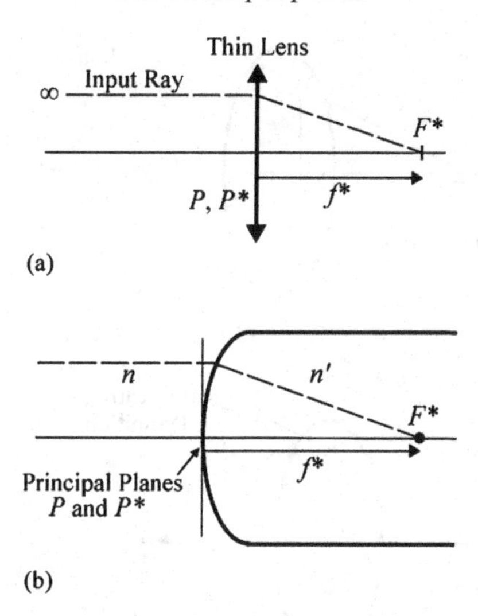

Figure 7.1 Principal planes for: (a) a thin lens, and (b) an SRS.

pass through either focal point in one space must, in the conjugate space of the lens, emerge parallel to the optical axis. This fact is what enables one to locate both the focal points and principal planes.

Consider a ray originating at the front focal point (F), as shown in Figure 7.2(a), which therefore emerges parallel to the optical axis in image space. From the opposite side of the lens, the emerging ray appears to be singularly refracted at an imaginary surface P, instead of twice refracted like the real ray, as shown. This imaginary surface is a unique location in which the ray completely changes to its parallel direction. The ray is really broken into three segments by the lens: two external to the thick lens, and one internal to the lens. The two external segments can be extended to an intersection point as illustrated in Figure 7.2(a). The principal plane's axial location is determined by this intersection, indicated as P. The principal plane is perpendicular to the optical axis and is located at the principal point (P). Similarly, the back principal plane can be located by a ray parallel to the optical axis, propagating in object space, which focuses to the back focal point (F^*), as shown in Figure 7.2(b). Parallel rays in object space can be thought of as refracting once at the back principal plane (P^*) instead of twice, as shown by the paraxial ray trace.

The principal points (P, P^*) are at the intersections of these planes with the optical axis. The obvious situation exists that the principal points and the lens vertices (the intersections of the lens surface with the optical axis) do not generally coincide. The distance between the front focal point (F) and principal plane (P) is the front focal length (f). The distance from the back principal plane (P^*) to the back focal point (F^*) is the back focal length (f^*).

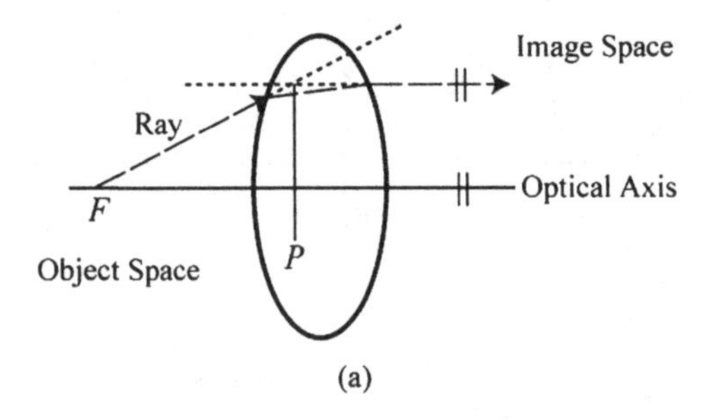

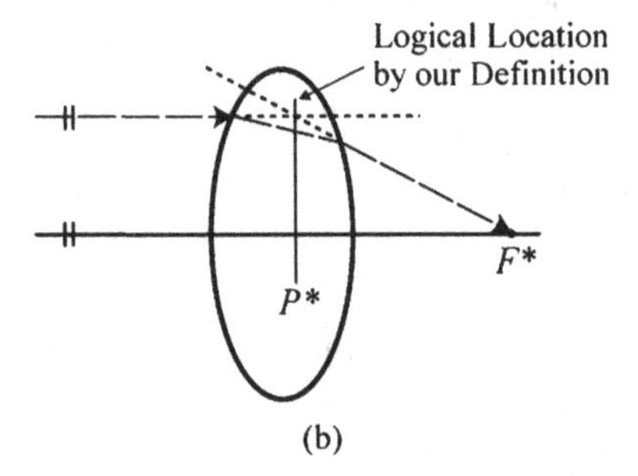

Figure 7.2 Incoming and emerging ray paths: (a) incoming ray from the front focal point emerges parallel to the optical axis. (b) Incoming ray parallel to the axis goes through the back focal point.

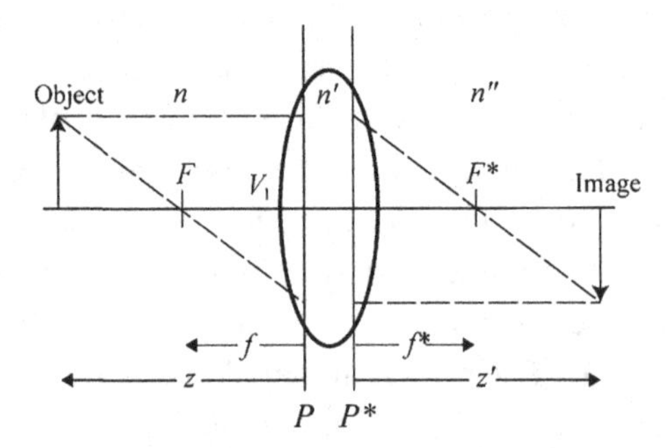

Figure 7.3 Thick lens Gaussian optics.

These principal planes represent a lens, so the object and image distances (z, z') are measured from these principal planes, as shown in Figure 7.3. Having now established the location of the principal planes, think of a simple lens system as equivalent to two principal planes with some separation. The principal planes have a one-to-one correspondence between each point on each plane, or a transverse magnification of 1 ($M_t = 1$). In addition, the thick lens (Figure 7.4(b)) can be thought of as being a thin lens with a hiatus, or empty space, between P and P^*. Thus, all the thin lens equations apply if the distances are measured from the principal points P and P^*.

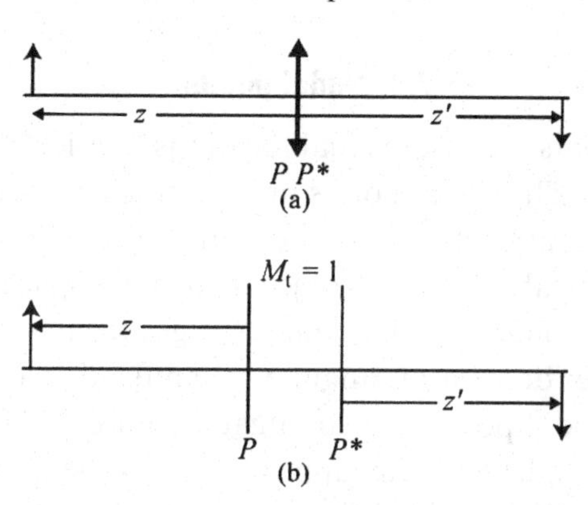

Figure 7.4 Lens principal planes: (a) thin lens; (b) thick lens.

The thick lens can be treated as an equivalent thin lens with a hiatus between P and P^*. For a thin lens, the principal plane separation distance ($\overline{PP^*}$) is zero (Figure 7.4(a)). For a thick lens, the front and back principal planes are separated by a distance. As in the case of the thin lens, a one-to-one mapping occurs between the principal planes ($M_t = 1$) in a thick lens as well.

7.2 Focal points

If parallel light rays (plane waves) are incident on a lens, the lens focuses them to the back focal point F^*. If the lens has a positive optical power, the back focal point (F^*) is to the right. If it is a negative optical power lens, the back focal point (F^*) is to the left. Thus the plane wave can be made to converge or diverge. If light is collimated by a lens, or made into a plane wave in image space, its origin in object space must be from the front focal point (F). These two focal points (F and F^*) are two more cardinal points of a lens system. These two points are not conjugate. Their conjugates are at infinity.

The focal lengths are measured from the principal points to the focal points. The front focal length (f) is the distance $\overline{PF}$, and the back focal length (f^*) is the distance $\overline{P^*F^*}$, as shown in Figure 7.3. The effective focal length (*efl*) of a lens is defined with respect to air ($n = 1$), and therefore is related to the optical power

$$\phi = 1/efl. \tag{7.1}$$

With the additional caveat that

$$\phi = -n/f = n''/f^*. \tag{7.2}$$

If $n \neq n''$ then the front and back focal lengths are not equal. Your eye is an example of such a system.

7.3 Nodal points

The (two) nodal points are also virtual locations in a lens; however, they are not principal points. The nodal points are locations within a lens which give unit angular magnification for a ray from an object. This ray subtends the same angle to the optical axis in both object and image spaces. Any ray from an object that passes through a nodal point emerges from a second nodal point with the same angle, or with an angular magnification of 1 from object to image space. The nodal points are conjugate in an optical system. The axial points at which the undeviated ray appear to cross the optical axis are called the nodal points (N, N^*), as shown in Figure 7.5.

The nodal points are typically colocated with the principal points. This is because the object and image spaces are typically in the same medium or in media with the same index of refraction. However, if the indices of refraction are not the same in the object and the image space, the nodal points shift with respect to the principal points, as shown in Figure 7.6. In this case, the nodal points are displaced relative to the principal points. The equivalent Gaussian diagram is shown in Figure 7.6. Figure 7.7 shows two parallel rays in object space without the thick lens superimposed. One ray is traversing the focal point, while the second ray is passing through the front nodal point (N) in object space, and is thus called a nodal ray. By the definition, this nodal ray in

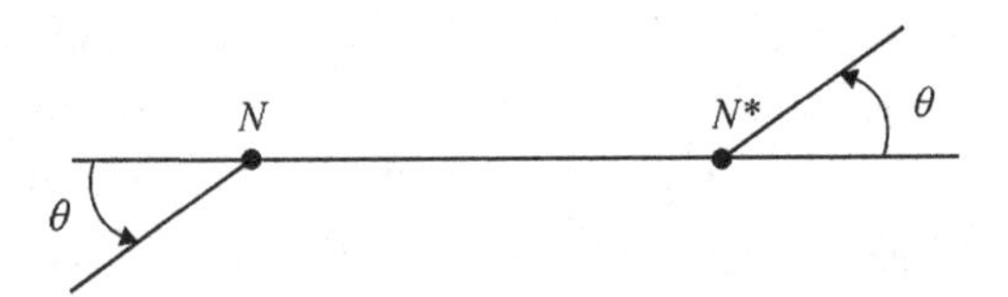

Figure 7.5 Nodal points with a ray angular magnification of 1.

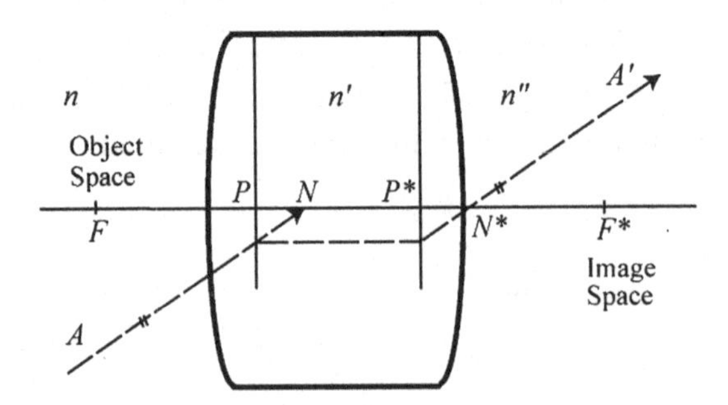

Figure 7.6 Nodal points for the general case of object and image spaces having different indices of refraction.

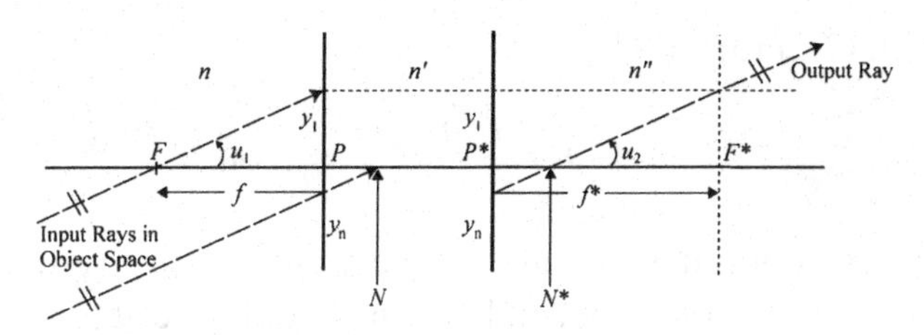

Figure 7.7 Location of nodal points when object space and image space are different media.

image space subtends the same angle through the rear nodal point (N^*), as it does through the front nodal point (see Figure 7.7).

Since the principal planes have a transverse magnification of 1 ($M_t = 1$), any ray incident on the front principal plane (P) exits at the same height (y) on the rear principal plane (P^*). For these two rays, shown in Figure 7.7, their y height is the same on the front and back principal planes (P, P^*). In addition, the focal ray in object space must exit the image space parallel to the axis. The rays through the nodal points form equal angles with the optical axis in object and image space. The intersection in image space of these two rays defines the back focal plane at F^*.

Since

$$u_1 = u_2 \tag{7.3}$$

and y_n at P is equal to y_n at P^*, triangle Py_nN is identical to $P^*y_nN^*$. Therefore

$$\overline{PN} = \overline{P^*N^*}. \tag{7.4}$$

From Figure 7.7, the values of the nodal ray angle in each space are:

$$\tan u_1 = u_1 = \frac{y_1}{-f} \qquad \text{and} \qquad \tan u_2 = u_2 = \frac{y_1}{f^* - \overline{P^*N^*}}. \tag{7.5}$$

Therefore, from Equation (7.3):

$$-f = f^* - \overline{P^*N^*}. \tag{7.6}$$

Then the distance from the principal to the nodal points is

$$\overline{P^*N^*} = f^* + f = \overline{PN}. \tag{7.7}$$

Recall Equation (7.2) for f and f^*:

$$f = -\frac{n}{n''} f^*. \tag{7.8}$$

If the index of refraction has the same value in object space and image space ($n = n''$), then $f = -f^*$ and the nodal and principal points are at the same location:

$$\overline{P^*N^*} = 0 = \overline{PN}. \tag{7.9}$$

Similarly, for the general case of $n \neq n''$, from the identical triangles (Py_nN) in Figure 7.7, the separations of principal points and nodal points are equal:

$$\overline{P^*N^*} = \overline{PN}. \tag{7.10}$$

From Equations (7.7) and (7.8), the distance between the principal points and nodal points is

$$\overline{P^*N^*} = f^* \left(\frac{n'' - n}{n''} \right). \tag{7.11}$$

So, if the refractive index in image space is greater than that in object space ($n'' > n$) the nodal point is located toward image space (n'') relative to P. Consider an air–glass–water transfer through a lens. N and N^* are moved toward the optical space with the higher index of refraction, in this case, toward the water. Rewriting Equation (7.11):

$$\overline{P^*N^*} = f^* - \frac{nf^*}{n''}.$$

If $n'' > n$, the nodal points are located toward the image relative to the principal points (positive lens). If $n'' < n$ then the nodal points are located towards the object relative to the principal points (negative direction).

7.4 Determining cardinal points

Figure 7.8 shows the most general case of a thick lens located between media of different refractive indices. The figure shows all six cardinal points. The cardinal points are separated in the layout because the refractive index is different in each of the three spaces: n for object space, n' for lens space, and n'' for image space.

The six cardinal points provide a quick description of an optical system's characteristics (Table 7.1). The method and equations necessary to find their locations, relative to the vertices, are presented in this section.

Table 7.1. *Six cardinal points*

Focal points	F, F^*
Principal points	P, P^* ($M_t = 1$)
Nodal points	N, N^* ($M_\alpha = 1$)

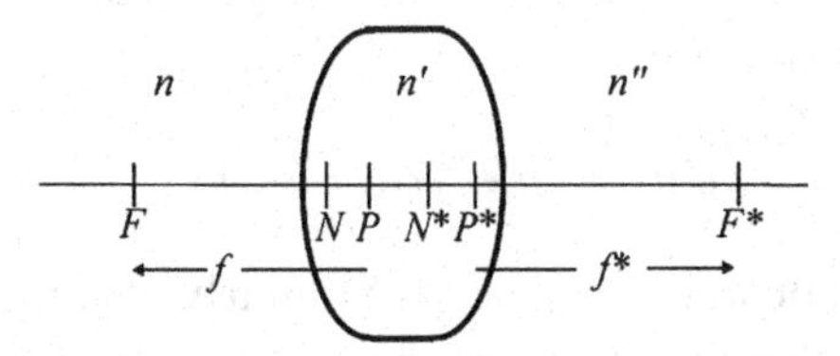

Figure 7.8 Gaussian optics illustrating the six cardinal points for the most general case with different indices of refraction for each space ($n > n''$).

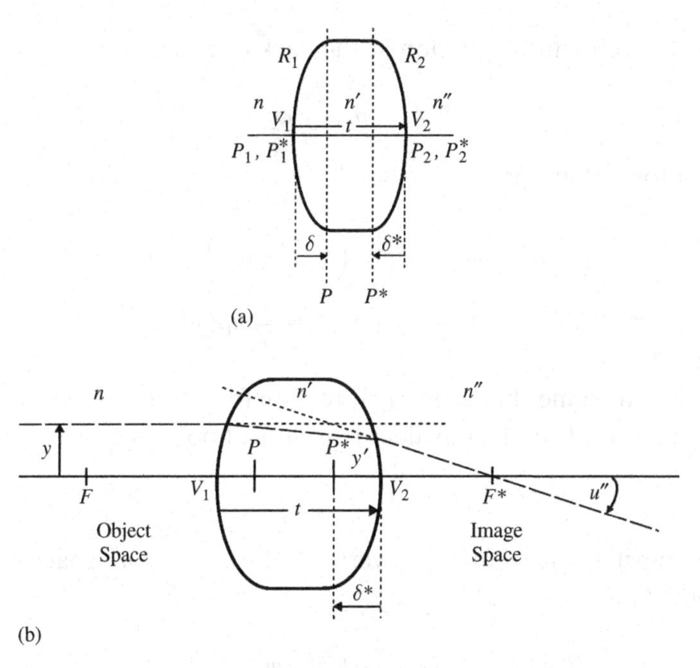

Figure 7.9 (a) Cardinal point locations for a single thick lens. (b) Location of rear principal plane relative to the rear vertex of a single thick lens.

Figure 7.9(a) shows the physical layout of a single thick lens. In order to locate the equivalent principal planes P, P^* for this thick lens, the vertices of the lens may be used as the reference points. Recalling our definition of parallel and focal rays, a second layout can be drawn for this lens (Figure 7.9(b)). When a ray from infinity is propagated through the lens, the extension of the incident ray (in object space) and the extension of the focal ray (in image space) intersect, as shown in the dotted lines in Figure 7.9(b). The plane that contains this intersection point is the back principal plane. The distance from the rear vertex (V_2) to that plane will be defined as δ^*.

The assumption is that all the refraction that takes place in the lens occurs at this place, so using the refraction equation (Equation (5.20)) for the actual paraxial ray in object space,

$$n'u' = nu - y\phi_1 = -y\phi_1, \tag{7.12}$$

where ϕ_1 is the optical power of surface 1. Similarly, using the transfer equation between the front and back surfaces of the lenses for the paraxial ray:

$$y' = y + n'u'\frac{t}{n'} = y - y\phi_1\frac{t}{n'}. \tag{7.13}$$

Reapplying the refraction equation at the back (second) surface:

$$n''u'' = n' - y'\phi_2. \tag{7.14}$$

Substituting for $n'u'$ and y':

$$\begin{aligned} n''u'' &= -y\phi_1 - \left(y - y\phi_1\frac{t}{n'}\right)\phi_2 \\ &= -y\left(\phi_1 + \phi_2 - \frac{t}{n'}\phi_1\phi_2\right). \end{aligned} \tag{7.15}$$

In order to get the same change in ray direction for a single optical power, one can write, in terms of total equivalent lens optical power ϕ,

$$n''u'' = -y\phi. \tag{7.16}$$

Using the transfer equation from vertex V_2 to P^* in n'' space (shown in Figure 7.9(b)):

$$y' = y - \frac{\delta^*}{n''}n''u''. \tag{7.17}$$

Rearranging Equation (7.17) to solve for δ^*/n'', while substituting for y' (from Equations (7.13)) and $n''u''$ (from Equation (7.16):

$$-\frac{\delta^*}{n''} = \frac{y' - y}{n''u''} = \frac{\left(y - y\phi_1\frac{t}{n'}\right) - y}{n''u''} = \frac{y\phi_1\frac{t}{n'}}{y\phi}. \tag{7.18}$$

Rearranging terms to get the distance to the back principal point from vertex V_2:

$$\frac{\delta^*}{n''} = -\frac{t}{n'}\frac{\phi_1}{\phi}. \tag{7.19}$$

The negative sign is used because the derivation was from right to left. Therefore,

$$\delta^* = -\frac{t}{n'}\frac{\phi_1}{\phi} \cdot n''. \tag{7.20}$$

Similarly, for the distance from the front vertex to the front principal plane, δ in Figure 7.9:

$$\frac{\delta}{n} = \frac{t}{n'}\frac{\phi_2}{\phi}, \tag{7.21}$$

or

$$\delta = \frac{t}{n'}\frac{\phi_2}{\phi} \cdot n. \tag{7.22}$$

Using Equations (7.20) and (7.22) for δ and δ^* for a given thick lens or lens system, the location of the principal planes can be determined. Once the principal planes are located, the object/image distances and focal lengths are measured from these planes. The distance between the principal planes $\overline{PP^*}$ for a thick lens in air ($n = n'' = 1$) is

$$\overline{PP^*} = t - \delta + \delta^* = \frac{t}{n'}(n' - 1) - \frac{\phi_1\phi_2 t^2}{n'^2\phi}. \tag{7.23}$$

Example 7.1

For the thick meniscus lens shown below, find the principal planes, P, P^*, and nodal points N, N^*.

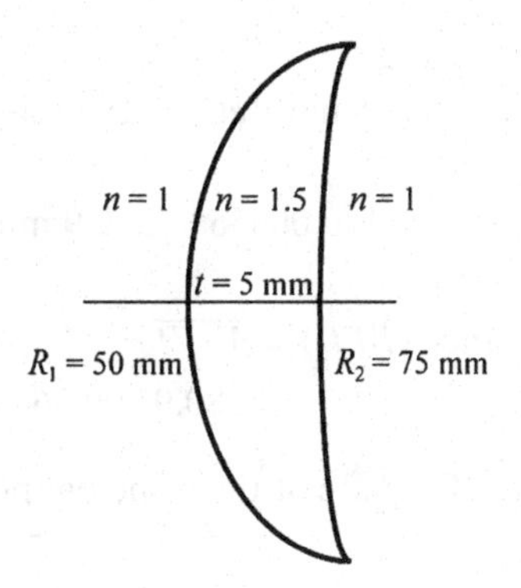

$$\phi_1 = \frac{1.5 - 1}{50} = 0.01\ \text{mm}^{-1},$$

$$\phi_2 = \frac{1 - 1.5}{75} = -0.006\,667\ \text{mm}^{-1}.$$

$$\begin{aligned}\phi &= \phi_1 + \phi_2 - \phi_1\phi_2\frac{t}{n'} \\ &= 0.01 - 0.006\,667 - (0.01)(-0.006\,667)\left(\frac{5}{1.5}\right) \\ &= 0.0033 + 0.000\,223\,33 = 0.003\,56\ \text{mm}^{-1},\end{aligned}$$

$$\delta = n\frac{t}{n'}\frac{\phi_2}{\phi} = 1\left(\frac{5}{1.5}\right)\left(\frac{-0.006\,667}{0.003\,56}\right) = -6.25\ \text{mm},$$

$$\delta^* = -n''\frac{d}{n'}\frac{\phi_1}{\phi} = -1\left(\frac{5}{1.5}\right)\left(\frac{0.01}{0.003\,56}\right) = -9.38\ \text{mm}.$$

Let us layout the thick lens and the principal planes. δ is measured from the first surface to the front principal plane, and δ^* is measured from the second surface to the rear principal plane.

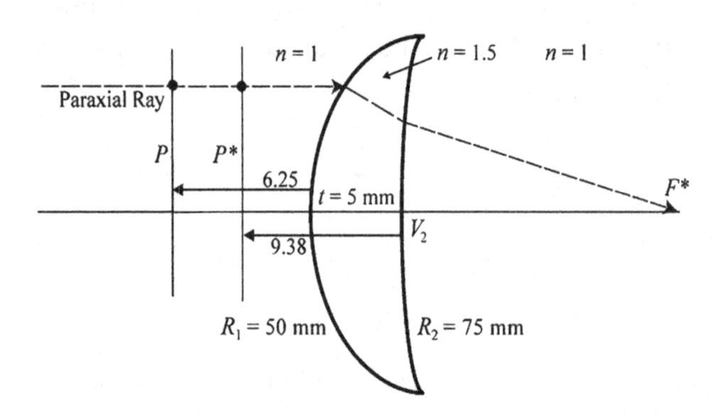

$$\text{Effective focal length}(efl) = \overline{P^*F^*}(\text{in air}).$$

$$f^* = efl = 1/0.00356 = 280.9\ \text{mm}.$$

$$\begin{aligned}\text{Back focal distance } (BFD) &= \overline{V_2F^*} = f^* + \delta^* \\ &= 280.9 - 9.36 = 271.54\ \text{mm}.\end{aligned}$$

Where are the nodal points? They are at the principal points, since $n = n''$!

Table 7.2. *Glossary of terms*

Significant points on the optical axis		
Principal points	$P \quad P^*$	Conjugate ($M_t = 1$)
Nodal points	$N \quad N^*$	Conjugate ($M_\alpha = 1$)
Focal points	$F \quad F^*$	Not conjugate
Vertices	$V_1 \quad V_2$	Not conjugate
Object/Image points	$O \quad I$	Conjugate
Significant directed distances		
Front/rear focal lengths	$\overline{PF} = f$, $\overline{P^*F^*} = f^*$	
Front /back focal distances	$\overline{V_1F} \quad \overline{V_2F^*}$	
Object/images distances	$\overline{PO} \quad \overline{P^*I}$	
Vertex thickness	$\overline{V_1V_2}$	
Principal point separation	$\overline{PP^*}$	
Nodal point separation	$\overline{PN} = f + f^* = \overline{P^*N^*}$	

Table 7.3. *Gaussian properties of a thin lens*

Thin lens in air (thickness (t) $\rightarrow 0$) t = lens thickness	
$\phi = (n'_g - 1)(C_1 - C_2)$	$\overline{PN^*} = 0; f^* = \dfrac{1}{\phi}$
$\delta \Rightarrow 0,\ \delta^* \Rightarrow 0$	$\overline{PP^*} = 0$

The most important terms and their definitions are listed in Table 7.2. A summary of the Gaussian properties and cardinal points are given in Table 7.3 for a thin lens, in Table 7.4 for a single refractive surface, and in Table 7.5 for a thick lens.

The locations of the principal planes (colocated with nodal points since these lenses are in air) are shown in Figure 7.10 for various shapes of lens. It is worth emphasizing that the principal planes can be located outside the physical structure of glass such as in the case of meniscus-shaped lenses for both positive and negative lenses. In plano-convex and plano-concave lenses, one of the principal planes lies at the vertex of the lens.

7.5 Thick lens combinations

A combination of thick lenses can also be reduced to a set of cardinal points and an effective focal length, just as in the case with a single thick lens. Likewise, a triplet can be reduced by combining two adjacent thick lenses into an equivalent

Table 7.4. *Gaussian properties of a SRS*

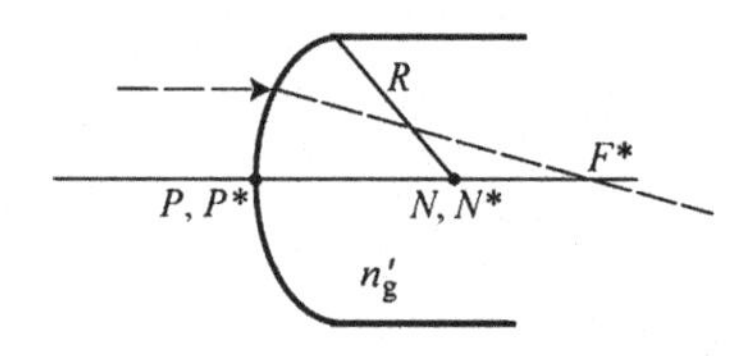

$\phi = (n'_g - n)C$
P, P^* at vertex
N, N^* at center of curvature (C)
Since:

$$f + f^* = -\frac{n}{\phi} + \frac{n'_g}{\phi} = -\frac{nR}{n'_g - n} + \frac{n'_g R}{n'_g - n} = R$$

Table 7.5. *Gaussian properties of a thick lens*

Thick lens in air

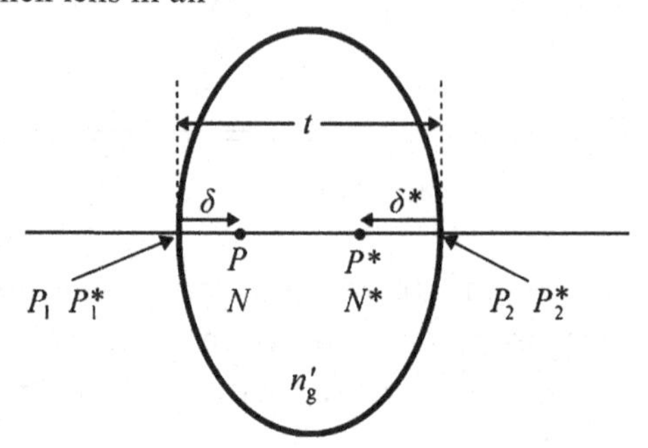

$n = n'' = 1$

$n_g{}' = \text{glass index}$

$\phi_1 = (n'_g - 1)C_1$

$\phi_2 = (1 - n'_g)C_2 = -(n'_g - 1)C_2$

$$\phi = (n'_g - 1)\left[C_1 - C_2 + \frac{(n'_g - 1)}{n'_g} t C_1 C_2\right]$$

$$\delta = \frac{t}{n'_g}\frac{\phi_2}{\phi}; \quad \delta^* = -\frac{t}{n'_g}\frac{\phi_1}{\phi}$$

$$\overline{PP^*} = \frac{n'_g - 1}{n'_g} t - \left(\frac{t}{n'_g}\right)^2 \frac{\phi_1 \phi_2}{\phi}$$

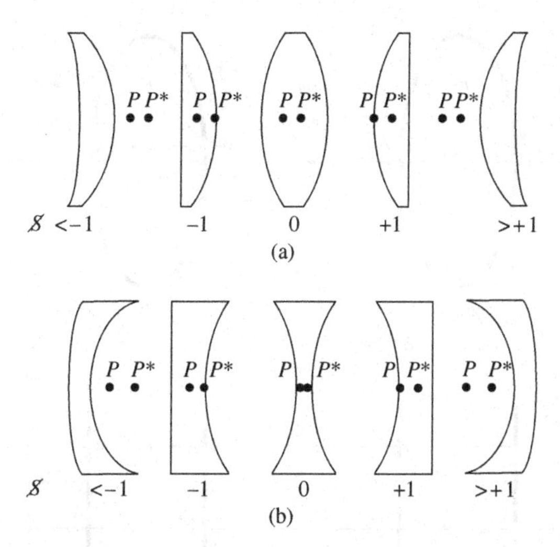

Figure 7.10 Shape factors and the location of principal planes (pts) in air of thick lenses: (a) positive lens; (b) negative lens.

lens (defined by equivalent cardinal points and the effective focal length) and then repeating that combination process for that equivalent lens and the last original lens. This process of reducing a group of lenses to a set of cardinal points and an effective focal length is called "Gaussian reduction." In most cases, the cardinal points are measured relative to the front vertex of the first lens.

7.5.1 Gaussian reduction of two thick lenses

Consider two thick lenses separated by an axial distance of t with an index n_2 in the medium between the lenses. Furthermore, imagine the most general case where the index of refraction is different for all spaces, as shown in Figure 7.11(a). However, for most real world problems, air separates the lenses; thus, the indices n_0, n_2 and n_4 are all 1.

To begin our discussion of Gaussian reduction, we will turn to a more detailed illustration of this two thick lens system (with paraxial approximation), shown in Figure 7.11(b).

P_x and $P^*{}_x$ $(x = 1, 2, 3)$ are the principal planes for each surface of optical power (two per lens). To reduce each lens, labeled in the diagram by subscripts and superscripts a and b, to an equivalent single power and set of cardinal points via the use of Gullstrand's equation and the shift equation for principal planes δ and δ^*, the following equations are used:

$$\phi_a = {}^a\phi_1 + {}^a\phi_2 - {}^a\phi_1{}^a\phi_2\frac{t_a}{n_1}, \tag{7.24}$$

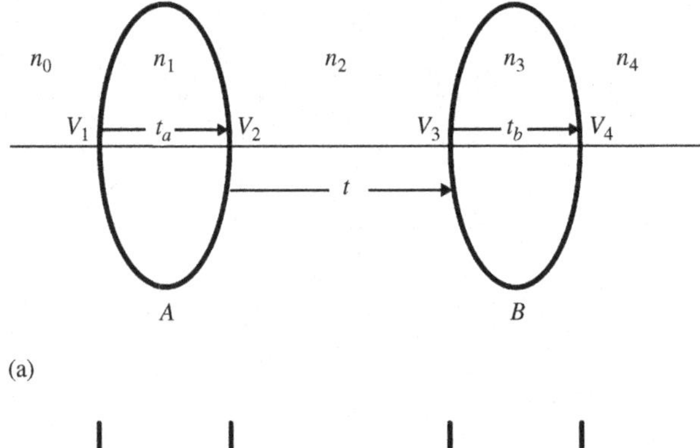

(a)

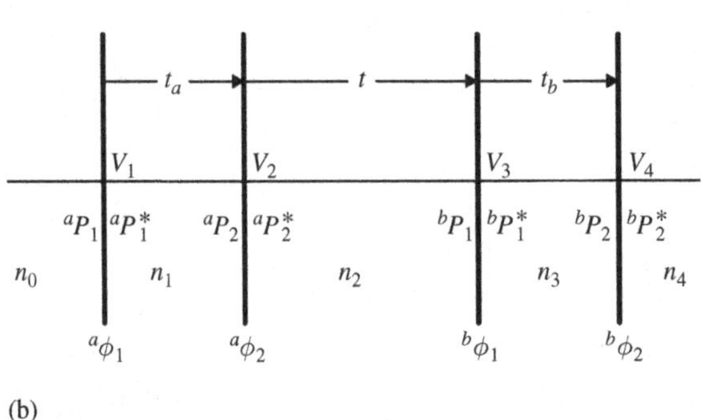

(b)

Figure 7.11 (a) Gaussian reduction of two thick lenses in three media. (b) The corresponding principal planes used for Gaussian reduction.

$$ {}^a\delta = n_0 \frac{{}^a\phi_2}{\phi_a} \frac{t_a}{n_1}, \tag{7.25} $$

$$ {}^a\delta^* = -n_2 \frac{{}^a\phi_1}{\phi_a} \frac{t_a}{n_1}, \tag{7.26} $$

$$ \phi_b = {}^b\phi_1 + {}^b\phi_2 - {}^b\phi_1 {}^b\phi_2 \frac{t_b}{n_3}, \tag{7.27} $$

$$ {}^b\delta_1 = n_2 \frac{{}^b\phi_2}{\phi_b} \frac{t_b}{n_3}, \tag{7.28} $$

$$ {}^b\delta_2^* = -n_4 \frac{{}^b\phi_1}{\phi_b} \frac{t_b}{n_3}, \tag{7.29} $$

$$ t_{ab} = t + {}^b\delta_1 - {}^a\delta_2^*. \tag{7.30} $$

Figure 7.12 illustrates the effect of the Gaussian reduction after this first step in the process. The vertices of the lens are indicated with V_n. The corresponding principal plane for each lens is located. Also shown is t_{ab}, the separation between the back principal plane ${}^aP_{12}^*$ of lens a and the front principal plane ${}^b\mathrm{P}_{12}$.

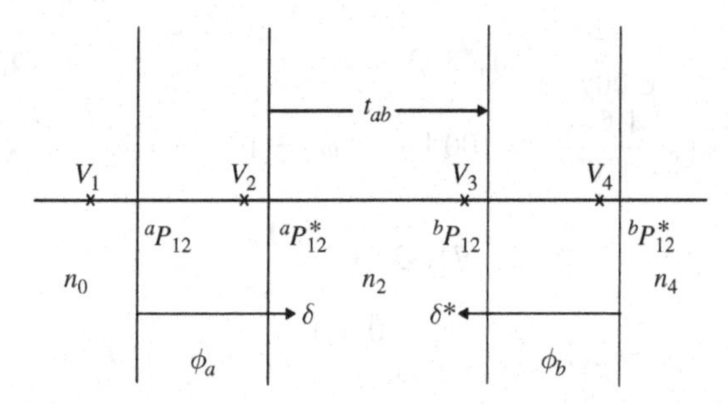

Figure 7.12 Thick lens system after the first step of Gaussian reduction.

The final Gaussian reduction step is to solve for the equivalent optical power and set of principal planes that will represent these two thick lenses. The layout should have distances measured relative to the vertex of the first lens V_1. Solving for the total optical power ϕ of the two thick lenses:

$$\phi = \phi_a + \phi_b - \phi_a \phi_b \frac{t_{ab}}{n_2}, \tag{7.31}$$

$$\delta = n_0 \frac{\phi_b}{\phi} \frac{t_{ab}}{n_2}, \tag{7.32}$$

$$\delta^* = -n_4 \frac{\phi_a}{\phi} \frac{t_{ab}}{n_2}. \tag{7.33}$$

As shown in Figure 7.12, δ is measured from $^aP_{12}$, and δ^* is measured from $^bP^*_{12}$. The effective focal length is just $1/\phi$. Using this process and these equations we find the equivalent of two thick lenses, presented as the most general case. Thus, this approach may be applied to a two thick lens system with any parameters (n_1, n_2, n_3, t, $^a\phi_1$, $^b\phi_1$, etc.).

Example 7.2

Calculate the locations of the cardinal points of an equivalent system to the two thick lens combination shown below. Reference to vertex 1 (V_1).

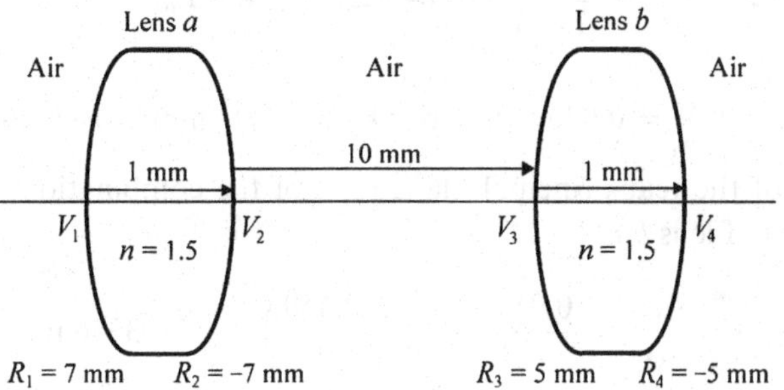

$$^a\phi_1 = \frac{1.5-1}{0.007} = 71.43\text{ D},\quad ^a\phi_2 = \frac{1-1.5}{-0.007} = 71.43\text{ D},$$
$$^b\phi_1 = \frac{1.5-1}{0.005} = 100\text{ D},\quad ^b\phi_2 = 100\text{ D}.$$

$$\phi_a = 2(71.43) - (71.43)^2\left(\frac{0.001}{1.5}\right) = 139.459\text{ D},$$
$$\phi_b = 2(100) - 100^2\left(\frac{0.001}{1.5}\right) = 193.33\text{ D}.$$

$$\delta_a = \frac{^a\phi_2}{\phi_a}\frac{t}{n} = 0.000\,341\,5\text{ m from } V_1,$$

$$\delta_a^* = \frac{-^a\phi_1}{\phi_a}\frac{t}{n} = -0.000\,341\,5\text{ m from } V_2,$$

$$\delta_a^* + \overline{V_1V_2} = 0.000\,658\,5\text{ m from } V_1,$$

$$\delta_b = \frac{^b\phi_2}{\phi_b}\frac{t}{n} = 0.000\,344\,8\text{ m from } V_3,$$

$$\delta_b^* = \frac{-^b\phi_1}{\phi_b}\frac{t}{n} = -0.000\,344\,8\text{ m from } V_4,$$

$$\delta_b^* + \overline{V_3V_4} = 0.000\,655\text{ m from } V_3.$$

Now combine the two thick lenses to get the equivalent system's cardinal points:

$$t = \overline{P_a^*P_b} = 10\text{ mm} + 0.3415\text{ mm} + 0.3448\text{ mm} = 0.010\,686\text{ m}.$$

$$\phi_T = \phi_a + \phi_b - \phi_a\phi_b\left(\frac{t}{n}\right) = 44.668\text{ D},\quad t = 0.010686\text{ m and } n = 1.0.$$

The location of the front principal plane (P) of the combination from the rear principal plane of lens b is

$$\delta_T = \frac{t\phi_b}{n\phi_T} = 0.046\,253\text{ m from } P_a,$$

$$1\text{ mm} + \delta_a^* + \delta_T = 0.04691\text{ m from } V_1,\quad 46.91\text{ mm} = \overline{V_1P}\text{ and }\overline{V_1N},$$

The location of the rear principal plane (P^*) of the combination from the rear principal plane of lens b is

$$\delta_T^* = \frac{-t\phi_a}{n\phi_T} = \frac{0.010\,686}{1}\left(\frac{139.459}{44.668}\right) = -0.03336\text{ m},$$

$$\overline{V_1P^*} = \overline{V_1N^*} = 0.011\,344\,8 + (-0.033\,35) = -0.022\,015 \text{ m from } V_1P^*,$$

$$f^* = \frac{1}{\phi} = \frac{1}{44.668} = 0.02239 \text{ m},$$
$$f^* + \overline{V_1P^*} = \overline{V_1F^*} = 0.022\,39 - 0.022\,015 \rightarrow V_1F^* = 0.000\,375 \text{ m},$$

$$f = \frac{-1}{\phi} = \frac{-1}{44.668} = -0.02239 \text{ m},$$
$$f + \overline{V_1P} = -0.02239 + 0.04691 = 0.02452 \text{ m from } V_1.$$

7.5.2 Gaussian reduction of three thick lenses

The Cooke triplet shown in Figure 7.13(a) is probably the most widely evaluated and interesting lens. Its uniqueness lies in the fact that all third-order aberrations can be corrected by adjusting the many available variables (six radii, stop positions and thicknesses). Typically, correcting for all third-order ray aberrations is not done (see Chapter 11). What is often required during design and analysis of such a triplet is the reduction of the lenses to an equivalent set of principal planes and focal lengths, as shown in Figure 7.13(b).

During our reduction of this Cooke triplet, we will assume that the three lenses are all in air. We will follow the same procedure that was followed in the

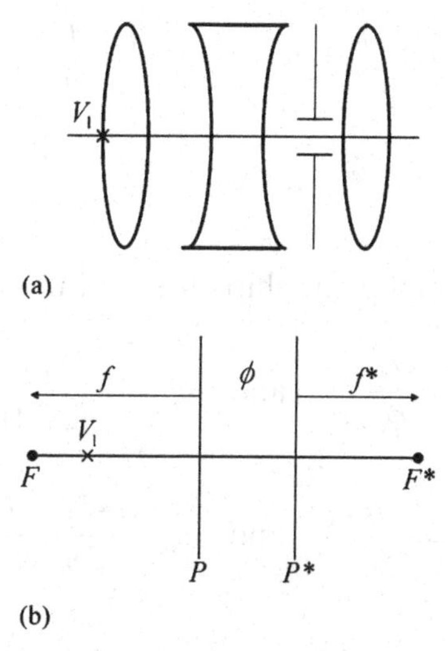

Figure 7.13 (a) Cooke triplet, sometimes called a Taylor triplet (after the optical designer who invented it). (b) Reduced equivalent lens.

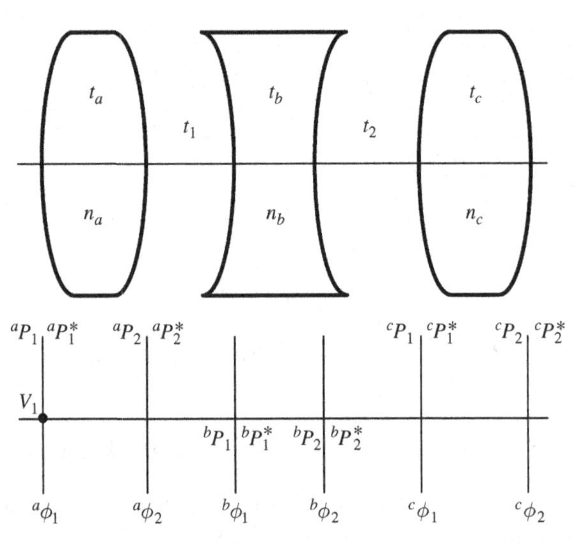

Figure 7.14 Triplet surfaces and corresponding principal planes.

case of two thick lenses, defining the principal planes for each surface, along with the optical power of each surface, as shown in Figure 7.14.

Combining the optical power of both curved surfaces for each lens gives each lens' optical power:

$$\phi_a = {}^a\phi_1 + {}^a\phi_2 - {}^a\phi_1{}^a\phi_2\frac{t_a}{n_a}, \tag{7.34}$$

$$\phi_b = {}^b\phi_1 + {}^b\phi_2 - {}^b\phi_1{}^b\phi_2\frac{t_b}{n_b}, \tag{7.35}$$

$$\phi_c = {}^c\phi_1 + {}^c\phi_2 - {}^c\phi_1{}^c\phi_2\frac{t_c}{n_c}. \tag{7.36}$$

The corresponding principal plane shifts for the three lenses are

$$\delta_a = \frac{{}^a\phi_2}{\phi_a}\frac{t_a}{n_a} \quad \text{and} \quad \delta_a^* = \frac{-{}^a\phi_1}{\phi_a}\frac{t_a}{n_a}, \tag{7.37}$$

$$\delta_b = \frac{{}^b\phi_2}{\phi_b}\frac{t_b}{n_b} \quad \text{and} \quad \delta_b^* = \frac{-{}^b\phi_1}{\phi_b}\frac{t_b}{n_b}, \tag{7.38}$$

$$\delta_c = \frac{{}^c\phi_2}{\phi_c}\frac{t_c}{n_c} \quad \text{and} \quad \delta_c^* = \frac{-{}^c\phi_1}{\phi_c}\frac{t_c}{n_c}. \tag{7.39}$$

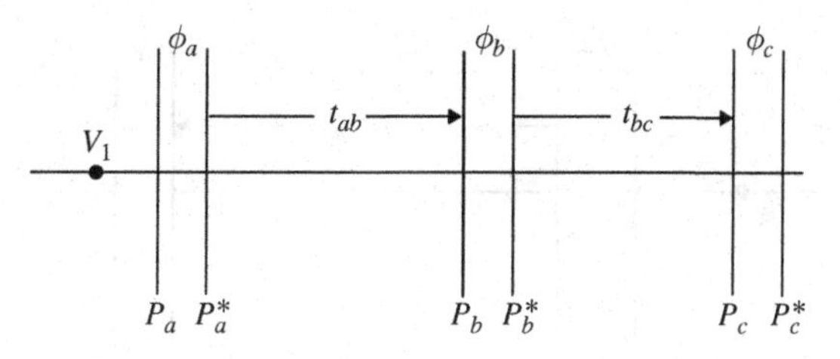

Figure 7.15 Equivalent optical powers of each lens of the triplet with the corresponding separations between the principal planes.

The reduction process, thus far, is shown in Figure 7.15. In Figure 7.15, the spacings between the principal planes, (P_a^*, P_b) and (P_b^*, P_c), respectively, are:

$$t_{ab} = t_1 - \delta_a^* + \delta_b, \tag{7.40}$$

$$t_{bc} = t_2 - \delta_b^* + \delta_c. \tag{7.41}$$

At this point in the Gaussian reduction, there are two approaches for the next step: should we combine lenses a and b into an equivalent, or reduce lenses b and c? Recall that we can only combine adjacent lenses. It is not correct to combine lenses a and c. For this discussion, we will arbitrarily decide to combine a and b and find their equivalent. The optical powers and corresponding shifts of the principal plane are

$$\phi_{ab} = \phi_a + \phi_b - \phi_a \phi_b t_{ab}, \tag{7.42}$$

$$\delta_{ab} = \frac{\phi_b}{\phi_{ab}} \frac{t_{ab}}{1}, \tag{7.43}$$

$$\delta_{ab}^* = -\frac{\phi_a}{\phi_{ab}} \frac{t_{ab}}{1}. \tag{7.44}$$

This reduces the diagram to two sets of principal planes and two optical powers, as shown in Figure 7.16, where the separation between P_{ab}^* and P_c is

$$t_{abc} = t_{bc} - \delta_{ab}^*, \tag{7.45}$$

the final reduction to the equivalent power is

$$\phi = \phi_{ab} + \phi_c - \phi_{ab} \phi_c t_{abc} \tag{7.46}$$

and

$$\delta = \frac{\phi_c}{\phi} t_{abc}, \tag{7.47}$$

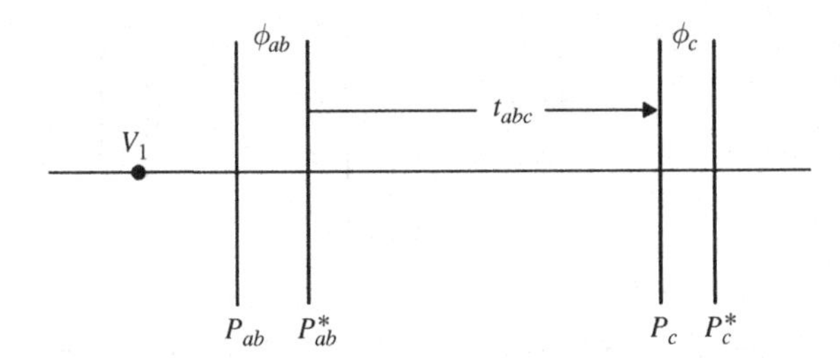

Figure 7.16 Principal plane location of lenses *a* and *b* combined, separated from principal planes of lens *c* of triplet.

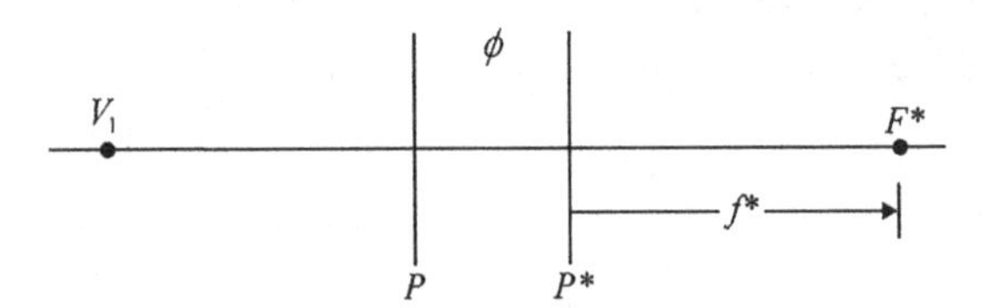

Figure 7.17 Equivalent lens layout.

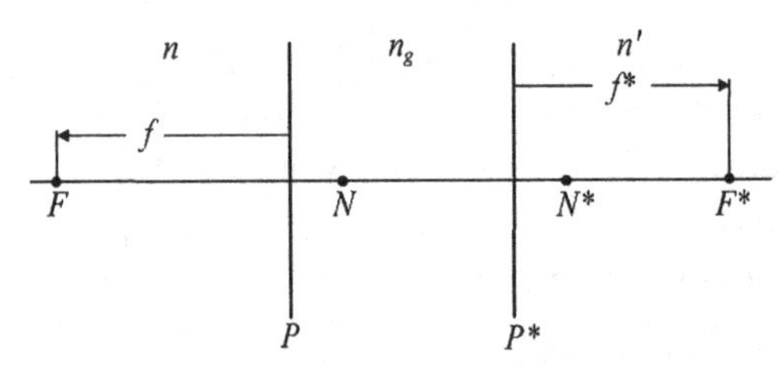

Figure 7.18 The image space (n') is different from the object space (n) so the nodal points are not at the principal points: for the case shown $n' > n$.

$$\delta^* = -\frac{\phi_{ab}}{\phi} t_{abc}, \tag{7.48}$$

as shown in Figure 7.17.

Also the effective focal length is

$$efl = 1/\phi. \tag{7.49}$$

The Cooke triplet has thus been reduced to a set of equivalent principal planes and an effective focal length. This is the focal length that is quoted for the three-element system. Generally, any multi-element optical system can be reduced to give an effective focal length, one example being the objective lens for cameras.

It should be pointed out again that if the indices of refraction of the object and image spaces are unequal, then the nodal points will shift, as shown in Figure 7.18 (see Equation (7.11)):

$$\phi = \frac{-n}{f} = \frac{n'}{f^*} = \frac{1}{efl}, \; n' \neq n, \; n' > n. \tag{7.50}$$

Example 7.3

Find the equivalent power, principal plane, and effective focal length for the cemented doublet shown.

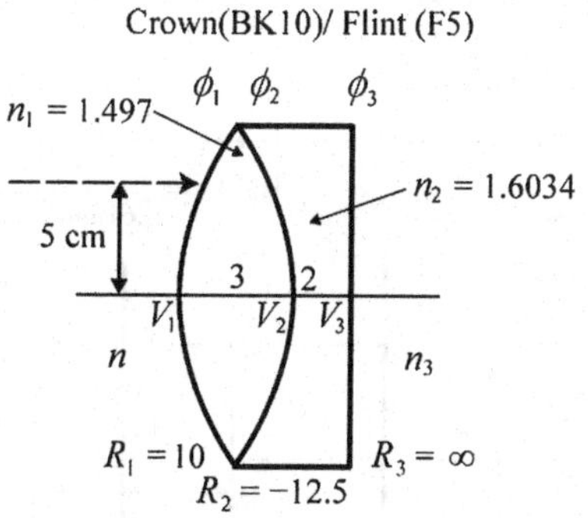

Crown(BK10)/ Flint (F5)

$n_1 = 1.497$, $n_2 = 1.6034$, $R_1 = 10$, $R_2 = -12.5$, $R_3 = \infty$, 5 cm

$$\phi_1 = \frac{1.497 - 1}{10} = 0.0497,$$

$$\phi_2 = \frac{1.6034 - 1.497}{-12.5} - 0.008512,$$

$$\phi_3 = 0.$$

$$\phi_{12} = \phi_1 + \phi_2 - \phi_1\phi_2\frac{t}{n} = 0.0497 - 0.008\,512 + \frac{(0.0497)(0.008\,512)3}{1.497}$$

$$\rightarrow \phi_{12} = 4.204(10^{-2})\ \text{mm}^{-1}.$$

$$\frac{\delta_1}{n} = \frac{\phi_2}{\phi_{12}}\frac{t}{n_1} = -\frac{0.008\,512}{4.203\,57(10^{-2})} \times \frac{3}{1.497} = -0.4058\ \text{mm};$$

$$\rightarrow \delta_1 = -0.4058.$$

$$\frac{\delta_1^*}{n_2} = -\frac{\phi_1}{\phi_{12}}\frac{t}{n_1} = -\frac{0.0497}{4.20357(10^{-2})} \times \frac{3}{1.497} = -2.3694\,\text{mm}$$

$$\rightarrow \delta_1^* = (-2.3694)(1.6034) = -3.799\ \text{mm}.$$

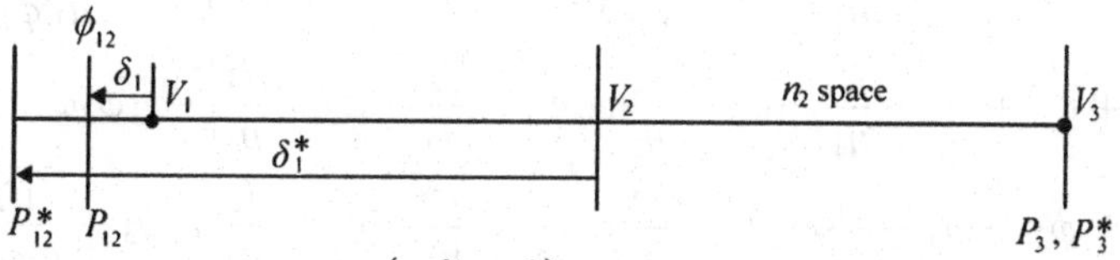

ϕ_{12}, δ_1, V_1, V_2, n_2 space, V_3, δ_1^*, P_{12}^*, P_{12}, P_3, P_3^*

$$\phi_{123} = \phi_{12} + \phi_3 - \phi_{12}\phi_3\left(\frac{-\delta_1^* + 2}{n_2}\right) = \phi_{12}, \phi_3 = 0 \rightarrow f_{\text{eff}} = 23.8\ \text{mm},$$

$$\delta = \frac{\phi_3}{\phi_{123}}\left(\frac{-\delta_1^* + 2}{n_1}\right) = 0,$$

$$\delta^* = -\frac{\phi_{12}}{\phi_{123}}\left(\frac{-\delta_1^* + 2}{n_2}\right) = -\frac{\phi_{12}}{\phi_{123}}\left(\frac{5.799}{1.6034}\right) = -3.61 \text{ mm}.$$

Effective focal length (*efl*) $= \overline{P^*F^*}$. In the last image refractive index is 23.8 mm.

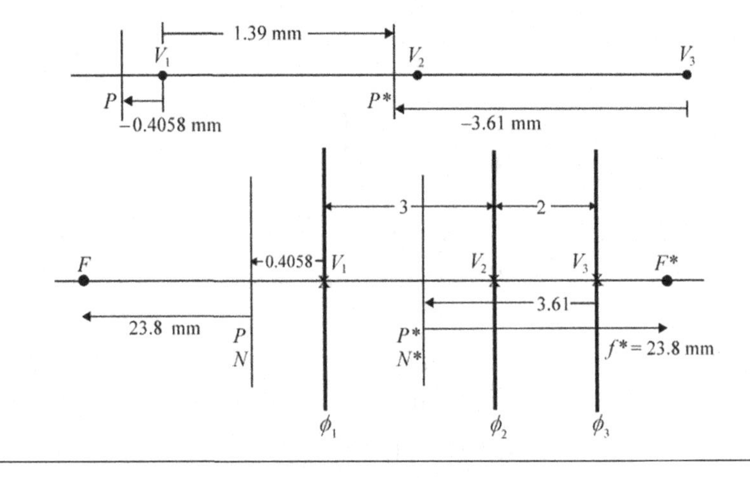

7.5.3 Gaussian reduction of an achromat

The equation for the combination of three surfaces of optical power for a thick lens (e.g. a cemented doublet) is the same as that for three thin lenses (Equation (6.66)), but one must allow for differences in the index of refraction between media with differing optical powers, as shown in the following derivation:

$$\begin{aligned}
\phi_{123} &= \phi_{12} + \phi_3 - \frac{t_1}{n_2}\phi_{12}\phi_3 = \phi_{12} + \phi_3 - \frac{(t_2 - \delta^*)}{n_2}\phi_{12}\phi_3 \\
&= \phi_1 + \phi_2 + \phi_3 - \frac{t_1}{n_1}\phi_1\phi_2 - \frac{(t_2 - \delta^*)}{n_2}\phi_{12}\phi_3 \\
&= \phi_1 + \phi_2 + \phi_3 - \frac{t_1}{n_1}\phi_1\phi_2 - \frac{t_2}{n_2}\phi_{12}\phi_3 + \frac{\delta^*}{n_2}\phi_{12}\phi_3 \\
&= \phi_1 + \phi_2 + \phi_3 - \frac{t_1}{n_1}\phi_1\phi_2 - \frac{t_2}{n_2}\left(\phi_1 + \phi_2 - \frac{t_1}{n_1}\phi_1\phi_2\right)\phi_3 - \frac{t_1}{n_1}\frac{\phi_1}{\phi_{12}}\frac{1}{n_2}\phi_{12}\phi_3 \\
&= \phi_1 + \phi_2 + \phi_3 - \frac{t_1}{n_1}\phi_1\phi_2 - \frac{t_2}{n_2}\phi_1\phi_3 - \frac{t_2}{n_2}\phi_2\phi_3 + \frac{t_1 t_2}{n_1 n_2}\phi_1\phi_2\phi_3 - \frac{t_1}{n_1}\phi_1\phi_3 \\
&= \phi_1 + \phi_2 + \phi_3 - \frac{t_1}{n_1}\phi_1\phi_2 - \frac{t_2}{n_2}\phi_1\phi_3 - \frac{t_2}{n_2}\phi_2\phi_3 - \frac{t_1}{n_1}\phi_1\phi_3 + \frac{t_1 t_2}{n_1 n_2}\phi_1\phi_2\phi_3 \\
&= \phi_1 + \phi_2 + \phi_3 - \frac{t_1}{n_1}\phi_1\phi_2 - \frac{t_1}{n_1}\phi_1\phi_3 - \frac{t_2}{n_2}\phi_1\phi_3 - \frac{t_2\phi_2\phi_3}{n_2} + \frac{t_1 t_2}{n_1 n_2}\phi_1\phi_2\phi_3 \\
&= \phi_1 + \phi_2 + \phi_3 - \frac{t_1\phi_1}{n_1}(\phi_2 + \phi_3) - \frac{t_2\phi_3}{n_2}(\phi_1 + \phi_2) + \frac{t_1 t_2}{n_1 n_2}\phi_1\phi_2\phi_3.
\end{aligned} \tag{7.51}$$

Problems

7.1 Derive the expression for the distance (δ) from the front vertex V_1 to the front principal plane (P).

7.2 Prove (using equations) that the one principal point of a plano-convex lens is always on the curved surface.

7.3 For a convex-plano lens in air, show that the back principal plane (P^*) is always t/n from vertex 2.

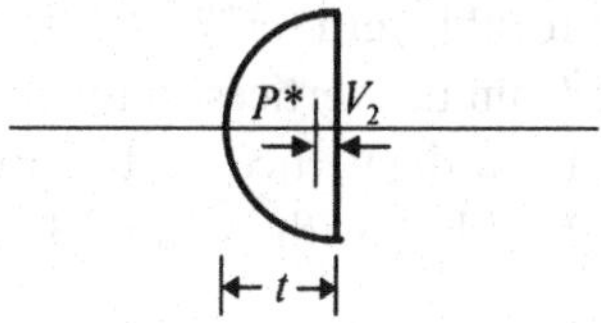

7.4 An equi-convex thick lens has $R_1 = -R_2 = 4$ cm with $n = 1.5$. What lens thickness would produce a zero optical power lens? Can you make a general statement when $R_1 = -R_2$?

7.5 An equi-concave negative lens has an axial thickness of 0.2 cm, with $|R| = 15$ cm. What is the edge thickness of a 2.5 cm diameter lens?

7.6 A thick lens has $R_1 = R_2 = 3$ cm, $n = 1.5$ with a thickness of 0.5 cm ($t = 0.5$ cm). Determine the power and location of the cardinal points in air.

7.7 A thick lens has two concentric surfaces $R_1 = 20$ cm, $R_2 = 15$ cm, with a thickness of 5 cm. It separates air from water.

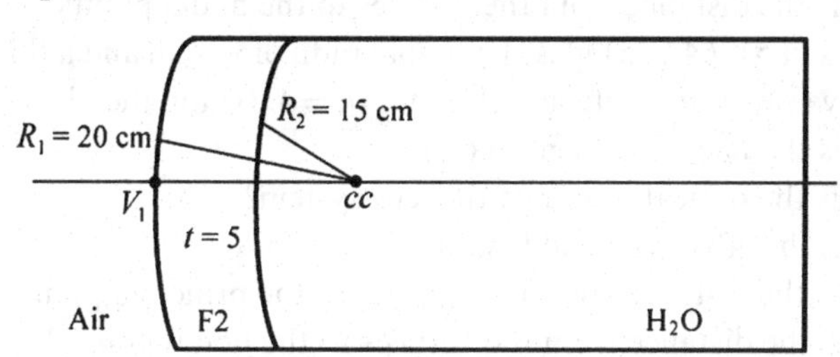

(a) What is the optical power (diopters)?

(b) Make a sketch of the cardinal points relative to the front vertex.

7.8 You have two lenses. Lens A is a negative meniscus and Lens B is a positive meniscus. The lenses have the prescriptions shown in Table 7.6.

Table 7.6.

Lens A	Lens B
$R_{a1} = 7$ cm	$R_{b1} = 3$ cm
$R_{a2} = 3$ cm	$R_{b2} = 7$ cm
$t = 0.5$ cm	$t = 0.5$ cm
$n_a = 1.45$	$n_b = 1.55$

(a) What is the optical power of each lens?
(b) Determine their cardinal points in air and sketch the lenses.
(c) Determine the optical power of an in-contact combination of these lenses in the following two cases:
  (1) lens A to lens B,
  (2) lens B to lens A.

7.9 An equi-convex thick lens with radii of 9 cm is made of 755276.479 glass with an axial thickness of 3 cm.
(a) What is the effective focal length (*efl*)?
(b) What is the distance from the vertices to the focal points?
(c) What is the distance from the vertices to the principal points?

7.10 A lens made of 517642.251 glass with radii of 3 cm and 5 cm has a thickness of 2 cm.
(a) What is the effective focal length?
(b) What are the locations of the principal points relative to the vertices?
(c) What is the effective focal length if the radii are interchanged?

7.11 A thick lens made of 755276.479 glass, $R_1 = 4$ cm, and $R_2 = -2$ cm, is 3 cm thick. This lens is placed at the end of a tank containing a transparent liquid with a refractive index of 1.42. R_2 is in contact with the liquid.
(a) What is the front focal length? What is the back focal length?
(b) What is the distance from the vertices to the focal points?
(c) What is the distance from the vertices to the principal points?
(d) What is the distance from the vertices to the nodal points?

7.12 A thick lens of 517642.251 glass has equal radii of +5 cm and a thickness of 2 cm. The concave side is in contact with water ($n = 4/3$), and the object space is in air.
(a) What is the back focal length (f^*)?
(b) What is the optical power of the lens system?
(c) What is the effective focal length?
(d) What is the distance from the vertices to the principal points?
(e) What is the distance from the vertices to the nodal points?

7.13 Determine the Gaussian properties of the eye using the values shown in Table 7.7 (i.e. locate and label the cardinal points relative to the cornea vertex). (All dimensions are in millimeters).

Table 7.7.

Surface	Radius	Thickness	Refractive index
1	7.80	0.55	1.3771
2	6.50	3.05	1.3374
3	11.03	4.00	1.4200
4	−5.72		1.3360

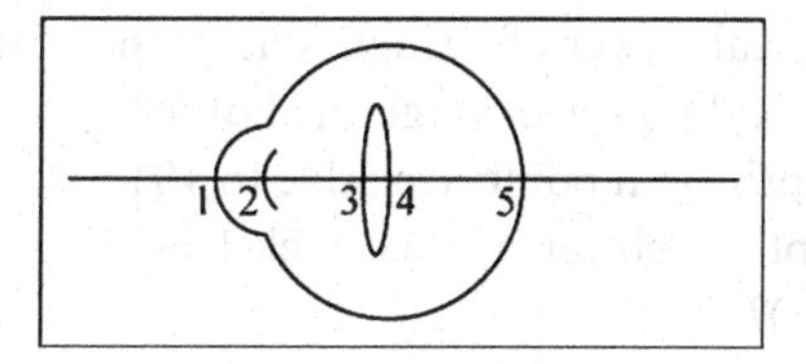

Then, fill in the correct location for the cardinal points of an eye in the diagram shown below

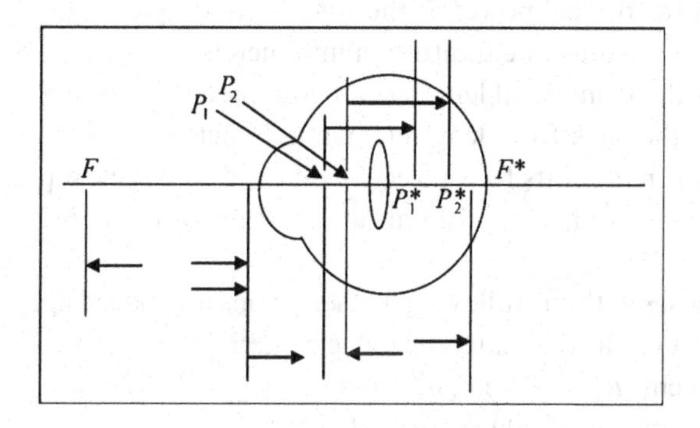

and answer the following questions to the nearest tenth (0.1).

(a) What is the total optical power (diopters)?

(b) What is the front focal length (f)?

(c) What is the front focal distance?

(d) What is the cornea to front principal plane distance?

(e) What is the cornea to rear principal plane distance?

(f) What is the distance between the front principal plane and the front nodal point?

(g) What is the rear focal length?

(h) What is the size on the retina of a 4 mm object at 20 ft from the eye?

(i) What is the back focal distance?

7.14 A thick lens of 4 mm with radii of 4 cm and 6 cm has index of refraction of 1.65 and is located in the side of a fish tank. The water is on the $R_1 = 4$ cm side. A fish is 15 cm away from the lens.

(a) Find the image location (relative to V_1).

(b) Where are the nodal points located relative to the vertices?

7.15 Consider the bottom of a cola bottle with an outside radius of 4 in and an inside radius of 3.75 in, made of glass with a refractive index of 1.5.

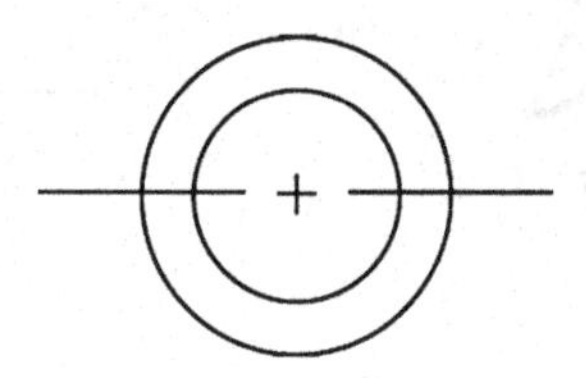

(a) What is the optical power of a single side of the bottle?
(b) What is the optical power through the bottle?
(c) Where are the principal points (relative to V_1)?
(d) What is the optical power of this thick lens (if the bottle were filled with water, $n = 1.3\bar{3}$)?

7.16 A glass with a refractive index of 2 ($n = 2$) has an image formed in the glass, 25 mm to the right of the vertex, while the object is at infinity in front of the vertex.
(a) What is the optical power of the surface (in diopters)?
(b) What is its radius of curvature in millimeters?
(c) What is the front focal length (f) in millimeters?
(d) What is the back focal length (f^*) in millimeters?

7.17 Find the cardinal points for two equi-convex thin lenses, separated by 11 cm in air. One lens has an f^* of 10 cm, and the other is an $F/1$ lens with a 5 cm diameter.

7.18 Two thin lenses with the following prescriptions are placed against each other (in physical contact) at the matching –25 cm radii:
(1) $R_1 = 20$ cm, $R_2 = -25$ cm, $n_1 = 1.4$,
(2) $R_1 = -25$ cm, $R_2 = 100$ cm, $n_1 = 1.8$.
Calculate the following:
(a) The individual optical powers and focal lengths of the two thin lenses;
(b) The combined optical power and focal length of the doublet.

7.19 For a marble of radius 1 cm, with a refractive index of 2, and an object 10 cm away (left of the marble):
(a) What is the optical power in diopters?
(b) What is the transverse magnification?
(c) Is the image real or virtual?

7.20 Calculate and show on a sketch the cardinal points of the following lens (A and B) configurations. Reference to vertex 1.

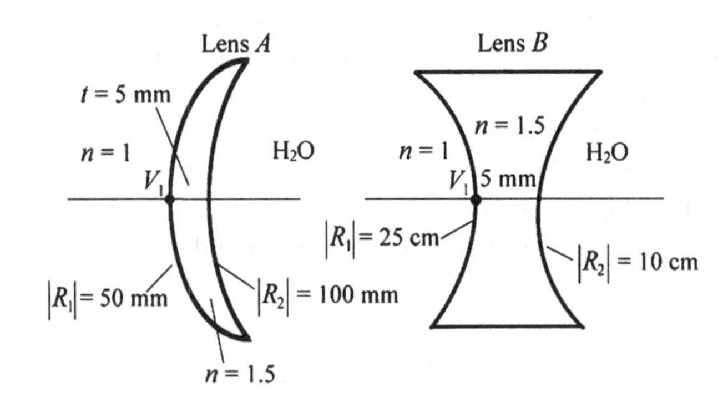

7.21 Calculate and show on a sketch the cardinal points of an equivalent system equaling a two thick lens combination, as shown below. Reference to vertex 1.

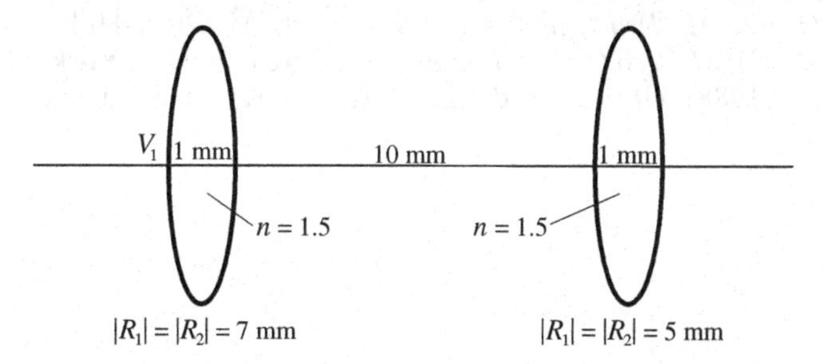

7.22 A +5.00 diopter lens forms a real image on a screen placed 100 cm away from the object. Find the two object distances possible. What is the transverse magnification for each position?

7.23 A 3 mm thick lens has a front radius of 8 mm and a back radius of –6 mm and a refractive index of 1.5. It images from air to water ($n = 1.3\bar{3}$).
   (a) What is the optical power of this lens (in diopters)?
   (b) If the image is in the water, where is an object measuring 1 mm high and 25 cm away imaged?
   (c) Sketch a layout of the cardinal points.

7.24 An equi-convex thick lens is made of 777444.333 glass. Calculate the radius of curvature for the lens's surfaces with a thickness of 2 mm that is necessary to give an optical power of 4.5 diopters for *d* light.

7.25 Find the cardinal points for the two-thick-lens system shown as a telephoto lens below.

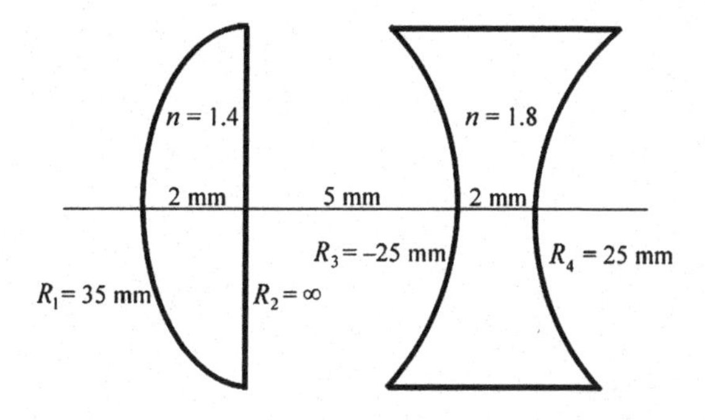

Bibliography

Bass, M. (1995). *Handbook of Optics*, Vol. I. New York: McGraw-Hill.

Kingslake, R. (1978). *Lens Design Fundamentals.* New York: Academic Press.

Mouroulis, P. and MacDonald, J. (1997). *Geometrical Optics and Optical Design.* New York: Oxford University Press.

Sears, F. W. (1958). *Optics*, Reading, MA: Addison-Wesley.
Shannon, R. R. (1997). *The Art and Science of Optical Design*. Cambridge, UK: Cambridge University Press.
Smith, W. J. (1992). *Modern Lens Design*. New York: McGraw-Hill.
Smith, W. J. (2000). *Modern Optical Engineering*, third edn. New York: McGraw-Hill.
Welford, W. T. (1988). *Optics*, third edn. Oxford, UK: Oxford University Press.

8

Mirrors

In Chapter 2 we introduced the concept of a plane mirror and its effect on the handedness of an image. An effect of Snell's law provides the law of reflection: the angle of incidence is equal to the angle of reflection, along with a sign change relative to the normal of the surface (see Equation (2.14)). Rays from an object or any point on the object are reflected according to Snell's law in the plane of incidence. The plane of incidence is the plane composed of the incident ray and the surface normal, as shown in Figure 8.1, in which the plane of the paper contains the ray and the normal (η).

As discussed in Chapter 2, the image of point P is located as far behind the mirror as the point is in front of the mirror. For an extended object made up of a continuum of points, as shown in Figure 8.2, the image is located by tracing rays backward in the plane of incidence. The image of the arrow has been inverted upon reflection, and point A' is below point B' on this image. What we have been doing is ray tracing in the plane of incidence. If we look at the object directly, we see a different orientation in the plane of incidence than we do if we look at the object via the mirror. An observer looking at the object and image, as shown in Figure 8.3, sees an inverted image. To the observer at position (1), B lies to the right of A. Now for the observer at position (2), B' appears to the left of A'; the image is left-handed. This is because the observer has changed his or her point of view. Consider our stick man shown in Figure 8.4. Although he is inverted, the left and right hand have the same right to left orientation as the object.

The orientation of the image after reflection is exactly as was discussed in Section 2.5, using the letter F as a simple example, and tracing through the mirror systems with the left vertical line of F in the plane of incidence. This procedure is the same as was discussed in Chapter 2 for deviating prisms.

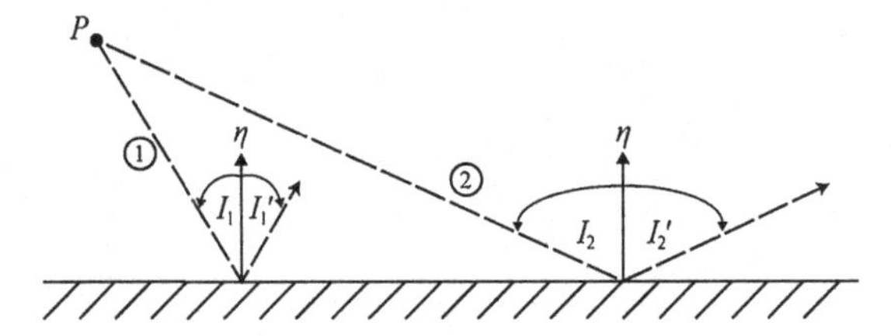

Figure 8.1 Point source reflection in the plane of incidence.

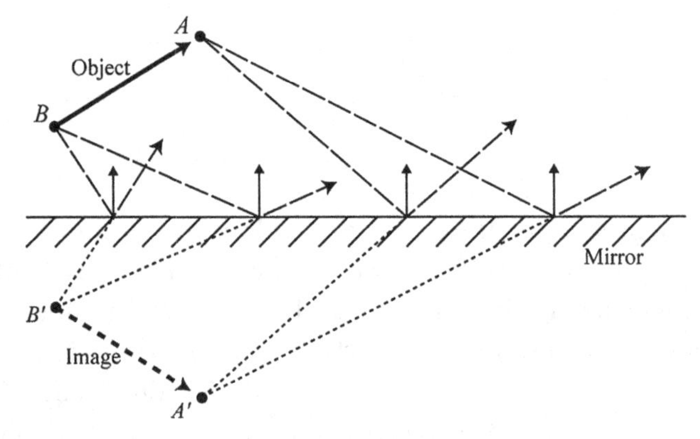

Figure 8.2 Projection of a line object through a plane mirror.

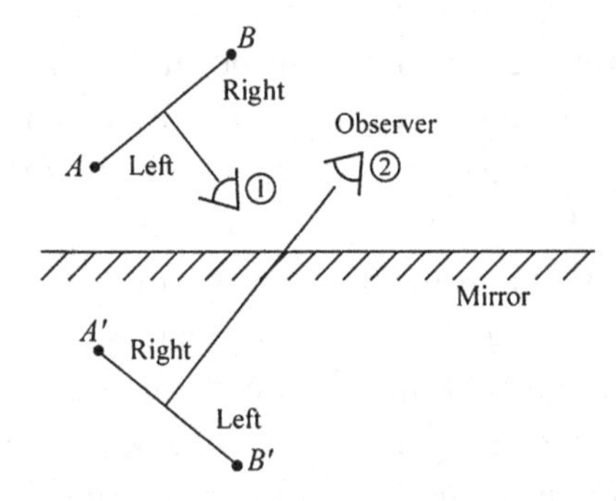

Figure 8.3 Observing a line object directly and via a plane mirror.

8.1 Plane mirrors

The simplest and by far the most common optical element is the flat or plane mirror. There are many different types of mirror, such as dressing mirrors, two-way mirrors, and rear view mirrors in cars. The applications and uses of plane mirrors are many and varied. In scientific systems, flat mirrors are most often used to reduce the overall length of an optical system by folding it or to correct the parity of the image. The key reasons for using a plane mirror are:

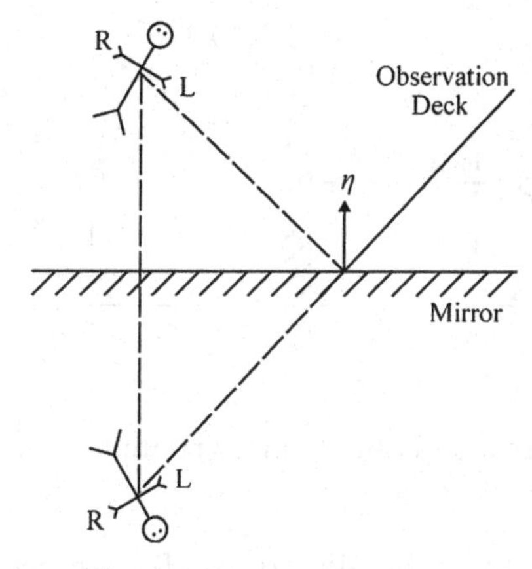

Figure 8.4 Prospective reflection from plane mirror.

- to fold the optical system to get extra distance via virtual space;
- to correct handedness;
- to produce multiple virtual images.

The apparent position of the image is the same distance behind the mirror as the actual object is in front of the mirror. The image is the same size as the object, and is called a virtual image (i.e. the rays of light from the object do not actually go to the image, but only appear to, as the extensions of the reflected light rays seem to intersect behind the mirror).

Glass mirrors date from the Middle Ages. They were made in large quantities in Venice, Italy, from the sixteenth century. Mirrors made from Murano glass featured a back covered with a thin coating of tin mixed with mercury. After 1840, a thin coating of silver was generally used. Using mercury in an enclosed vessel to form a flat surface is a technique that is still employed today to produce a reference surface. More recently, aluminum has been introduced as the reflecting material, because it is almost as efficient a reflecting material as silver, but is more resistant to oxidation. The aluminum is placed on either the front or back surface of a glass plate. Placing it on the back surface is far more common, because this arrangement offers protection against scratches and other damage to the surface. In this case, however, the reflectivity may not be as high and a shadowing may occur, since the glass surface has about a 4% reflection. The advantage is that the glass surface can be readily cleaned without damaging or destroying the aluminum. Typically, front surface mirrors are used for scientific purposes because infrared light does not transmit though normal glass, so a back surface mirror would not reflect efficiently. Since light is defined according to our convention as traveling from left to right, once a mirror is put into an optical system, the ray path is reversed or

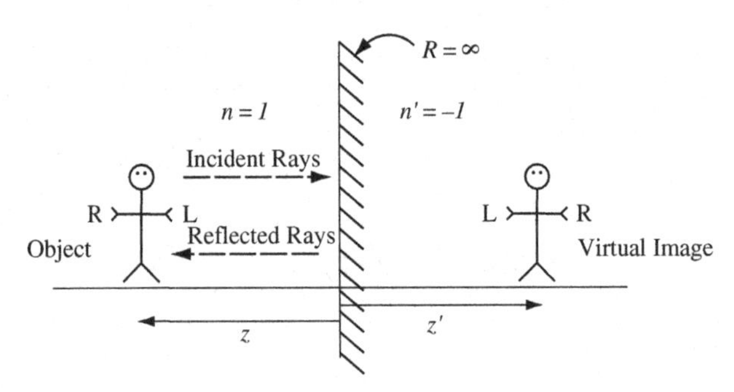

Figure 8.5 Plane mirror and object–image relationship.

goes from right to left (negative direction). To compensate for that negative distance, the sign of the index of refraction is changed. Light traveling from right to left carries a negative index of refraction. In air the index would be −1. With this concept in mind, consider an object in front of a plane mirror, as shown in Figure 8.5. Using the Gaussian equation (6.3) for object–image relationship:

$$\frac{n'}{z'} = \frac{n}{z} + \phi. \tag{8.1}$$

substituting for the parameters n, n', and the optical power as shown in Figure 8.5:

$$\frac{-1}{z'} = \frac{1}{z} + \frac{-1 - 1}{\infty}. \tag{8.2}$$

This proves that the image distance is equal to the object distance in magnitude but is opposite in sign.

$$z' = -z. \tag{8.3}$$

It also demonstrates that the transverse magnification of the vertical image is

$$\begin{aligned} M_{\mathrm{t}} &= \frac{y_i}{y_0} = \frac{z'/n'}{z/n} = \frac{-z/-1}{z/1} \\ &= 1 \end{aligned} \tag{8.4}$$

Thus, the image is upright but the handedness has been changed because there has been one reflection. Note that the reverted image is bilaterally symmetric about the plane of incidence, which, in most cases, is the vertical plane; and thus, the image is not flipped top to bottom but is flipped left to right.

For the cases studied so far, the convergence or divergence of the rays determines whether the object or image is real or virtual. In Table 8.1 objects

Table 8.1. *Ray convergence/divergence vs. real or virtual objects/images*

Objects			Images		
Emitted rays are diverging	Emitted rays from object are converging	Emitted rays are parallel to the optical axis	Refracted rays are converging	Refracted rays are diverging	Refracted rays are parallel to optical axis
Real object	Virtual object	Real or virtual	Real	Virtual	Real or virtual
$z(-)$	$z(+)$	$z = \pm\infty$	$z'(+)$	$z'(+)$	$z' = \pm\infty$

and images are identified as "real" or "virtual" depending on whether their rays (emitted or refracted) are diverging, converging, or parallel to the optical axis.

8.2 Spherical mirrors

Spherical mirrors, like lenses, have a center of curvature that is at an equal distance from any point on the spherical surface. The optical axis is a line from the center of curvature to the vertex of the spherical mirror, similar to the lens except simpler in concept. In the lens case, the optical axis is determined by two surfaces' centers of curvature and two vertices.

The vast majority of concave spherical mirrors are used in cosmetic applications, e.g. shaving or beautification. However, in the scientific community, spherical mirrors are often used because of the very large diameters available. Typically, refracting optical element lenses are no larger than about 1 m in diameter. Mirrors, on the other hand, can be made with diameters as large as 8.5 m, and there are plans for even larger (32 m) telescope mirrors.

In a convex spherical mirror, the vertex of the mirror is nearer to the object than the edges, i.e. the mirror bulges toward the object. The image formed is always smaller than the object and is always erect. It is never real, because the reflected rays diverge outward from the face of the mirror and are not brought to a focus. The image, therefore, is determined by the apparent extension of the rays behind the mirror as in the case of the plane mirror. Figure 8.6 shows the layout of both a concave and a convex mirror.

The optical power of a spherical mirror surface is similar to that of a lens, except that the index of refraction, after reflection from the mirror, is the negative of the index of refraction in front of the mirror ($n' = -n$). From the expression for optical power,

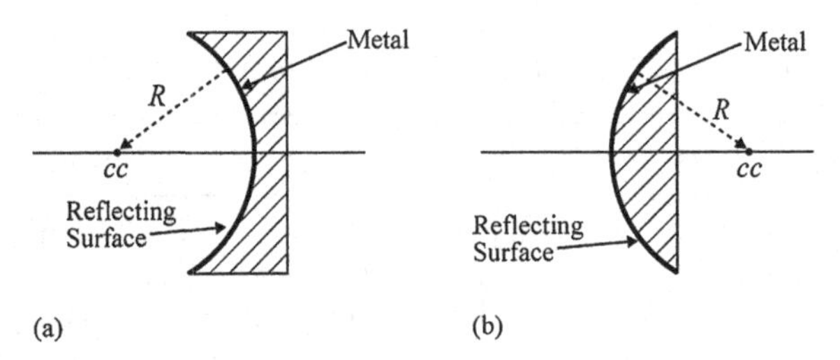

Figure 8.6 (a) Concave mirror and (b) convex mirror.

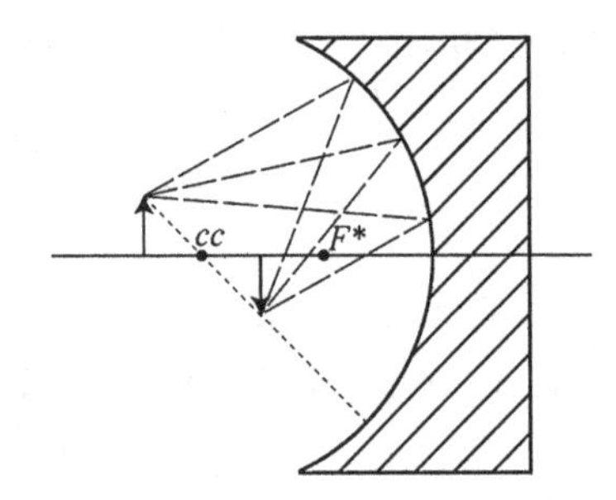

Figure 8.7 Rays sketched from the object point to the image point.

$$\phi = \frac{n' - n}{R} = \frac{-n - n}{R} = \frac{-2n}{R}. \tag{8.5}$$

For a spherical mirror, if R is a positive value, then the power is negative, if R is a negative value, then the power is positive.

Spherical mirrors are used because they offer several advantages over refractive elements. Some advantages are:

(a) Mirrors can be made to any size.
(b) Since the refractive index is not a function of wavelength, mirrors have no chromatic dispersion.
(c) Mirrors are more compact for use in an optical system layout.
(d) Mirrors produce about 4 times the optical power of an equivalent spherical surface made of glass ($n \approx 1.5$).
(e) A mirror acts as a thin lens.

Spherical mirrors have a few disadvantages which limit their use:

(a) The surface precision, or accuracy, must be much better than that of a refracting surface.
(b) The spherical mirror surface has a much reduced field of view (see Figure 8.7) than corresponding refractive optics for the same optical power.

As shown in Figure 8.8, rays incident on a spherical surface are reflected such that the angle of incidence equals the angle of reflection. However, as we

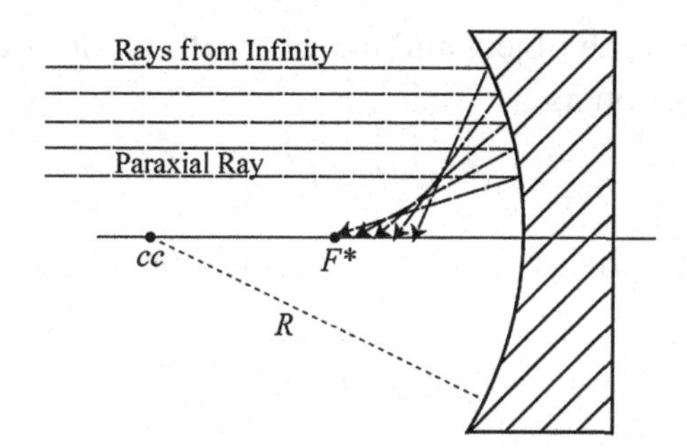

Figure 8.8 Spherical mirrors with rays from infinity.

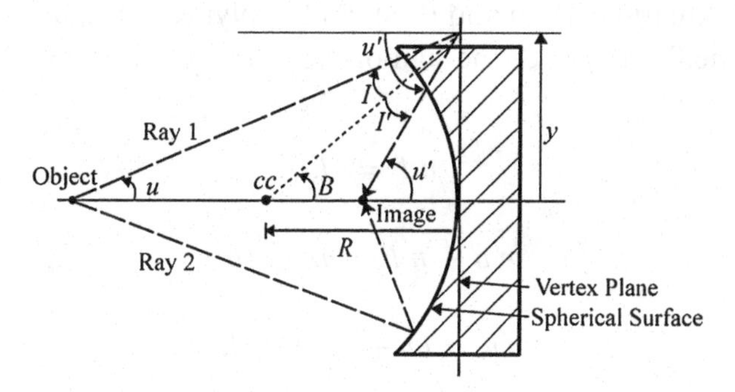

Figure 8.9 Paraxial case for a spherical mirror.

move the rays further from the axis, note that the reflected ray crosses the axis closer to the vertex. This is what is known as a spherical aberration, and this produces a blur of a point, which is increased in size compared with the ideal point. As with a lens system, we need to constrain ourselves to rays close to the axis, or paraxial rays only. Restricting ourselves to the paraxial domain causes all rays to converge to a point at the focal point, F^*. As shown in Figure 8.8, the normal to the surface is through the center of curvature (*cc*), and the ray path, after reflection, goes back to the focal point F^*.

Zooming in on the region near the axis causes the spherical surface to appear as a plane, and therefore, we avoid sag, as shown in Figure 8.9. In this restricted analysis, all reflection would take place at the vertex plane. Therefore, the reflection is really taking place on the tangent to the sphere; however, we project the rays back to the vertex plane, as shown Figure 8.9. The angle of incidence (I) and the angle of reflection (I') are relative to the normal of the spherical surface, shown as a dashed line. We will assume for this discussion that the reflection takes place at the vertex plane. In this case, a second ray (Ray 2) converges to the same point as Ray 1. In fact, all paraxial rays from object point will be imaged to image point.

A relation between the object and image locations can be derived by recalling the law of reflection as:

$$-I = I'. \tag{8.6}$$

From Figure 8.9, the angles the ray makes with the optical axis are

$$u = B + I \tag{8.7}$$

and

$$u' = B + I'. \tag{8.8}$$

Rearranging Equations (8.7) and (8.8), and applying the small angle approximation for Snell's law, since the sign of the index of refraction changes upon reflection:

$$n'I' = nI, \tag{8.9}$$

$$n'u' - n'B = nu - nB \tag{8.10}$$

$$n'u' = nu + (n' - n)B, \tag{8.11}$$

where B is the angle from the center of curvature, as shown in Figure 8.9:

$$B = -y/R, \tag{8.12}$$

which leads to the refraction equation, Equation (5.20), for paraxial optics:

$$n'u' = nu - \frac{n' - n}{R}y. \tag{8.13}$$

Therefore, the optical power of this spherical surface, as concluded earlier, is

$$\phi = \frac{n' - n}{R}. \tag{8.14}$$

For this case in air ($n = 1$), the optical power for the mirror is

$$\phi = \frac{-1 - 1}{R} = \frac{-2}{R}. \tag{8.15}$$

Recalling Equation (7.2) with the appropriate indices and solving for front and back focal lengths (f) and (f^*):

$$\phi = \frac{-n}{f} = \frac{n'}{f^*}, \tag{8.16}$$

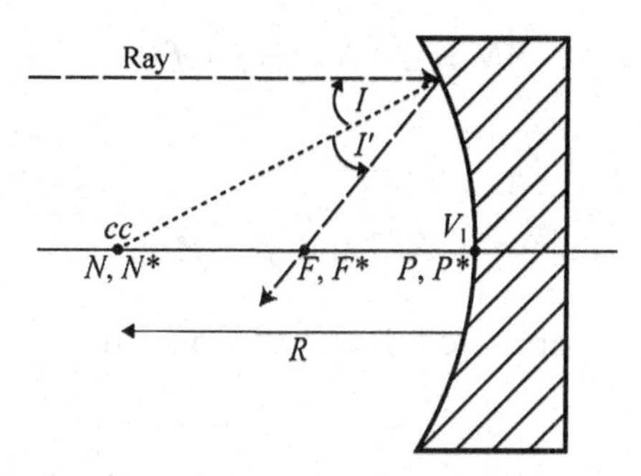

Figure 8.10 Single reflecting spherical concave surface in air.

$$f = \frac{R}{2}, \tag{8.17}$$

$$f^* = \frac{R}{2}. \tag{8.18}$$

Therefore, the front and back focal points lie on top of each other but in different optical spaces. This sometimes becomes confusing when they are drawn on top of one another. For a single reflective surface, object space and image space coincide. An interesting fact is that, if the spherical surface is in water or is the back side of a glass substitute such as a mangin mirror (rear surface of a spherical glass surface is metalized), the optical power is changed:

$$\phi = \frac{-2n}{R}. \tag{8.19}$$

but the front and back focal lengths are not changed. In summary, for a concave spherical mirror there are four main points to remember, as illustrated in Figure 8.10:

(a) The focal lengths are equal to the radius divided by 2 ($R/2$).
(b) The concave spherical mirror is a thin lens with the principal planes, (P, P^*) at the vertex.
(c) All media, after a reflection in the system, are treated as having negative indices of refraction.
(d) The nodal points N and N^* lie at the center of curvature since $n' \neq n$.

The last point is not completely obvious, since any ray through the center of curvature in Figure 8.10 is reflected back on itself in order to have an angular magnification of 1. Another way of resolving the nodal point location is via Equation (7.7):

$$\overline{PN} = \overline{P*N*} = f + f*,$$

where

$$f* = R/2 \text{ and } f = R/2.$$

The nodal points are separated from the principal points by the radius distance:

$$\overline{PN} = R. \tag{8.20}$$

Example 8.1

For the given system, find z' for a 5 mm object at the *cc* of a concave mirror with radius of curvature of 400 mm.

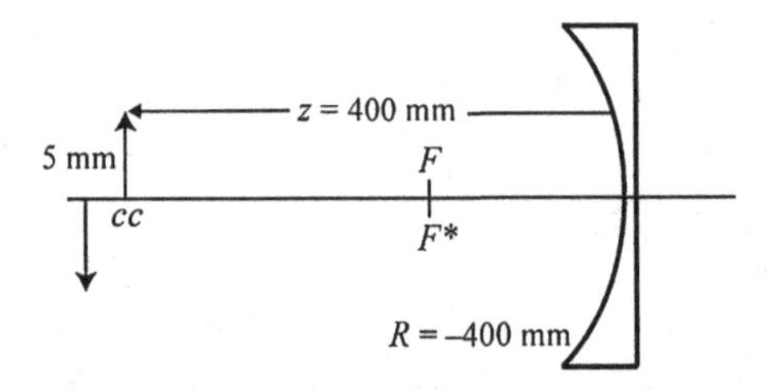

$$\phi = \frac{-2}{R} = \frac{-2}{-400} = \frac{1}{200} \quad \text{or} \quad \phi = -\frac{n}{f} = \frac{n'}{f*} = \frac{-1}{-200} = \frac{1}{200},$$

$$\phi = \frac{1}{200} = 5\,\text{diopters}, \; f = f* = -200,$$

$$\frac{n'}{z'} = \frac{n}{z} + \phi \rightarrow \frac{-1}{z'} - \frac{1}{-400} = \frac{-2(1)}{-400} \rightarrow \frac{-1}{z'} = \frac{1}{400} \rightarrow z' = -400\,\text{mm}.$$

Note: The space where $n' = -1$ is a different space than object space, but is located in the same *physical space* as the object.

Transverse magnification:

$$M_\text{t} = \frac{Y_\text{i}}{Y_\text{o}} = \frac{nz'}{n'z} \rightarrow M_\text{t} = \frac{1(-400)}{-400(-1)} = -1.$$

The image height is –5 mm, inverted. The magnification is 1.

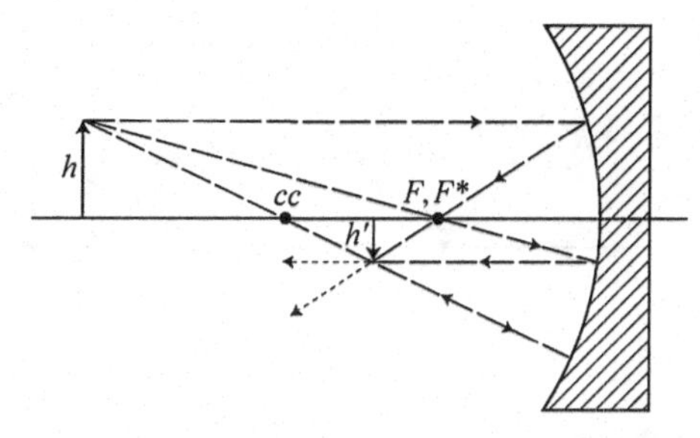

Figure 8.11 Using selected rays to find the image location.

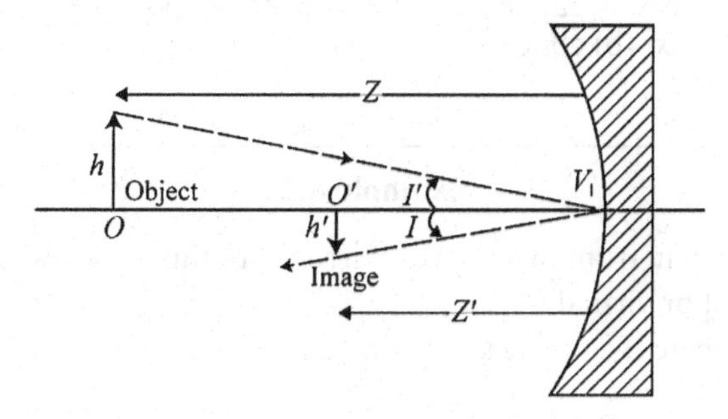

Figure 8.12 Image magnification via the chief ray.

The object–image relationship can be determined by ray tracing some selected rays, as shown in Figure 8.11, for a positive optical power spherical mirror.

A ray from the top of the object parallel to the axis is reflected through the back focal point (F^*). A ray through the front focal point (F) becomes parallel to the optical axis, and a ray through the center of curvature is reflected back on itself, as shown in Figure 8.11.

The magnification, h'/h, can be determined by applying similar triangles to the chief ray that comes from the top of the object to the vertex and reflects back to the top of the image, as shown in Figure 8.12. Since the angles are equal at the vertex, the triangles, V_1Oh and $V_1O'h'$, are similar, so the transverse magnification is

$$M_{\text{t}} = \frac{\overline{O'h'}}{\overline{Oh}} = \frac{z'/n'}{z/n} = \frac{-z'}{z}. \tag{8.21}$$

This is the same as the definition of transverse magnification for lenses.

A convex spherical mirror has a negative optical power, as stated earlier. Since the radius has a positive value, the expression for power is negative, and both focal points lie on the positive side of the mirror at $R/2$, as shown in Figure 8.13. The nodal points are again separated from the principal points by the radius of curvature, $(f^* + f)$.

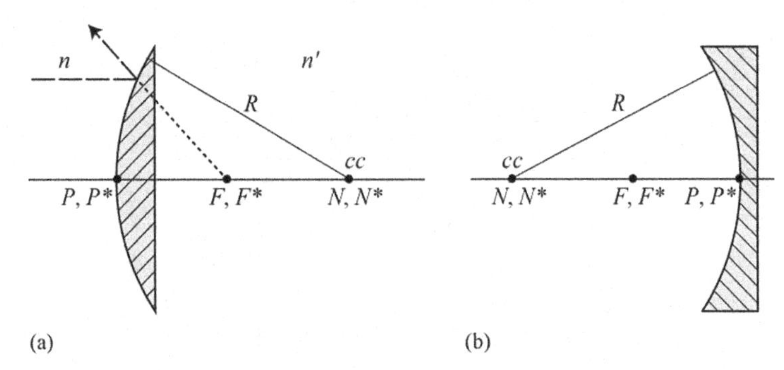

Figure 8.13 Convex negative power and concave positive power spherical mirrors: (a) convex; (b) concave.

Example 8.2

If an object is 60 cm in front of a convex mirror of radius 20 cm, where is the image location? Is it real or virtual?

Ray trace to locate the image approximately:

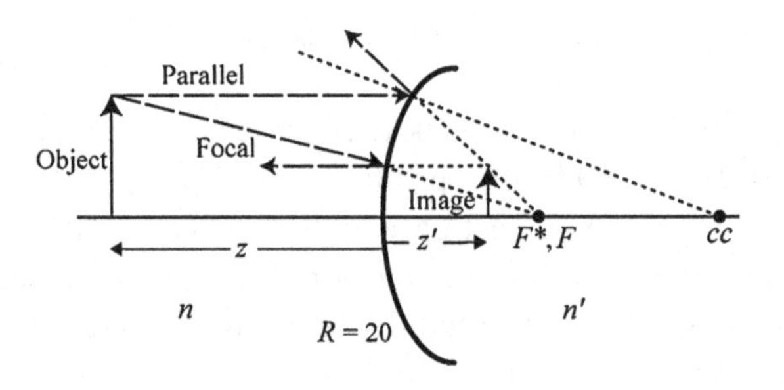

Gaussian solution:

$$\frac{n'}{z'} = \frac{n}{z} + \phi;$$

$$\frac{n'}{z'} = \frac{n}{z} + \frac{n'-n}{R} = \frac{n}{z} - \frac{1}{f_2} = \frac{1}{-60} + \frac{-1-1}{20} = \frac{-1}{60} - \frac{1}{10} = \frac{-7}{60}$$

$$\frac{-1}{z'} = -\frac{7}{60} \Rightarrow z' = \frac{60}{7} \text{ cm; virtual.}$$

8.2.1 Catadioptic systems

If a lens has both refractive and reflective elements, then the lens is said to be catadioptic. For a lens that has both refractive and reflective elements, the determination of the image location can become complicated since, after a reflection of light, the indices of refraction are negative for all the rays until another reflection occurs. This also requires the handedness of the image to be considered in this type of system. If a refractive element follows a mirror, the refractive index of the refracting lens must be considered negative. The easiest way to explain this situation is by the example shown below.

Example 8.3

What is the total power of the system below, which contains a spherical mirror ($R = -25$ cm), and a lens (refracting the rays) placed 10 cm in front of the mirror?

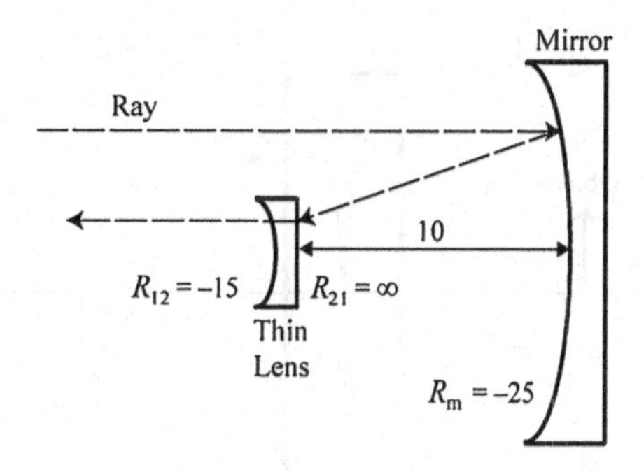

Mirror optical power: $\phi_{\mathrm{m}} = \dfrac{-2}{R_{\mathrm{m}}} = \dfrac{-2}{-25} =$ positive power.

Lens optical power of first surface: $\phi_1 = \dfrac{-1.5 - (-1)}{\infty} = 0.$

Lens optical power of

second surface: $\phi_2 = \dfrac{n' - n}{R_2} = \dfrac{-1 - (-1.5)}{-15} = -\dfrac{1}{30}$ (negative).

$$\phi_t = \frac{2}{25} - \frac{1}{30} - \left(\frac{2}{25}\right)\left(\frac{-1}{30}\right)\left(\frac{-10}{-1}\right) = \frac{11}{150}.$$

For ease of layout and ray tracing, a system is often unfolded to eliminate the need to use negative refractive indices and negative distances. If one thinks about reduced distance (length/refractive index), the negative signs cancel when using reduced length. The unfolding process requires all media, after reflection, to have both positive refractive indices and positive distances. Since t/n is usually used in equations, the negative signs cancel out.

8.2.2 Unfolding mirror systems

Since a mirror is a perfect thin lens equivalent, its thickness is zero in the unfolded layout. So the layout of the lenses does not need to be adjusted for path length. The advantages of unfolding an optical system are that the light propagates in the positive direction, and the layout is what is expected from previous work.

Example 8.4

A 5 mm object is 400 mm in front of a concave mirror with a radius of 400 mm. What are the image location and image height?

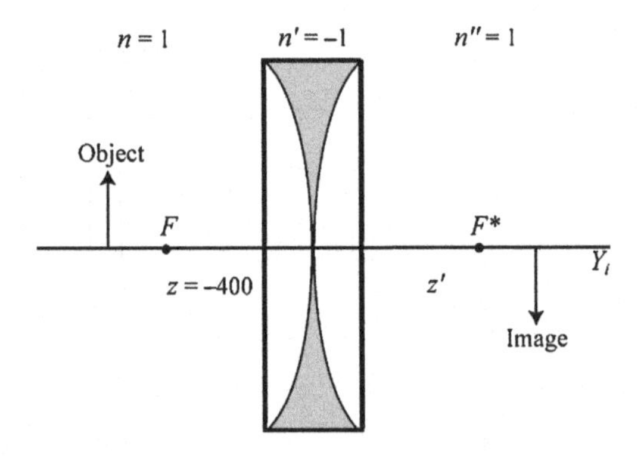

Use unfolded layout

$$\frac{n''}{z'} - \frac{n}{z} = \phi = \frac{n' - n}{R} = \frac{-1 - 1}{-400} = \frac{1}{200}$$

$$\frac{1}{z'} - \frac{1}{-400} = \frac{1}{200} \Rightarrow z' = 400$$

$$M_t = \frac{z'/n}{z/n} = \frac{400/1}{-400/1} = -1$$

$$Y_i = -5\,\text{mm}.$$

Image is –5 mm!!

As indicated by the drawing of the unfolded mirror, the optical element really is a thin lens. For a convex mirror (see Figure 8.14(a)), the corresponding unfolded mirror is shown in Figure 8.14(b). The optical power from Equation (8.5) with a positive radius of curvature is negative (if $R = 20$ mm, $f^* = 10$ mm).

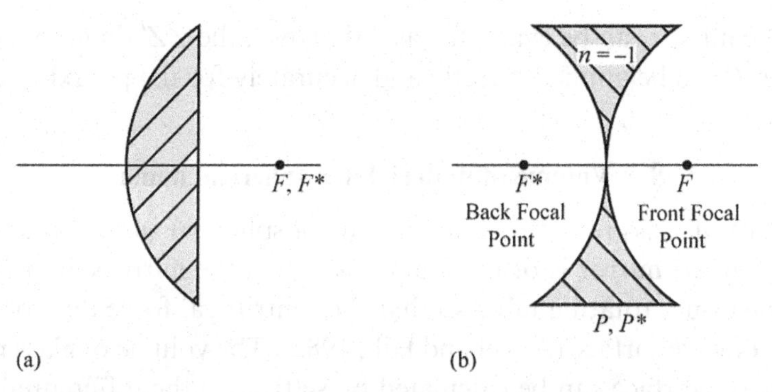

Figure 8.14 (a) Standard layout of a negative convex mirror; (b) unfolded negative convex mirror layout.

Example 8.5

If an object is 60 cm in front of a convex mirror of radius (+)20 cm, where is the image in the unfolded system?

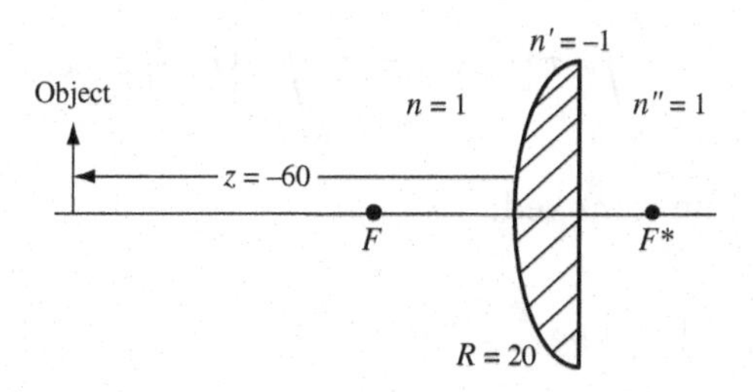

$$\frac{n''}{z'} = \frac{1}{z} + \phi = \frac{1}{-60} + \frac{-1-1}{20} = -\frac{7}{60}.$$

$z' = -\frac{60}{7}$ in the unfolded layout. In the real layout, it is to the right of the mirror's vertex.

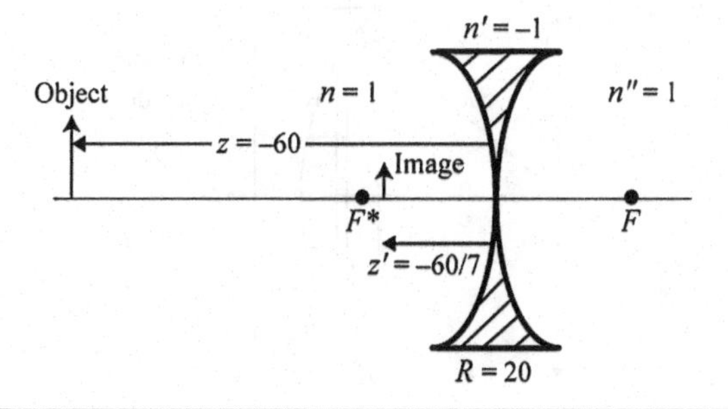

Since the mirror can be evaluated as a thin lens, the ZZ' diagram discussed in Chapter 6 can be applied directly and accurately for the paraxial case.

8.3 Volume of material in a spherical dome

The amount of glass removed when making a spherical surface is sometimes substantial if the mirror is of the 8 m class. Often the mirrors are made with molten glass on a rotating table so that the centrifugal force due to spinning creates a concave surface (Angel and Hill, 1982). The volume of glass removed for a concave surface can be calculated by setting up the differential volume element, as shown in Figure 8.15, for a rotationally symmetric mirror around the z axis. The volume is calculated for a given sag, h in this case.

The differential volume for a disk in Figure 8.15 is

$$\mathrm{d}v = \pi y^2 \mathrm{d}z, \tag{8.22}$$

and the integration should be taken from $R - h$ to R. So the total volume is

$$v = \int_{R-h}^{R} \pi\left[y^2\right]\mathrm{d}z = \pi \int_{R-h}^{R} \left[R^2 - z^2\right]\mathrm{d}z. \tag{8.23}$$

Solving for the volume removed:

$$\begin{aligned} V &= \int_{R-h}^{R} \pi\left[R^2 - z^2\right]\mathrm{d}z \\ &= \pi\left[R^2 z - \frac{z^3}{3}\right]_{R-h}^{R} \\ &= \pi\left[R^3 - \frac{R^3}{3}\right] - \pi\left[R^2(R-h) - \frac{(R-h)^3}{3}\right], \end{aligned}$$

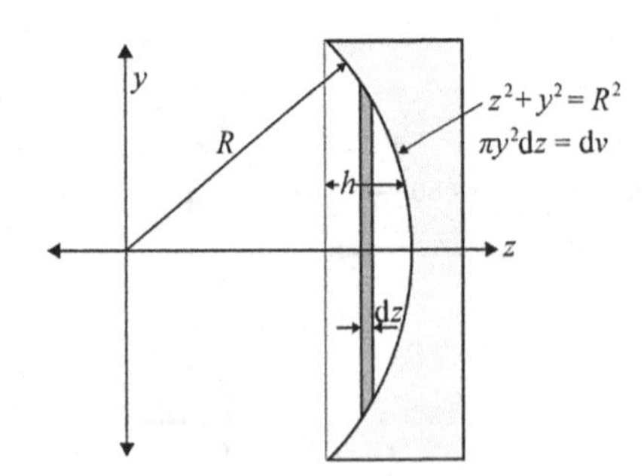

Figure 8.15 Volume calculation of a spherical dome.

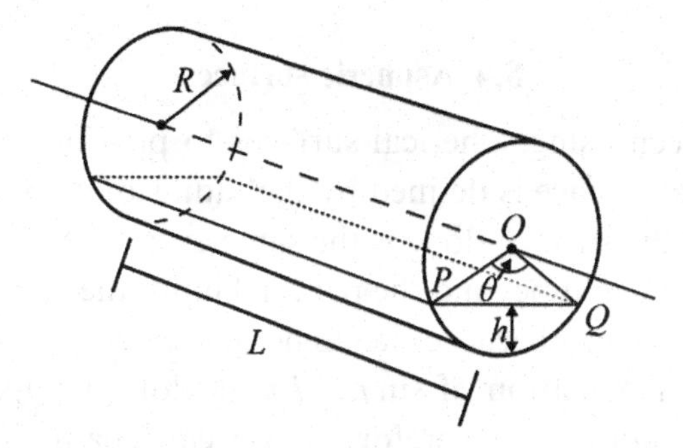

Figure 8.16 Volume of a section of a cylinder.

where

$$\frac{(R-h)^3}{3} = \frac{(R^3 - 3hR^2 + 3h^2R - h^3)}{3},$$

$$V = \pi\left[R^3 - \frac{R^3}{3}\right] - \pi\left[R^3 - hR^2 - \frac{(R^3 - 3hR^2 + 3h^2R - h^3)}{3}\right]$$

$$= \frac{\pi(3h^2R - h^3)}{3}$$

$$= \frac{\pi h^2(3R - h)}{3}. \tag{8.24}$$

This is the volume of a dome cap of height h for a sphere of radius R. The volume of glass in a cylinder lens can also be calculated, as shown below.

The area of a sector POQ is

$$A_s = \frac{\theta}{360}\pi R^2.$$

The area of triangle POQ in Figure 8.16 is calculated using the equation for a triangle:

$$A_T = (R-h)(2Rh - h^2)^{1/2}. \tag{8.25}$$

Therefore, the area of a sector of a circle with height h is just the difference between these areas. To compute the volume of a cylindrical lens, the area times the length (L) is required.

$$V = \left[\left(\frac{\theta}{360}\right)\pi R^2 - (R-h)(2Rh - h^2)^{1/2}\right]L. \tag{8.26}$$

This is the volume of a cylindrical lens of thickness h and length, L.

8.4 Aspheric surfaces

So far, we have been using spherical surfaces to provide optical power to a system. A spherical surface is defined by the radius of curvature, and has the characteristic that the surface slope is the same everywhere on the surface. As we increase the ray height beyond the paraxial limit, the rays no longer follow simple linear expressions, because the angle of incidence has increased to the point that the approximation of $\sin I \cong I$ is no longer applicable. This was demonstrated in Figure 8.8; therefore, if we can change the surface slope beyond the paraxial ray, we can make this approximation ($\sin I \cong I$) extend to larger clear apertures. Such a surface is an aspheric. As the name implies, the surface's local curvature varies with distance from the optical axis in a rotationally symmetric surface. So, as the distance from the optical axis changes, the slope of the surface also changes. The obvious choices for these surfaces are paraboloids, ellipsoids, and hyperboloids of revolution, which are symmetrical about the optical axis. These surfaces are called conics: in the paraxial domain they differ only slightly from a sphere, but outside the paraxial domain they are very different; thus, causing rays at large angles and distances to also focus to the focal point.

Recall that the equation of a spheroid at the origin, with the optical axis along the z direction is

$$x^2 + y^2 + (z - R)^2 = R^2. \tag{8.27}$$

The sag away from the x–y plane in one dimension is

$$z = R - \sqrt{R^2 - y^2}. \tag{8.28}$$

The evolution of the conics is shown in Figure 8.17, where the two foci for a spheroid are at the center of the sphere.

For the ellipse, the two foci are separated by some distance, as shown. If the foci are further separated, taking one focus to infinity, we have a parabola. In

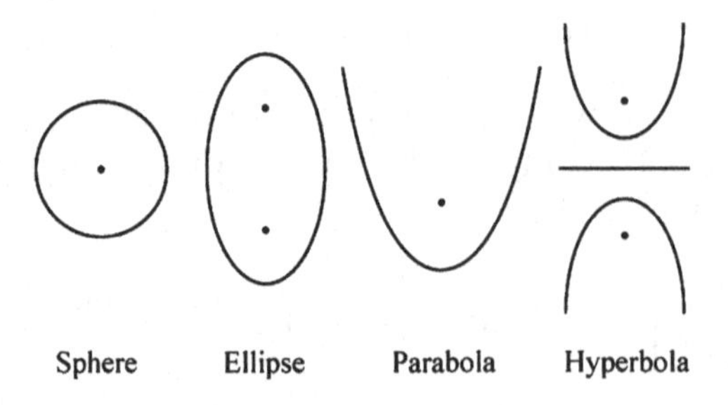

Figure 8.17 Conic shapes as the foci shift.

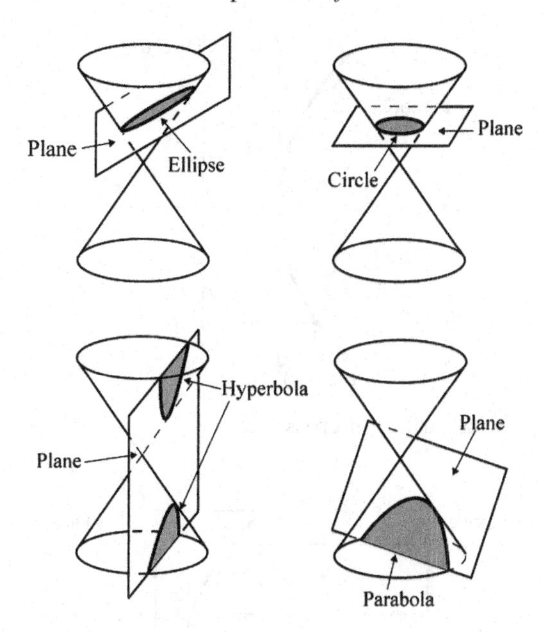

Figure 8.18 Conic sections.

the case of a hyperbola, one focus disappears into positive infinity and reappears at negative infinity.

These curves are called conic sections, because each is obtained by the intersection of a plane and a right circular cone. Figure 8.18 illustrates how planes intersect with cones to produce the four types of conic sections.

8.4.1 Paraboloid mirror

A paraboloid is a surface generated by a point moving in such a way that its distance from a fixed point (F) is equal to its distance from a fixed line (called a directrix). As shown in two dimensions in Figure 8.19, since the paraboloid is symmetrical about the z axis, it is a parabola in the y–z plane.

By the definition of a parabola, the line segments in Figure 8.19 are equal:

$$\overline{FS} = \overline{SM}. \tag{8.29}$$

From geometry and trigonometry, the value of the line segments is:

$$\sqrt{(-z-f)^2 + (-y)^2} = f - z. \tag{8.30}$$

Squaring,

$$\begin{aligned} z^2 + 2fz + f^2 + y^2 &= f^2 - 2fz + z^2 \\ y^2 &= -4fz. \end{aligned} \tag{8.31}$$

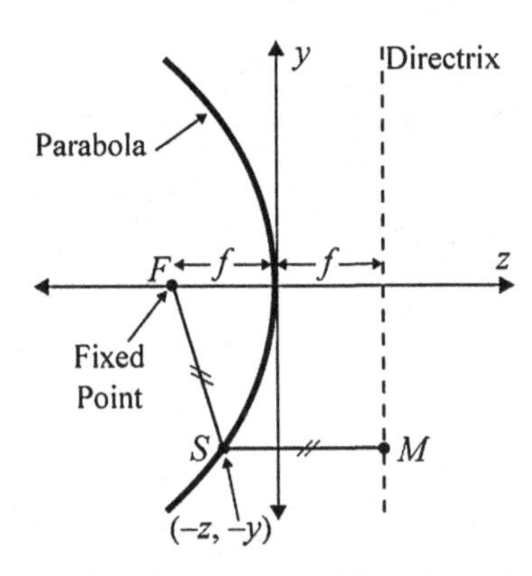

Figure 8.19 Parabola surface in cross section.

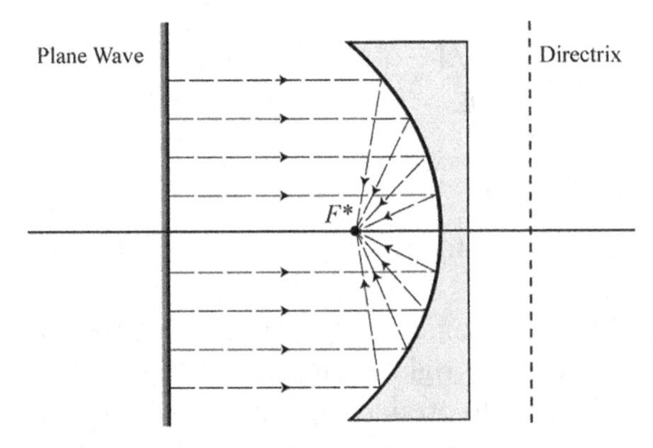

Figure 8.20 Plane wave is parallel to the directrix.

Equation (8.31) is the equation of a parabola in the y–z plane. The sag of the parabolic surface is:

$$z = \frac{y^2}{-4f}. \tag{8.32}$$

Recall the relationship (Equation (5.15)) between the focal length and radius of a sphere on the axis

$$z = \frac{y^2}{2R}. \tag{8.33}$$

This was an approximation of the sag of the sphere. This sag equation was an approximation for a spherical surface, but it is exact for a parabola beyond the paraxial region.

Consider a plane wave front, as shown in Figure 8.20, which is parallel to the directrix of the parabola. The optical path length is the same for all rays to the

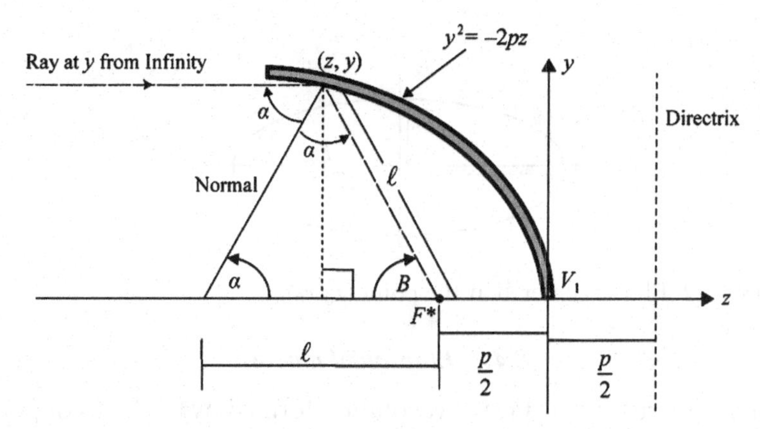

Figure 8.21 Any ray parallel to the axis converges to point F^*. Note: triangle $[\alpha,\alpha,B]$ is an isosceles triangle.

focal points, because the distance from the parabola to the directrix is equal to the distance from the parabola to point F^*, the focal point.

To extend this further, the law of reflection must be obeyed. So, as shown in Figure 8.21, the angle of incidence, α, is equal to the angle of reflection. From the right-hand triangle shown in Figure 8.21:

$$\begin{aligned}
\ell^2 &= y^2 + \left(z + \frac{p}{2}\right)^2 \\
&= y^2 + z^2 + pz + \frac{p^2}{4} \\
&= -2pz + z^2 + pz + \frac{p^2}{4} = z^2 - pz + \frac{p^2}{4} \\
&= \left(z - \frac{p}{2}\right)^2 \\
\ell &= z - \frac{p}{2} \\
z - \ell &= \frac{p}{2}
\end{aligned} \tag{8.34}$$

Since this is independent of ray height, y, a ray traveling from any point on a plane wave that impinges on a parabola will travel to a point $(z = p/2)$, and every ray will have equal optical path length (*OPL*). For any point on the parabola, the *OPL* $(z - 1)$ is a constant, shown to be $p/2$, which is the focal point. Therefore, if one is imaging a star with a large parabolic mirror (e.g. 4 m), all the starlight is focused to point F^* and there will not be any blur. Another parameter often quoted is the eccentricity of a conic. The eccentricity of a paraboloid is 1 $(\varepsilon = 1)$.

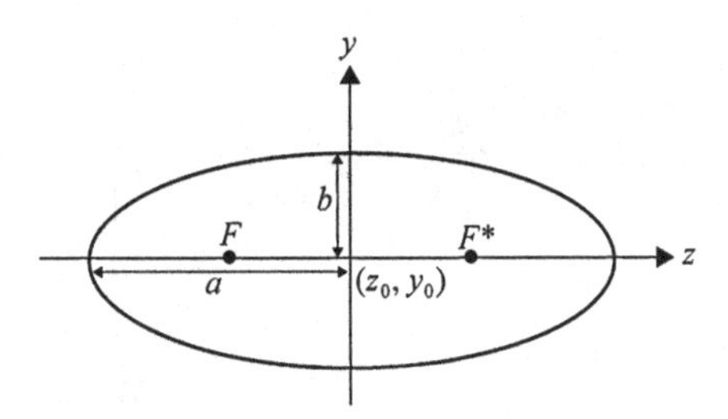

Figure 8.22 Ellipse centered at the point (z_0,y_0).

8.4.2 Ellipsoidal mirror

An elliptical mirror is a surface of revolution formed by the locus of points for which the total sums of the distances from two points are equal. Figure 8.22 shows an ellipse in two dimensions.

F and F^* are conjugate when the OPL is fixed between the two points. In Figure 8.22, the major axis is along z, and the minor axis is along y. The equation for an ellipse is

$$\frac{(z-z_0)^2}{a^2}+\frac{(y-y_0)^2}{b^2}=1, \tag{8.35}$$

where a is the semi-major axis and b is the semi-minor axis. The eccentricity is defined as

$$\varepsilon = f/a,$$

where

$$f=\pm\sqrt{a^2-b^2}. \tag{8.36}$$

The eccentricity of an ellipse varies between zero and 1 ($0<\varepsilon<1$).

There are several combinations for using ellipsoidal mirrors as concave or convex mirrors. Various choices are shown in Figure 8.23.

8.4.3 Hyperboloid mirror

A hyperboloid surface is formed from a locus of points such that, at any point on the surface, the difference of the lengths from two fixed points is a constant (fixed OPD in optics). As shown in Figure 8.24, the difference between the distances from F and F^* locates two curves in a two-dimensional layout. Therefore from the geometry of Figure 8.24 for the curve in the negative z region:

$$FQ - F^*Q = \text{constant}. \tag{8.37}$$

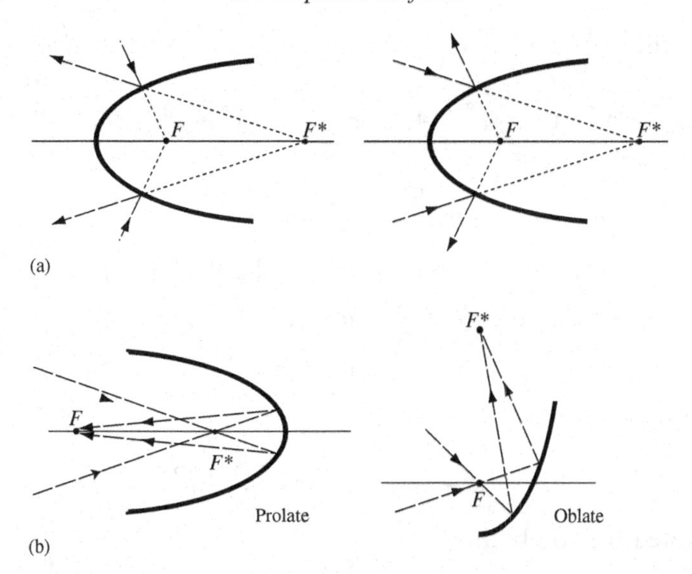

Figure 8.23 Ellipsoidal mirrors: (a) convex; (b) concave.

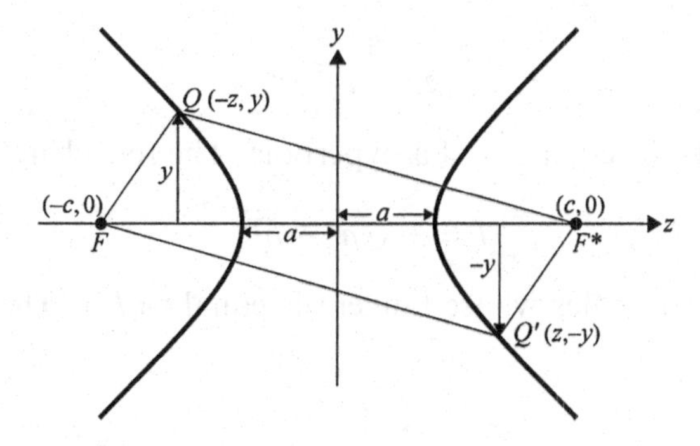

Figure 8.24 Hyperboloid geometry.

Additionally for the positive z region, the curve is such that

$$F^*Q' - FQ' = \text{constant}. \tag{8.38}$$

To form a hyperbola, the constants must be equal. We will assume bidirectional symmetry and choose a convenient constant of $\pm 2a$. The foci are at $\pm c$ on the z axis. The points Q and Q' are at $(-z, y)$ and $(z, -y)$. From the coordinate system shown in Figure 8.24 and the definition of a hyperbola:

$$FQ - QF^* = \pm 2a,$$

$$\sqrt{(z+c)^2 + y^2} - \sqrt{(c-z)^2 + y^2} = \pm 2a$$

$$\sqrt{(z+c)^2 + y^2} = \pm 2a + \sqrt{(c-z)^2 + y^2}.$$

Completing the square,

$$z^2 + 2cz + c^2 + y^2 = 4a^2 \pm 4a\sqrt{(c - z)^2 + y^2} + z^2 - 2cz + c^2 + y^2,$$

$$\begin{aligned} cz - a^2 &= \pm a\sqrt{(c - z)^2 + y^2} \\ c^2z^2 - 2ca^2z + a^4 &= a^2(z^2 - 2cz + c^2 + y^2) \\ c^2z^2 + a^4 &= a^2z^2 + c^2a^2 + a^2y^2 \\ (c^2 - a^2)z^2 - a^2y^2 &= a^2(c^2 - a^2), \end{aligned}$$

we then define

$$b^2 = c^2 - a^2 \quad \text{or} \quad c^2 = a^2 + b^2,$$

and substitute for c to obtain

$$b^2z^2 - a^2y^2 = a^2b^2,$$

$$\frac{z^2}{a^2} - \frac{y^2}{b^2} = 1. \tag{8.39}$$

Equation (8.39) is the equation of a hyperbola. The focal length is

$$f = \pm\sqrt{a^2 + b^2}, \tag{8.40}$$

with asymptotes at angles whose tangent is equal to b/a. The eccentricity is f/a ($\varepsilon > 1$).

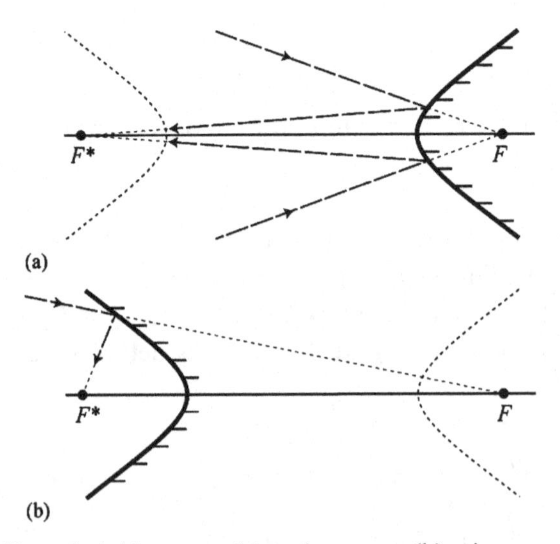

Figure 8.25 Hyperboloid convex (a) and concave (b) mirrors.

The hyperboloid mirror is very commonly used to displace the focal point while introducing very small aberrations. In the case of paraxial optics the directions of the rays are perfect shifts between one focal point and the second focal point. As shown in Figure 8.25, a hyperboloid mirror can be convex or concave. Rays that are directed toward one focus are reflected to the second focus.

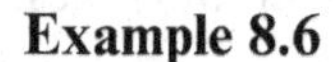

Example 8.6

For a hyperbola with $a=4$ and $b=3$, what is the focal length, f? What is the eccentricity (ε)?

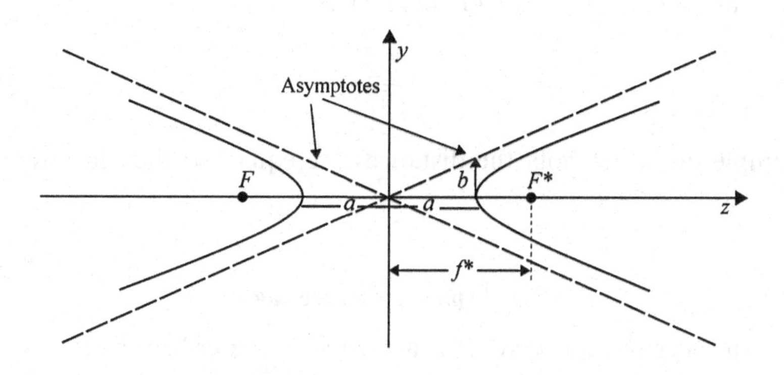

$$\frac{z^2}{a^2}-\frac{y^2}{b^2}=1 \quad \text{or} \quad \frac{z^2}{16}-\frac{y^2}{9}=1,$$
vertices at $\pm a = \pm 4$.

$$(f^*)^2 = a^2 + b^2,$$
$$f=\sqrt{16+9}=5.$$
$$\varepsilon=\frac{f}{a}=\frac{5}{4}.$$

In the case of Figure 8.25, the image at one focus is real while that at the other is virtual.

Conic curves or surfaces are characterized by their eccentricity (ε), which is a measure of how far the surface deviates from being a circle or sphere. A conic surface consists of a locus of points whose distance to a point F (the focus) equals the eccentricity times the distance to a line (directrix) which does not contain the point, F. The eccentricity, as defined, is shown in Figure 8.26. The

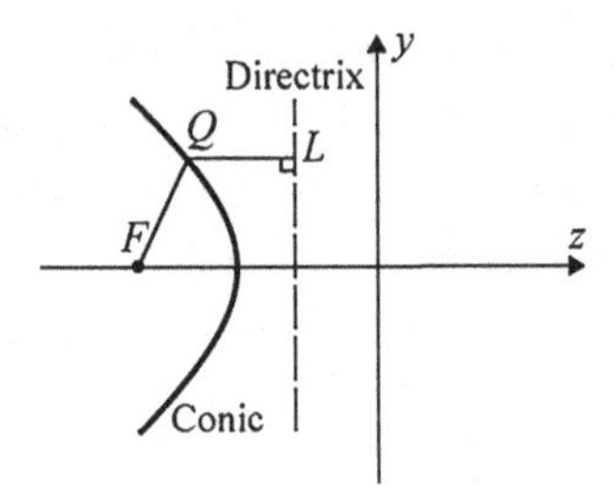

Figure 8.26 Eccentricity definition for conics.

eccentricity can be expressed as the ratio of the distance of the surface from the focus (F) and its distance from a line directrix (L):

$$\varepsilon = \frac{\overline{FQ}}{\overline{QL}}. \tag{8.41}$$

For example, in a parabola the distances are equal, so the eccentricity is 1 ($\varepsilon = 1$).

8.5 Aspheric surface sag

Aspheric surfaces in optical systems are typically described in terms of a sag equation that characterizes the surface for various zones in the y direction in a rotationally symmetric element. In the optics community, the sag is often expressed as a conic constant instead of the eccentricity (ε), which is the classical way of describing a conic in mathematical texts. The relationship between the conic constant (KK) and the eccentricity (ε) is

$$KK = -\varepsilon^2. \tag{8.42}$$

Table 8.2 shows the various values for the eccentricity and conic constant.

The expression used to describe a conic surface of revolution's sag from the vertex is

$$z = \frac{Cy^2}{1 + [1 - (1 + KK)C^2y^2]^{\frac{1}{2}}}, \tag{8.43}$$

where C is the base curvature of surface ($1/R$) and y is the zone height. This is the sag of an optical surface as a function of zone height, y. This equation works for all conic surfaces including a sphere. C is the reciprocal of the radius of curvature.

The relationship between the paraxial radius of curvature (R) and the vertex to foci distance (d_f) is

Table 8.2. *Values of the eccentricity and the conic constant for conic surfaces*

Conic	Eccentricity range	Conic constant range
Sphere	$\varepsilon=0$	$KK=0$
Ellipse	$0<\varepsilon<1$	$0>KK>-1$
Parabola	$\varepsilon=1$	$KK=-1$
Hyperbola	$\varepsilon>1$	$KK<-1$

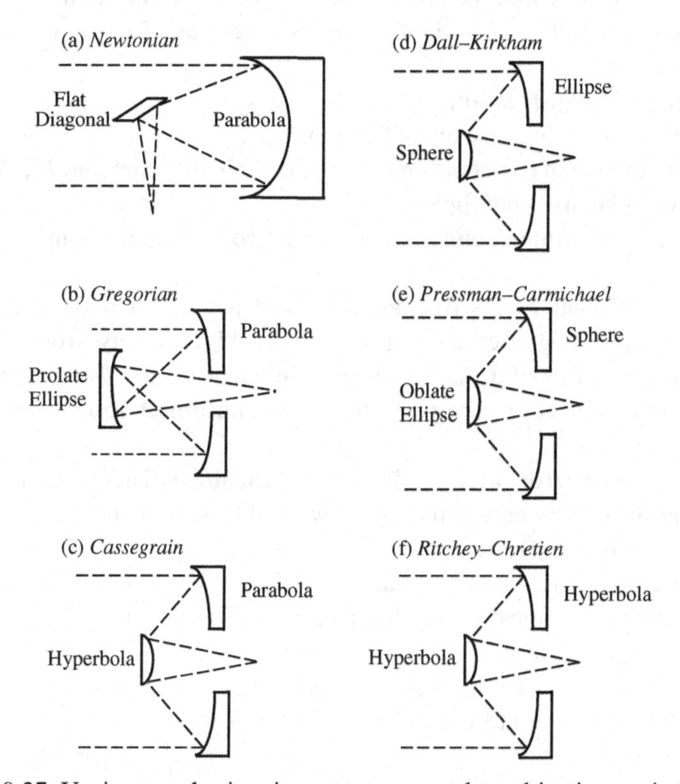

Figure 8.27 Various aspheric mirror systems used as objective optics.

$$d_{\mathrm{f}}=\frac{R}{KK+1}[1\pm\varepsilon]. \tag{8.44}$$

Aspheric mirrors are used in optical systems that are all reflective, such as the ones shown in Figure 8.27. All of these are two-mirror systems. The Newtonian telescope objective (Figure 8.27(a)) uses a single paraboloid mirror with optical power, plus a flat mirror with no optical power. The Gregorian telescope objective forms its images using paraboloid and ellipsoid mirrors. The Cassegrain system uses a parabola with its focal point colocated with the foci of one of the hyperbola's secondary mirror foci; while the Ritchey–Chretien system uses two hyperboloids with common foci.

Problems

8.1 What is the $(F/\#)_\infty$ of a 10 mm diameter concave mirror with a 25 mm radius of curvature? What would the $F/\#$ be if the mirror was a convex mirror?

8.2 A frosted light bulb, with a 2 in diameter, is imaged by a 2 in ball bearing which is 10 in away.
(a) Find the image location.
(b) What is the transverse magnification?

8.3 Calculate the sag of a spherical mirror with a 30 in diameter and $F/2$. What is the sag of a similar sized paraboloid?

8.4 Where are the cardinal points of a convex mirror of radius 30 cm in air? Make a sketch.

8.5 A mirror of what radius (include sign) will produce a real image twice the size (negative magnification) of the object 15 cm away from the mirror? What is the radius of a plano-convex thin lens with a refractive index of 1.5 that produces a real image twice the size of an object 15 cm away from the lens?

8.6 For a concave mirror, with a radius of –20 cm and a diameter of 4 cm:
(a) Find the optical power in diopters when it is used in air.
(b) What is the $(F/\#)_\infty$ in air?
(c) Make a sketch showing the cardinal points.
(d) What is the optical power (diopters) if the mirror is submerged in water ($n = 4/3$)?
(e) What are its focal lengths (f, f^*) in water?

8.7 A paraboloid mirror has the equation of $y^2 = -36z$.
(a) What is its *efl*?
(b) What is its sag, if its diameter is 60 cm?

8.8 Where are the cardinal points of a concave spherical mirror in air with radius of 30 cm? Make a sketch. Where are the cardinal points when this mirror is in water ($n = 1.33$)?

8.9 Design a 1 m spherical diameter mirror with an optical power of 2 diopters.
(a) What is the $(F/\#)_\infty$?
(b) What is the volume of glass removed?

8.10 A 2 mm object is located 20 cm in front of an equi-convex thick lens with $|R_1| = |R_2| = 25$ cm, thickness 1 cm and a refractive index of 1.5. The second surface is silvered. A thin lens with an effective focal length of 5 cm is located

5 cm to the left of the object. The layout is shown below. There are two images formed in the space to the left of the thin lens.

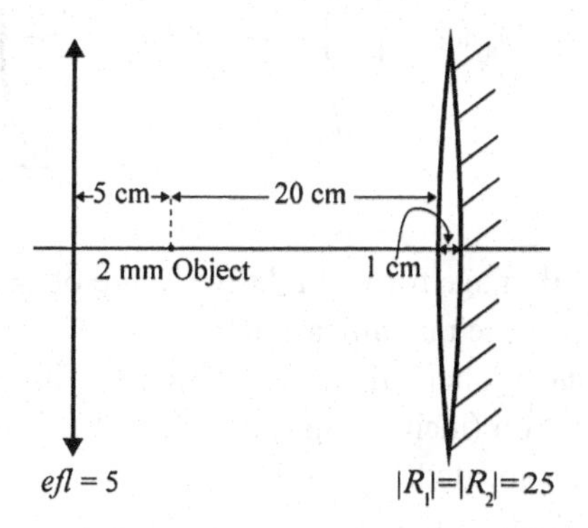

(a) Where are the final images relative to the thin lens?
(b) What is the image's magnification in each case?
(c) Are the images real or virtual?

8.11 A glass sphere ($n = 1.5$) with a radius of 5 cm is dipped in gold such that half of the sphere is covered with gold. An object is placed 15 cm in front of the glass sphere surface so that the gold surface acts as a concave mirror (see below).

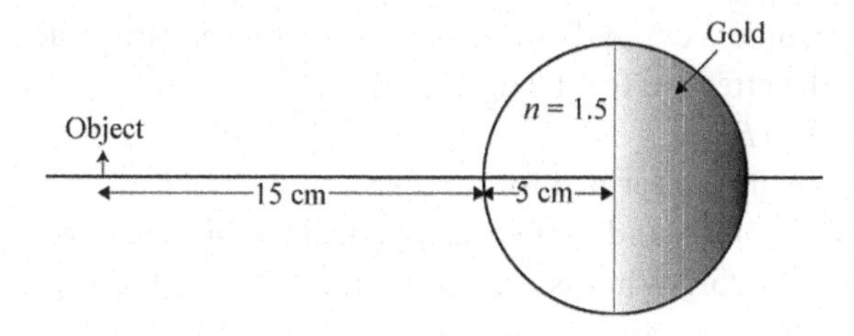

(a) Where is the image?
(b) What is the transverse magnification?
(c) Is the image real or virtual?

8.12 A concave spherical mirror has a radius of −40 cm.
(a) What is the optical power of the mirror in diopters?
(b) Where are the cardinal points?
(c) If a 3 cm high object is 20 cm in front of the mirror, where is the image located, and what is its size?
(d) If a 5 cm high object is 100 cm in front of the mirror, where is the image located and what is its transverse magnification?

8.13 An object is placed 20 cm in front of a thin lens with back focal length (f^*) = 10 cm. A concave mirror is located 30 cm beyond the lens with a radius of curvature of −20 cm.

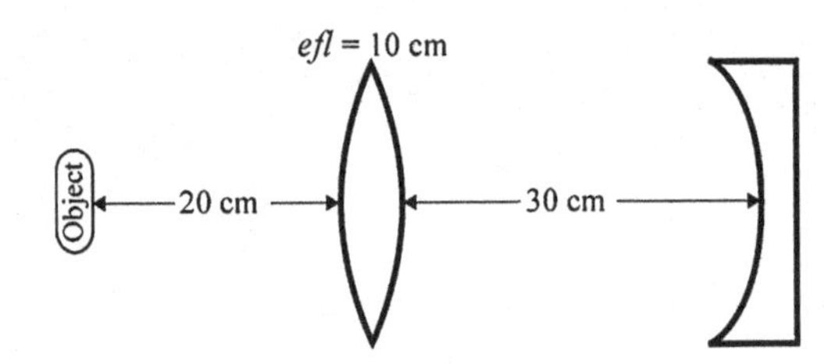

(a) Where is the final image formed relative to the object's position?
(b) What is the transverse magnification?

8.14 The radius of a concave spherical mirror is -40 cm. Find the image distances and transverse magnification (include sign) for each of the following locations if an object, 1.5 cm high, is located:
(a) 50 cm in front of the mirror;
(b) 40 cm in front of the mirror;
(c) 20 cm in front of the mirror;
(d) 10 cm in front of the mirror.

8.15 A concave spherical mirror has a radius of curvature of -35 cm and a diameter of 5 cm. For a 2 cm high object located 70 cm in front of it:
(a) Find the $(F/\#)_{\infty}$ and $(F/\#)_{\mathrm{W}}$.
(b) Find the transverse magnification.
(c) Find the optical power of the mirror (diopters).

8.16 A convex mirror is 3 cm in diameter and has a radius of curvature of 18 cm.
(a) What is the effective focal length (*efl*)?
(b) What is the $(F/\#)_{\infty}$?
(c) Where is an image formed for an object at negative infinity?

8.17 The sag for a parabola and that for a spherical surface are compared for a 6 inch diameter, $F/3$ mirror. What is the difference in sag at the edge? Which surface requires more glass removal?

8.18 Sketch the optical layout and label the surface type for the following:
(a) Cassegrainian telescope;
(b) Gregorian telescope;
(c) oblate ellipsoid.

8.19 For a paraboloid mirror with the equation $y^2 = -200z$ and a diameter of 50 cm,
(a) What is the sag?
(b) What is the focal length?

8.20 A concave mirror is used to focus the image of a spider (spinning its web) onto a nearby wall, 100 cm away from the spider, as shown below.

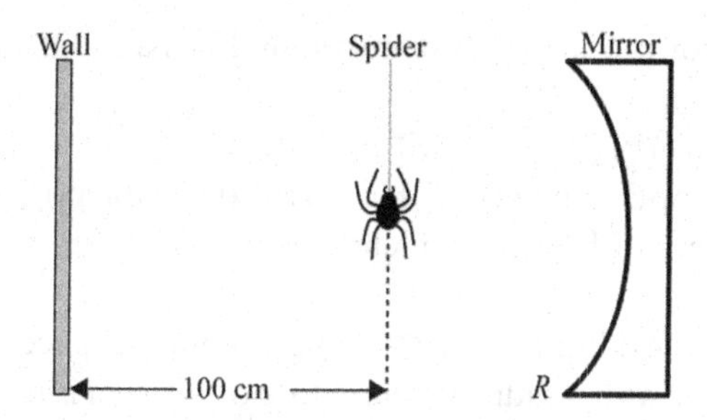

If the transverse magnification required is between −10 and −20, and you only have three mirrors at your disposal, $R_1 = -30$ cm, $R_2 = -10$ cm, and $R_3 = -2$ cm, which mirror would you use to cover that range of magnification? Make a sketch of the layout.

8.21 An equi-convex lens of refractive index 1.5, thickness 2 cm, and radii of 10 cm is aluminized (mirrored surface) on one side.

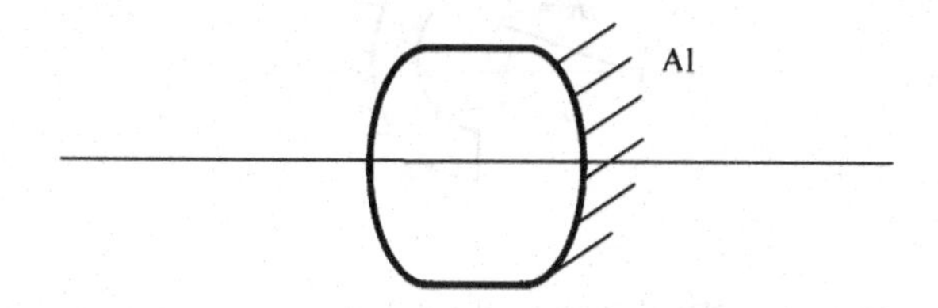

(a) What is its optical power (diopters)?
(b) Where are the cardinal points for the system?

8.22 A thin lens of refractive index 1.5 has radii of $R_1 = -5$ cm and $R_2 = -10$ cm. If the second surface is silvered, what is the optical power of the system?

8.23 A negative thin lens (optical power = −10 diopters) is located 10 cm behind a concave mirror with radius $R = -0.2$ m, as shown below.

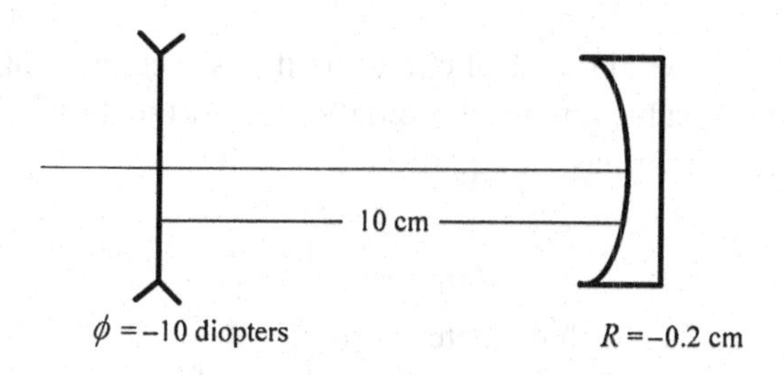

(a) What is the optical power of the system?
(b) Where are the cardinal points?
(c) Where is the final image formed for an object at infinity?

8.24 A concave mirror is 5 inches in diameter and has a radius of curvature of – 50 inches.
(a) What is the $(F/\#)_\infty$?
(b) What is the optical power of this mirror (in diopters)?
(c) If a 3 cm object is located 50 inches in front of the mirror, what is the size of the image?
(d) What is the working $F/\#$ for this object in image space?

8.25 A concave mirror with a radius of curvature of –16 cm is used in water ($n = 4/3$).
(a) What is the optical power in the water?
(b) What is the front focal length?
(c) What is the back focal length?
(d) Where are the nodal points?

8.26 A concentric lens of refractive index 1.5 with radii of –5 and –8 cm is aluminized on the 8 cm surface as shown below.

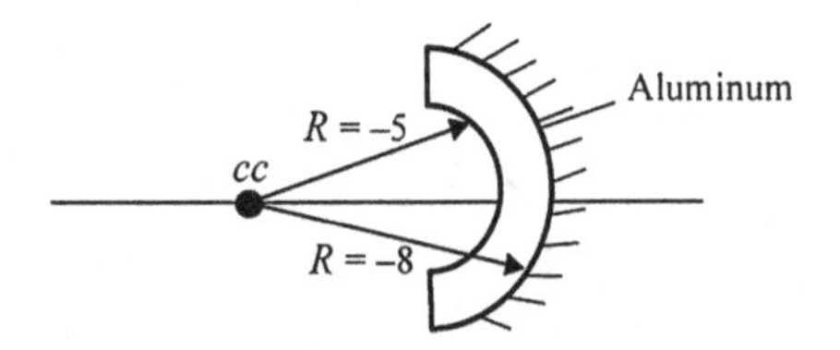

(a) What is the total optical power of the lens?
(b) Where is the back focal point located?

8.27 For a convex mirror with a radius of curvature of 15 cm and diameter of 5 cm:
(a) What is the $(F/\#)_\infty$?
(b) Where is the nodal point relative to the vertex?

8.28 An ellipsoidal mirror has the equation

$$\frac{z^2}{100} + \frac{y^2}{25} = 1.$$

(a) What is its paraxial radius of curvature if it is used in an prolate position?
(b) What is the y value on the ellipsoidal mirror at the focal points?

8.29 A hyperboloid mirror has the equation

$$\frac{z^2}{100} - \frac{y^2}{36} = 1.$$

What is the sag of a 3 inch diameter mirror?

8.30 A cocktail glass with a bowl shaped like a hemisphere with a diameter of 4 in is filled with a liquid to a depth of 1 in (maximum depth). What is the volume of this liquid?

8.31 A concave and a convex mirror are arranged such that they operate in a series, as shown below:

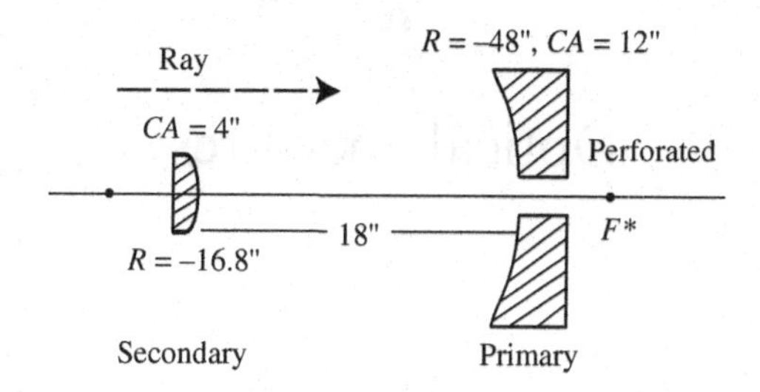

(a) What is the effective focal length?

(b) What are the aspheric surfaces you would use for this system to get maximum diffraction limited FOV?

(c) If this is a classical Cassegrain, what is the equation of the secondary mirror in order to have F^*, 3 in behind the primary vertex.

8.32 What are the general equations, in y–z coordinates, for the following when the vertex is at the origin of x, y, z:

(a) a circle;

(b) a parabola;

(c) an ellipse;

(d) a hyperbola.

8.33 There is a significant difference between an $F/1$ concave paraboloid mirror and an $F/1$ concave spherical mirror in the large, 8 m diameter, class. What is the difference in the volume of glass removed for a spheroid versus a paraboloid?

8.34 An 8 m concave mirror, $F/1.5$, is being fabricated.

(a) What is the volume of glass in (cubic inches) removed if the surface is spherical?

(b) What is the volume of glass in (cubic inches) removed if the surface is a paraboloid?

Bibliography

Angel, J. R. P. and Hill, J. (1982). Manufacture of large borosilicate glass honeycomb mirrors. *SPIE Proceedings on Advanced Technology Telescopes*, **332**, 298.

Anton, H. (1988). *Calculus with Analytic Geometry*, third edn. New York: Wiley.

Goble, L. W., Angel, J. R. P., Hill, J. M. and Mannery, E. J. (1989). Spincasting of a 3.5-m diameter $f/1.75$ mirror blank in borosilicate glass. *Proceedings of SPIE*, **966**, pp. 300–308.

Hecht, E. (1998). *Optics*, third edn. Reading, MA: Addison-Wesley.

Jenkins, F. A. and White, H. E. (1976). *Fundamentals of Optics*, fourth edn. New York: McGraw-Hill.

9
Optical apertures

Although an optical design goal is to collect as much light as possible, the amount of light which actually enters optical systems is a small fraction of the light which radiates off the object being viewed. The only usable light in an optical system is the light which enters the system at appropriate angles to reach the image location. Optical apertures limit the amount of light that enters the system, and also limit the area of the recorded image. Stops determine both the field of view and the illumination within the image. The field stop determines the field of view, which in turn determines how much of the object can be seen through the optical system. The field of view is calculated as the maximum angle from the optical axis at which light can enter and pass through the system. Image illumination is limited by aperture stops that constrict the bundle of light rays entering the system. Rays far from the optical axis, or those entering the system at too steep an angle, will not reach the image plane. Image illumination determines whether the image can be detected by an eye, a recording medium, or a two-dimensional detector array.

A stop (sometimes called a diaphragm) can be visualized as an opaque plate with a circular hole, akin to a metal washer or window in a building. There are two types of stops in optical systems, and they are named according to their functions within the system. The aperture stop (often shortened to Astop) limits the amount of radiation entering the system. It may be located at the front or in the middle of the optical system, but never at an image plane. The field stop limits the field of view, and may be located at the object plane, the image plane, or at any intermediate location. Figure 9.1 illustrates both an aperture stop and a field stop.

The aperture stop truncates the ray bundle at the upper and lower rim rays, as seen in Figure 9.1. Rays from the object beyond this bundle do not pass through the aperture stop and are lost from the image intensity, as indicated by

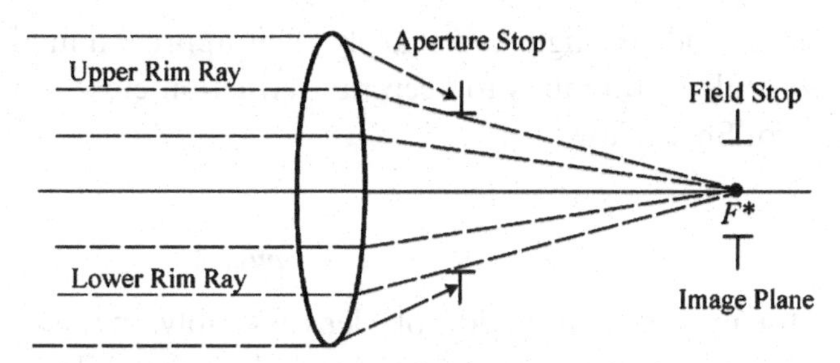

Figure 9.1 An aperture stop and a field stop.

the outermost rays shown by arrows in Figure 9.1. Each optical element in a rotationally symmetric optical system has a finite diameter, which is defined as the clear aperture (CA) of that element. Changing object or image positions can change the aperture or field stop locations as well as the field of view. This will be discussed in Section 9.1.3 in more detail.

9.1 Aperture stop

If an optical system has many elements, one element will cut off the light and create an edge outside of which rays cannot enter the optical system. The aperture stop is the limiting diameter of a lens, mirror, or baffle within a rotationally symmetric optical system. The light that forms the image is uniformly distributed across the aperture stop. Blocking a portion of the aperture stop does not cause blocking of a portion of the image; instead, the image is dimmed due to the loss of some energy from the object.

Larger aperture areas allow more light to be collected, making an image brighter. Since area depends on the square of the radius, the illuminance of the image depends on the square of the aperture stop's diameter. Larger apertures produce images with higher resolution. Image resolution is the degree to which an optical system can distinguish two closely spaced points in the image. Therefore, there are two reasons to design the aperture stop to be as large as possible: (1) greater image brightening, and (2) higher image spatial resolution.

Most optical systems, including telescopes, binoculars, and cameras, use the very first lens or mirror as the aperture stop. This is because it is expensive to make large optics, so it is common to concentrate the light near the first lens or mirror, enabling the light to go through smaller optics after entering the system. The light, however, is more difficult to handle when it is redirected at steep angles, so there is a design trade-off between concentrating light and directing it within optical systems. Groups of lenses are often used so that no single element bends the light too sharply. This approach minimizes aberrations. A good rule of thumb is to keep ray refraction close to the paraxial approximation of Snell's law.

9.1.1 Entrance and exit pupils

The aperture stop may be in the middle of a lens assembly. In this case, a person can only see the aperture stop by looking through the lenses. If looking from the object side, the person will see the entrance pupil; whereas, if looking from the image side, the person will see the exit pupil. The lenses between the aperture stop and the image or object make the size and location of the aperture stop appear different than it really is, due to the fact that the aperture stop is being imaged by those lenses. If the aperture stop is physically located in object space (there are no lenses between the object and the aperture stop), the aperture stop is the entrance pupil. If the aperture stop is the last element in the system in image space, the aperture stop is the exit pupil.

The pupils are the conjugate images of the aperture stop transferred by the lenses into object space or image space. The pupil location and size can be found by tracing rays from the aperture stop out of the system on either side. The pupils can also be located by tracking the conjugate image positions of the aperture stop through each lens. The effect of each lens is determined in proper sequence. First, the conjugate of the aperture stop by the closest lens is found. Then, that conjugate of the aperture stop is transferred to another conjugate position through the second lens, and then the third, and so on until the final lens is reached. The conjugate positions are determined by using the lens equation, $n'/z' - n/z = 1/f$. The magnification due to each lens is $M = (z'/n')/(z/n)$. The individual magnifications are multiplied together to determine the size of the pupil compared to the aperture stop.

The entrance and exit pupils are uniquely defined conjugate images to the aperture stop in object or image space, respectively. In Figure 9.2, the aperture stop is also the exit pupil because it is located in image space, indicated by ABC. Since the aperture stop is inside the focal length, the conjugate image is virtual and erect in object space, as shown by the dotted lines. This conjugate to the aperture stop in object space is the entrance pupil, indicated by $A'B'C'$. The diameter of the entrance pupil is determined by extension of the limiting rays in object space.

Figure 9.3 shows the aperture stop in front of the lens, and thus, the aperture stop becomes the entrance pupil. The exit pupil is the virtual image of the aperture stop. The size of the exit pupil is determined by the back projection of the upper and lower rim rays, as indicated in Figure 9.3. The central ray that

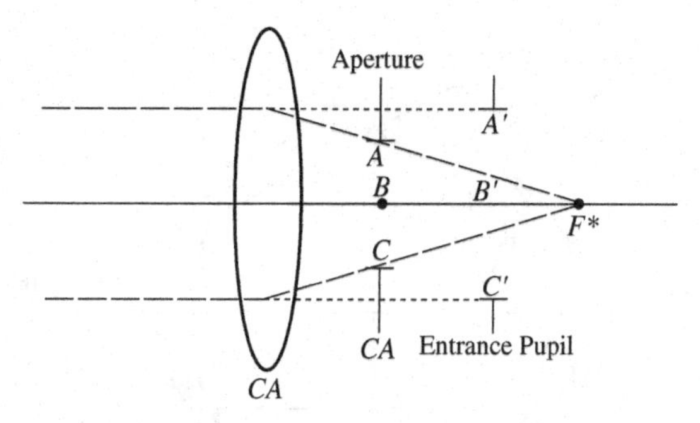

Figure 9.2 Entrance pupil of an optical system.

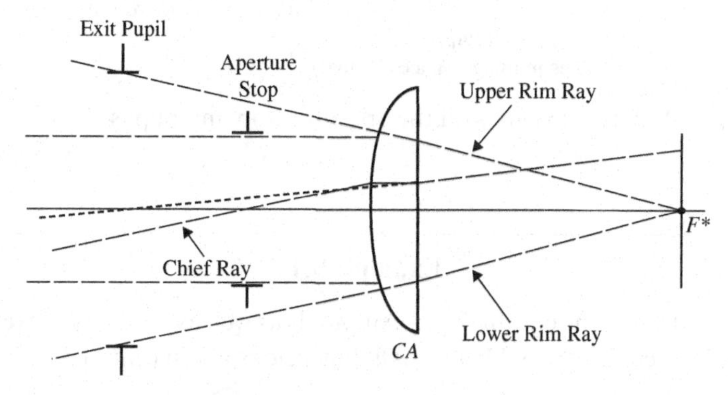

Figure 9.3 Exit pupil of an optical system.

determines the field of view is called the chief ray, and it goes through the center of the aperture stop.

Figure 9.4 shows an aperture stop between two lenses, forming what is known as a landscape lens. The aperture limits the rays in object space as shown. The image of the aperture stop in object space is indicated as A', and its image in image space is A''. The corresponding diameters are determined by the projection of the ray extensions in object and image space to these image locations, thus determining not only location but also size.

For the point object, the entrance pupil limits the bundle of rays which enters the optical system. For this case, the upper rim ray appears to go to A' (ray OA' in Figure 9.4). However, it is refracted by the lens and really goes through A, the edge of the aperture stop. The ray then goes through the second lens and forms the image I; however, the ray appears to have come from A'' to the image (ray $A''I$ in Figure 9.4).

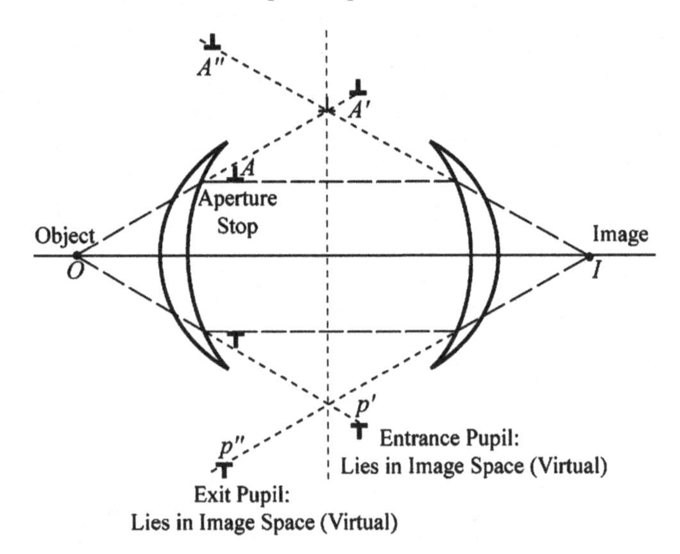

Figure 9.4 Landscape lens with the aperture stop and pupils.

Example 9.1

Find the location of the entrance and exit pupils as well as their respective sizes for the optical system illustrated below. All dimensions are in centimeters.

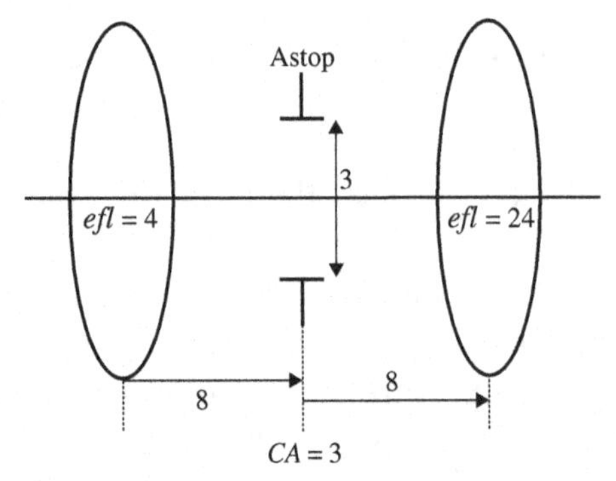

Step 1 Front half:
Entrance pupil:

$$\frac{n'}{z'} = \frac{n}{z} + \phi \rightarrow \frac{-1}{z'} = \frac{-1}{8} + \frac{1}{4}$$

(note: n and n' are negative because image of aperture is $R \rightarrow L$)

$$z' = -8 \rightarrow M_t = \frac{z'/n'}{z/n} = \frac{-8/-1}{8/-1} = -1 \rightarrow \text{size is 3 cm.}$$

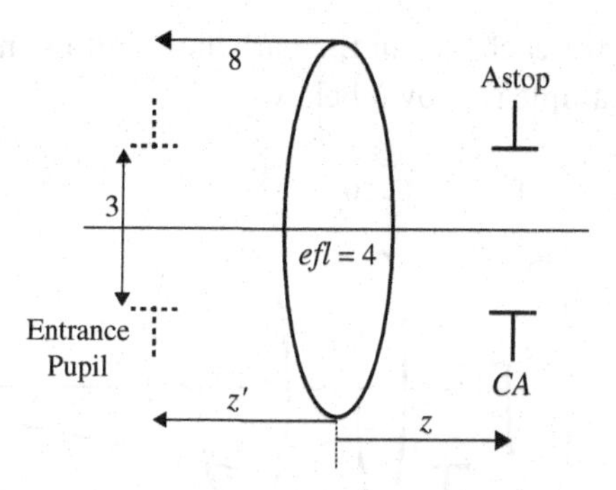

Step 2 Rear half:
Exit pupil:

positive 1 because $L \rightarrow R$

$$\frac{n'}{z'} = \frac{n}{z} + \phi = \frac{①}{z'} = \frac{1}{-8} + \frac{1}{24} = -\frac{2}{24} \rightarrow z' = -12$$

$$M_t = \frac{-12/1}{-8/1} = 1.5 \rightarrow \text{size is 4.5 cm.}$$

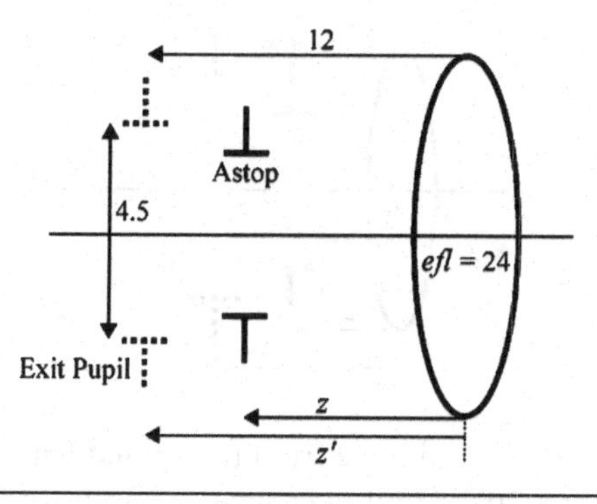

Example 9.2

In the system below, locate the entrance and exit pupils by finding the conjugate to the aperture stop throughout the system. All dimensions are in centimeters.

Use the lens formula $n'/z' - n/z = 1/f^*$ rearranged as $1/f^* + n/z = n'/z'$ repetitively for each lens in the system before the aperture stop to find the entrance pupil, and for each stop after the lens to find the exit pupil. The semi-diameter sizes of the pupils are determined by using the magnification formula

($M_t = (z'/n')/(z/n)$) for each lens in the path and multiplying all the magnifications together. An example is shown below.

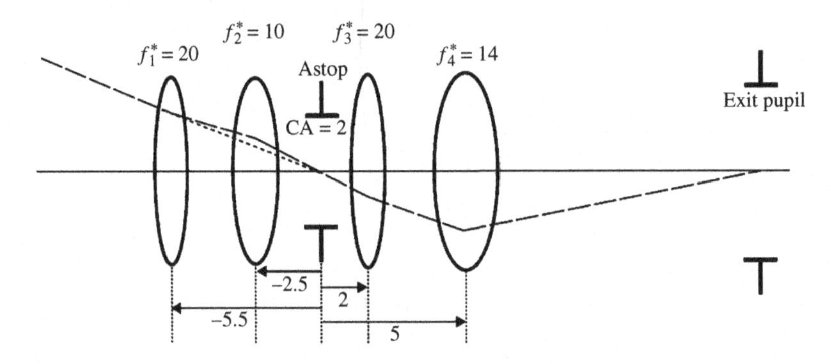

Find the entrance pupil by first finding the conjugate to the aperture stop from lens 2, $f^* = 10$. $-1/z' = -1/2.5 + 1/10 = -3/10$ so the image of the aperture is at a distance of 10/3 from lens 2, and the magnification $M_t M_t = -3.33/-2.5 = 1.33$ (upright virtual image).

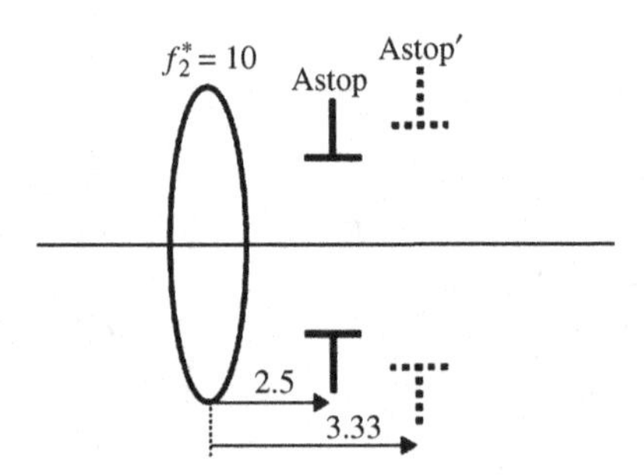

Now the virtual aperture conjugate from the second lens is imaged to a conjugate through lens1 ($f = 20$) to find the entrance pupil location:
$-1/z' = -1/6.33 + 1/20 = -0.1079$ and $M_t = 1.46$

The total magnification of the entrance pupil is (1.33)(1.46) = 1.95 times the radius of the aperture stop ($CA = 3.9$ cm).

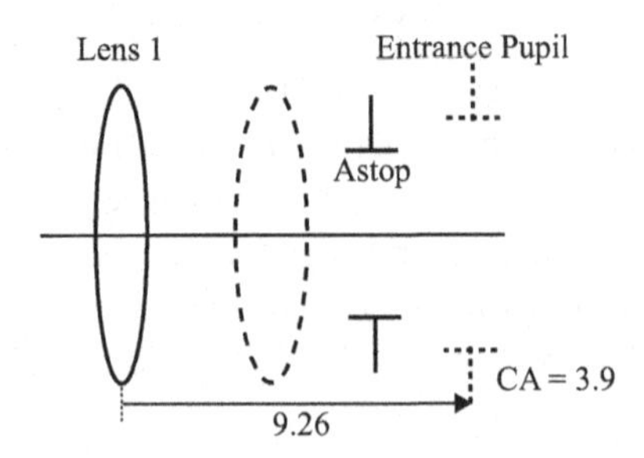

The entrance pupil location is 3.768 cm beyond the aperture stop. The same procedure is used to find the exit pupil at 22.1 cm in front of the aperture stop with $M_t = -4.70$ and $CA = 9.4$ cm.

The specifications for the entrance pupil size are typically given for optical instruments. For example, in a 10×50 pair of binoculars, both binocular entrance pupils are 50 mm in diameter, and the angular magnification is 10. The exit pupil would be 5 mm in this case. The exit pupil of a telescope/ microscope is the location where the eye is placed. Users of optics only care about the entrance and exit pupils, but optical systems designers are very mindful of the aperture stop in order to allow the most light into the system.

9.1.2 Telecentric pupil location

Telecentricity in a system is a special case in which the pupil is located at infinity. A system can be telecentric in object space, in image space or in both (a condition called double telecentricity). Telecentricity allows the object focus distance to vary without changing the magnification, so it is very useful for focusing in microscopes. When an optical system is telecentric in object space, the entrance pupil is located everywhere along the axis in object space between the lens system and negative infinity. When a system is telecentric in image space, the exit pupil is located everywhere along the axis in image space between the lens system and positive infinity.

As shown in Figure 9.5, the arrow-shaped object may be moved back and forth along the axis, causing it to go in and out of focus. The magnification will not change, because the ray shown at the top of the arrow (called the chief ray) is parallel to the axis in a telecentric system and goes through the center of the aperture stop.

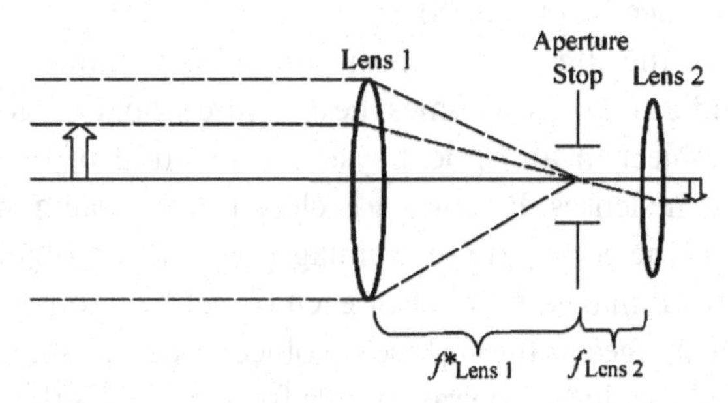

Figure 9.5 Telecentric system in object and image space (doubly telecentric).

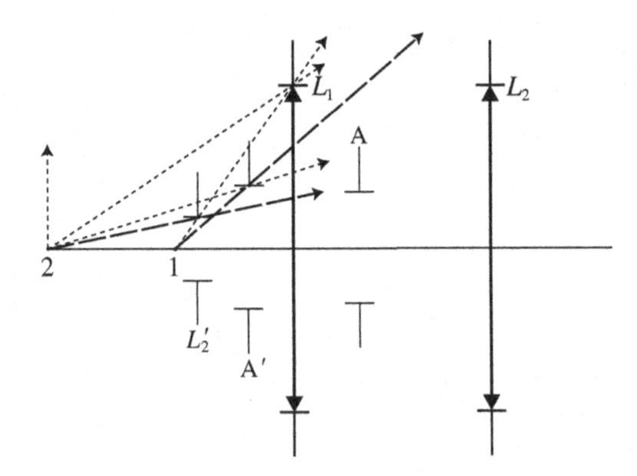

Figure 9.6 Determining which surface is the aperture stop in object space.

9.1.3 Aperture stop determination

The element that forms the aperture stop within a system depends upon the object location. As a result, the aperture stop may be one of the optical elements, a physical structure such as the optics housing, or a physical aperture stop. For an object located close to the optical system, the lenses on the far side of the system may become the aperture stop. The aperture stop may be found by tracing a sample ray starting from the object point on axis at an arbitrary angle and passing through the optical system. The element that is the aperture stop is the element at which the height of the sample ray is the highest percentage of the distance from the optical axis to that element's edge (clear aperture or CA). If the ray started at a slightly steeper angle, it would be clipped at the edge of the aperture stop and would not be at the edge of any other elements. For example, a sample ray trace may intersect the following five surfaces at 60%, 72%, 56%, 88%, and 87% of the height from the center of the surface to the edge of the lens element. In this example, the fourth element would serve as the aperture stop, since this ray is closest to the edge. The marginal ray is the one that goes through the edge of the aperture stop and is found by multiplying all angles and heights in the sample ray trace by (1.0/0.88).

Figure 9.6 shows the effect of object distance in determining the element that limits the ray bundle and thus becomes the aperture stop. Consider two different positions of the object in an optical system consisting of two lenses and an aperture between the lenses. Position 1 is close to the system and position 2 is farther away. L_2' is the conjugate to L_2 imaged by L_1 into object space, and A' is the conjugate to A imaged by L_1. For each object location, 1 or 2, draw lines to the edges of the images or the surfaces in object space, as shown in Figure 9.6. The entrance pupil that limits the ray bundle from point 1 is the conjugate to the

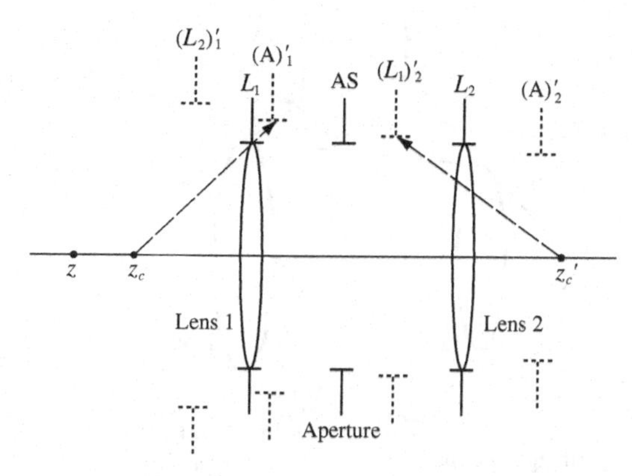

Figure 9.7 The entrance pupil, exit pupil, and aperture are all conjugates.

aperture stop (A'); thus, the aperture is the aperture stop for position 1. The entrance pupil formed by the image of lens 2 (L_2) limits the ray bundle originating from an object at point 2, so for this object, distance L_2 is the aperture stop. The dashed rays are the marginal rays for each object location, 1 or 2.

Figure 9.7 shows another system which consists of two lenses with an aperture between them. Any of these elements could be the aperture stop, depending on the location of the object. The stop or image of the stop that subtends the smallest angle from the axial point of the object is the entrance pupil (as discussed with Figure 9.6). The image formed by lens 2 of the aperture, A, is $(A)'_2$, and that of lens 1 is $(L_1)'_2$. Both are in image space.

Consider also the images in object space of the components shown in Figure 9.7. In object space, a ray that just passes by the edge of L_1 and image $(A)'_1$ locates a point z_c on the optical axis that is called the critical location. For objects beyond axial point z_c, the smallest ray angle from that axial object point, z, is to L_1, and determines the entrance pupil. In this case, L_1 subtends the smallest angle and, therefore, is the entrance pupil as well as the aperture stop of the system. In image space the exit pupil $(L_1)'_2$ is the image of the entrance pupil L_1. Notice that the exit pupil subtends the smallest angle from z'_c compared with the edge of lens 2 and the aperture stop image $(A)'_2$.

Figure 9.8 shows an example of an optical system comprising two lenses without an aperture stop, such that either lens can limit the bundle of rays. In this special situation, the image of the first lens by the second is $(L_1)'_2$, and the image of the second lens via the first is $(L_2)'_1$. Consider a moving point on the optical axis. There is a critical location z_c which exists such that an upper rim ray just passes through the edge of lens 1 and the edge of lens 2 simultaneously, as shown in Figure 9.8. In this unique location, the entrance pupil is the first

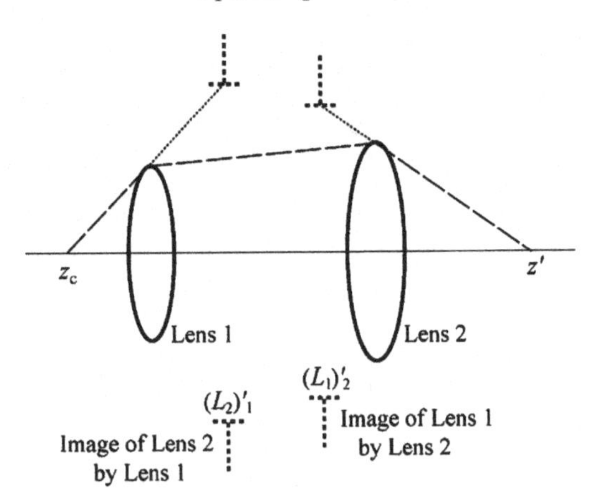

Figure 9.8 Two lenses without a given aperture stop; the aperture stop is object location dependent.

lens and the exit pupil is the second lens. If an object is to the left of this critical location z_c (farther away), the entrance pupil and aperture stop is lens 1 and the exit pupil is $(L_1)'_2$. If an object is to the right of z_c, lens 2 becomes the aperture stop and exit pupil, while the image of lens 2 through lens 1 becomes the entrance pupil, marked as $(L_2)'_1$ in Figure 9.8. For this particular, yet relatively common, situation the location of the object determines which element is the aperture stop.

9.2 Field stop

The field stop is the aperture in the optical system that limits the extent of the illuminated image plane and determines the extent of the field of view. The field of view is the angular extent of the object that can be seen by a viewer looking from the entrance pupil toward the object. The field stop limits the size and shape of the image, but not its brightness. It is located at the position of a real image having limited radial extent, and is typically, but not exclusively, in the image plane. Each point within the field stop location corresponds to a point in either the object or the image.

An opaque obstruction placed at the field stop does not change the overall brightness of the image, but shows up in the image as a dark shadow whose outline is the shape of the obstruction. The field stop is sometimes placed at the object plane; for example, the rectangular frame around slides in a slide projector. An example of the field stop in the image plane is the 36 mm × 24 mm rectangular dimension of the film in a 35 mm camera. The field stop can be circular, rectangular, or any other shape. The field stop may clip the edges of the image so that the edges will be occluded compared with the brightness at the center. Specks of dust on the front of a camera lens do not degrade the picture because they are near the aperture stop; however, if these specks were at the field stop, this dust would show up in the picture.

Windows are conjugate to the field stop in the same way that entrance pupils and exit pupils are conjugate to the aperture stop. The windows define the extent of the object that is viewable by the optical system in the same way that a window in a house limits the amount of an exterior object that may be viewed through the window. The exit window limits the edge extent of the image.

To increase the field of view, a field lens may be placed at an intermediate real image location within an optical system. The field lens allows a larger field of view by accepting parts of the image at larger angles than would be possible without it. It works by bending light at high angles back toward the optical axis, allowing rays to pass through the rest of the system. If the field lens has dust on its surface, dust will show up in the final image. To avoid image spots, the lenses are often placed slightly before or after the field stop. Alternatively, an air spaced doublet may be used to place the cardinal points of the lens at locations where there is no physical glass that can become dirty. Placing a positive (convex) lens at the field stop can dramatically increase the field of view by allowing a real image formed within the system to accept incoming light at higher angles that would otherwise be blocked by the outer edge of the system. This technique can sometimes increase the field of view by 25%.

9.2.1 Field of view

Field of view is defined as $\pm\theta$, where θ is the half angle from the optical axis, as shown in Figure 9.9. A larger field stop allows a larger field of view. The angular field of view is determined by the field stop size and distance to the exit pupil. Very short large diameter systems have the greatest field of view, and long thin systems have narrow fields of view.

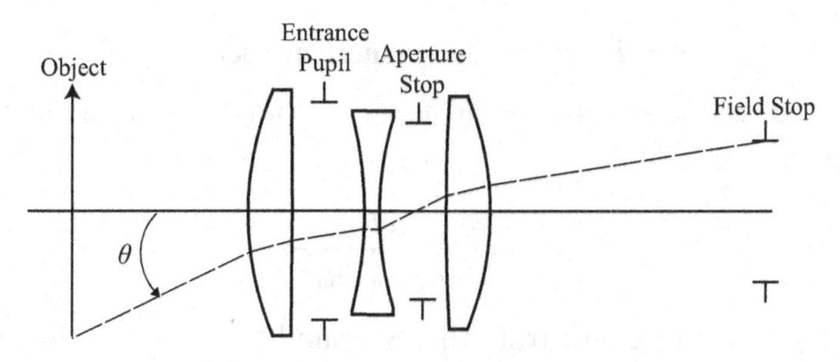

Figure 9.9 Field of view shown for a multi-lens system.

Table 9.1. *Typical field of view values*

Microscope eyepieces (standard)	41–45 degrees
Microscope eyepieces (expensive)	50–68 degrees
Microscope eyepieces area	3.3 × 4.4 mm
Rifle scopes at 1000 yards	8 yards

It is critical to understand clearly what is meant when given a specification for field of view, because the full angle, rather than the half angle, is sometimes provided. To avoid confusion, both the value for field of view and the angle used must be given explicitly.

Table 9.1 provides values of field of view as typically quoted, showing the necessity for defining terms as well as providing numbers. Sometimes angles are not used.

9.2.2 *Baffles and glare stops*

Baffles and glare stops prevent unwanted light from leaking into the system from undesirable locations, adding stray light to the desirable light. Baffles are placed as stops around the axis to block any stray light that might bounce off the walls within the system and find a way to the image, as shown in Figure 9.10. Baffles are typically placed at images of the aperture stop within the system, such as at entrance and exit pupil locations. Undesirable light may be of the wrong polarization or the wrong color. In these cases, color filters or polarizers are placed near the aperture stop or pupils, allowing all the light to pass through them, but preventing their images from showing up in the image. Filters and polarizing components are usually designed for collimated light, and should not be placed in areas of strong divergence or convergence.

9.3 *F*-number and numerical aperture

$F/\#$ is defined as the effective focal length divided by the diameter of the entrance pupil:

$$F/\# = \frac{efl}{D_{\text{entrancepupil}}}. \tag{9.1}$$

Alternatively, it is the length from the exit pupil to the image divided by the diameter of the exit pupil.

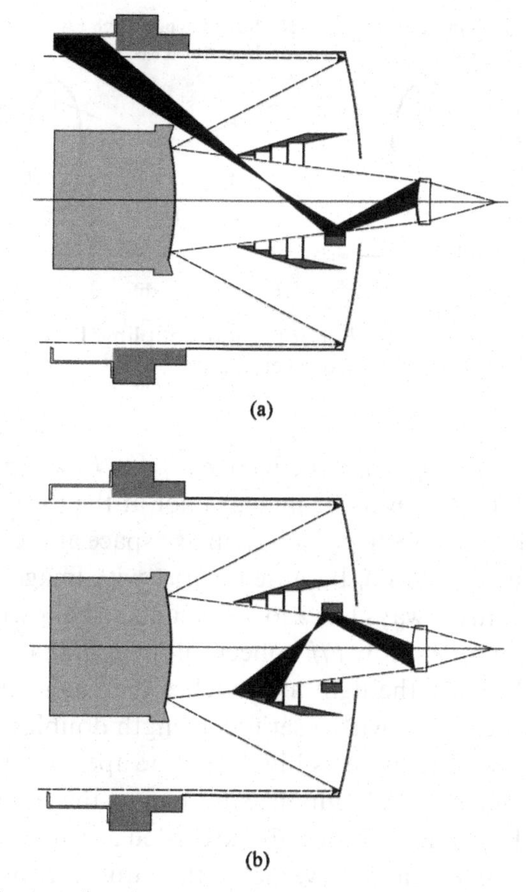

Figure 9.10 Cassegrain telescope with baffles: (a) wide angle stray radiation; (b) near-axis stray radiation.

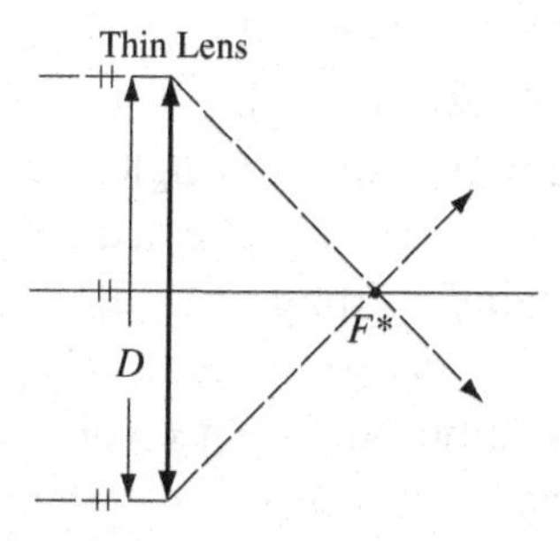

Figure 9.11 *F/#* configuration designation.

F/# is a measure of the angle at which the rays outside the lens are bent toward the optical axis. *F/#* varies from 1/2 for rays converging at a steep full angle of 45°, as shown in Figure 9.11, to very large values. Most systems work between *F/2* and *F/22*.

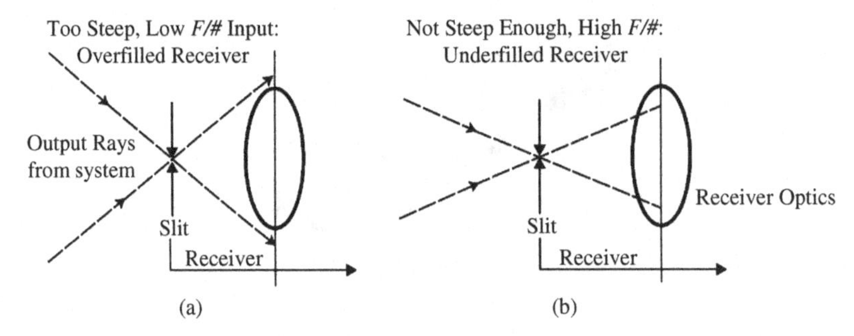

Figure 9.12 Need to match *F*/# for good coupling between systems: (a) overfilled receiver; (b) underfilled receiver.

The term working *F*/# ($(F/\#)_w$) or effective *F*/# ($(F/\#)_{eff}$) denotes the angular extent of the ray fan whenever the image is not at the focal point. $(F/\#)_w$ or $(F/\#)_{eff}$ tells us what the effective *F*/# is in image space at the image. This term is less commonly used than *F*/# because it varies with image location.

A steeply angled ray always has a low *F*/#, and a shallowly angled ray has a high *F*/#. A system with a low *F*/# collects more light that can be focused to a sharper point, because the light approaches the image at a higher angle. The light entering a system with a set focal length doubles in illuminance if the aperture area is made twice as big. Since the aperture area varies as the square of the diameter, the illuminance doubles as the *F*/# decreases by the square root of 2. For camera lenses, *F*/# are listed as values in a progression, where each successively higher value is the square root of 2 times the previous value. Common *F*-numbers in camera systems are *F*/1.4, *F*/2.0, *F*/2.8, *F*/4.0, *F*/5.6, *F*/11, *F*/16.8, *F*/22.

A lens is called a "fast" lens if it bends the rays steeply, so an *F*/2 lens is called fast while an *F*/11 lens is called slow. High *F*/# increases the depth of focus, but high *F*/# systems gather light over smaller angles, so less light enters a high *F*/# system. In photography, the trade-off between depth of focus and light gathering ability is adjusted by changing the *F*/# to fit every picture.

It is important to match *F*/# (see Figure 9.12) when combining optical systems, for example fiber optics and a spectrometer. The *F*/# of the second optical system should accept light rays at the same angle that they are sent by the first system. If that is not possible, the receiver for the second system should be underfilled. Spectrometers, for example, accept a ray bundle through a narrow slit. A spectrometer may be designated as *F*/4. If steeper rays enter the system, they will be lost around the outside of the collimating mirror, as shown in Figure 9.12(a). If less steep rays enter the system, they will pass through, but only a small section of the grating in the spectrometer will receive light, as shown in Figure 9.12(b) (recall the spectrometer resolving power). In this case, the spectrometer will be underfilled and the resolution will go down, as the spectrometer is not being used to its full potential light gathering $F/\#$.

It is important to remember that light at steep angles is much harder to direct than light at shallow angles. Sources of light such as bulbs often put out light in every direction, but the systems that use the light are designed such that all the light travels almost in line with the optical axis. In these cases, a condensing lens is used to bring the light from a bulb source in line with the optical axis of the system. For more information on illumination techniques, such as Koehler or critical illumination, see *Military Standardization Handbook of Optical Design (MIL-HDBK-141)* or Mouroulis and Macdonald (1997).

The numerical aperture is defined as the refractive index of the medium multiplied by the sine of the largest entrance ray angle with respect to the optical axis:

$$NA = n \sin u. \tag{9.2}$$

It is a figure of merit of the light collecting property of an optical system. The higher the numerical aperture, the greater the amount of light collected. The numerical aperture is typically defined in object space, and incorporates the maximum angle of object radiation through a medium with refractive index n that can be accepted by the optical system. The numerical aperture is always positive, regardless of whether the rays are divergent or convergent. More light can be collected at an effectively steeper angle if index matching oil is applied to the object. (See Figure 9.12, in which the acceptance angle changes due to different indices of refraction.) A low numerical aperture indicates a long depth of focus and long depth of field. The numerical aperture is related to $F/\#$ by:

$$NA = \frac{n}{\sqrt{4(F/\#)^2 + 1}}. \tag{9.3}$$

Snell's law tells us that $n \sin u$ in one medium is equal to $n' \sin u'$ in the next medium. The maximum ray angle that can enter the system defines the numerical aperture, $n \sin u$. In high refractive index materials, the maximum angle (u) is small because of Snell's law, but the numerical aperture remains constant. As shown in Figure 9.13, the ray angles are changed by going from a high index (n) to a low index (n') such that the numerical aperture remains constant.

Table 9.2. *Typical numerical aperture values*

Optical component		Typical numerical apertures
Microscope objective magnification	2 ×	0.055
	5 ×	0.14
	10 ×	0.28
	20 ×	0.41
	50 ×	0.55
	100 ×	0.7
Fiber optics		0.3–0.5
Fiber bundles		0.4

Table 9.3. *Angle, radians, NA (n = 1) and F/# equivalences*

Degrees	Radians	Numerical aperture	F/#
0	0.000	0.000	∞
5	0.087	0.087	5.7
10	0.174	0.174	2.8
15	0.262	0.259	1.9
20	0.349	0.342	1.4
25	0.436	0.423	1.1
30	0.524	0.500	0.9
35	0.611	0.574	0.7
40	0.698	0.643	0.6
45	0.785	0.707	0.5
50	0.872	0.766	
55	0.960	0.819	
60	1.047	0.866	
65	1.134	0.906	
70	1.222	0.940	
75	1.309	0.966	
80	1.396	0.985	

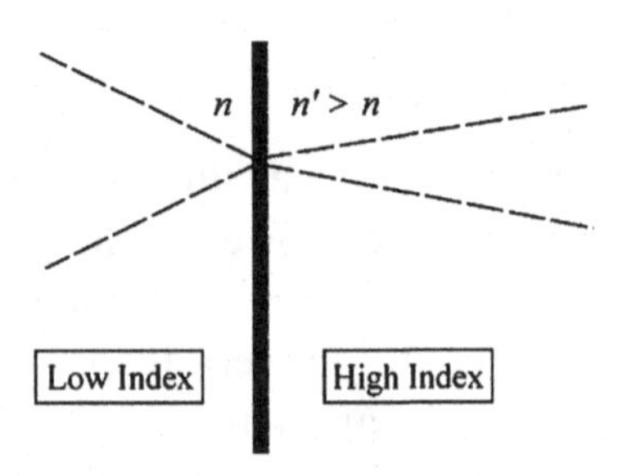

Figure 9.13 Numerical aperture remains the same when angles change due to the refractive index.

Numerical aperture varies directly with angle, rather than being inversely related like $F/\#$.

Numerical apertures for several optical components are provided in Table 9.2. Table 9.3 relates angles between 5° and 80° to radians, numerical aperture, and $F/\#$. Example 9.3 shows how to convert $F/\#$ to NA.

Example 9.3

To convert F/# to NA: $F/\# =$ focal length/diameter and $NA = n \sin u$ (n is the refractive index of the medium in which the angle is measured).

$F/\# = B/2A$, and by the definition of a sine, NA is defined as:

$NA = n\left(A/\sqrt{A^2 + B^2}\right)$

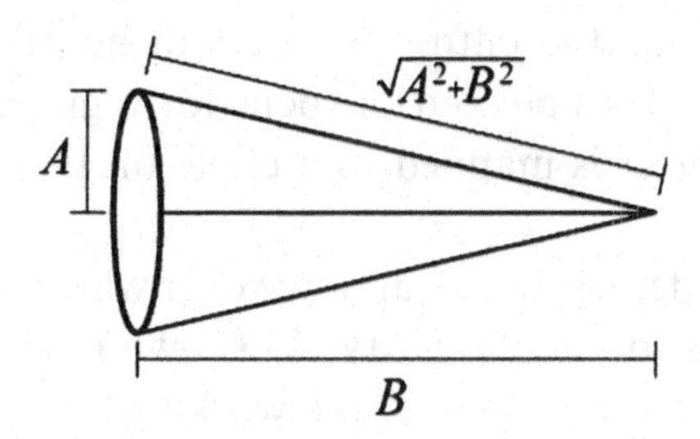

$$B^2 = 4A^2(F/\#)^2.$$

Set

$$A = 1$$

then

$$B^2 = 4(F/\#)^2$$

$$NA = \frac{n}{\sqrt{1 + 4(F/\#)^2}}, \quad F/\# = \frac{1}{2}\sqrt{\left(\frac{n}{NA}\right)^2 - 1}.$$

For small angles when $n = 1$: $\sin u = \tan u = u$, so

$$F/\# \approx \frac{1}{2u}, \quad NA \approx \frac{1}{2F/\#}.$$

9.4 Depth of focus and depth of field

The depth of field is the distance that the object can move longitudinally along the optical axis (z) while still in focus. This is also related to the change in image location called the depth of focus. Steeper (faster $F/\#$) ray bundles have a shorter depth of field. The depth of focus is the distance an image can be moved longitudinally along the optical axis without becoming noticeably blurred.

The depth of field is very important in photography when taking pictures in which there may be objects in both the foreground and the background. A large $F/\#$ is used if there is enough light available to prevent the close objects from being blurry while the picture contains in-focus distant objects. Some cameras have a depth of field preview button to show how well in focus the entire picture will be once the aperture stop is set. As the object is brought out of focus for a given conjugate distance, each point on the object is mapped to a circle on the image. This causes a blurred image.

The depth of focus defines the axial range over which an image plane can be translated without loss of image clarity. As shown in Figure 9.14, for a given acceptable blur size, b, depending on the working $F/\#$ for the corresponding image space, the depth of focus may vary widely. For a small $F/\#$ lens, the depth of focus is much smaller than for a large $F/\#$ lens.

The maximum blur size that governs the depth of focus is ultimately limited by diffraction. The diffraction-limited blur size diameter (b) is determined by (Jenkins and White, 1976):

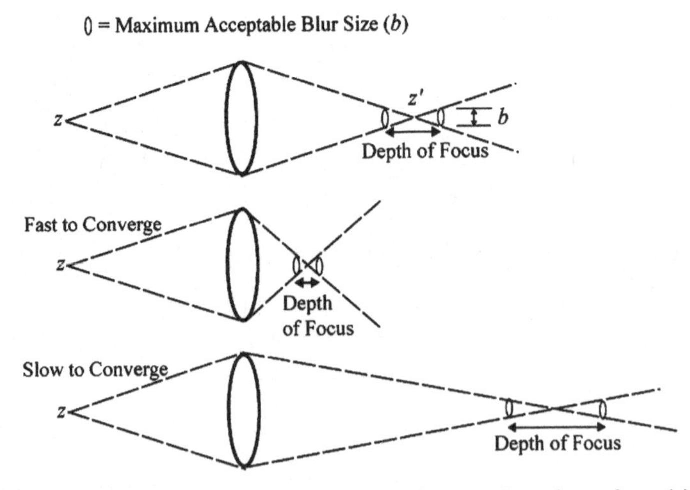

Figure 9.14 Depth of focus for a given blur size as a function of working $F/\#$.

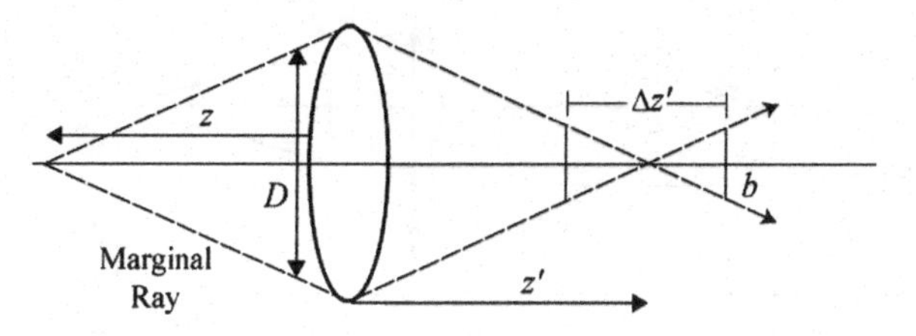

Figure 9.15 Image is a diffraction-limited blur

$$b = 2.44\lambda(F/\#)_{\mathrm{w}}. \tag{9.4}$$

Based on similar triangles, as shown in Figure 9.15, the depth of focus ($\Delta z'$) can be quantified, where $\Delta z'$ is the region of loss of image clarity due to diffraction:

$$\frac{D/2}{z'} = \frac{b/2}{\Delta z'/2}$$

or

$$\Delta z' = \frac{2bz'}{D}.$$

If the object is located at infinity and the system is diffraction limited, then the depth of focus is

$$\Delta z' = 2b(F/\#)_{\infty} = 4.88\lambda(F/\#)^2_{\infty}. \tag{9.6}$$

If the object is located at a finite distance z, as shown in Figure 9.15, then

$$z' = \left[\frac{1}{z} + \frac{1}{f^*}\right]^{-1}.$$

From the definition of $(F/\#)_{\mathrm{w}}$ in Equation (9.4),

$$\Delta z' = 2b(F/\#)_{\mathrm{w}} = 4.88\lambda(F/\#)^2_{\mathrm{w}}. \tag{9.7}$$

Recalling numerical aperture (NA) from Equation (9.3) and Example 9.3, the result can also be expressed as

$$\Delta z' = \frac{b}{NA} = \frac{1.22\lambda}{(NA)^2}. \tag{9.8}$$

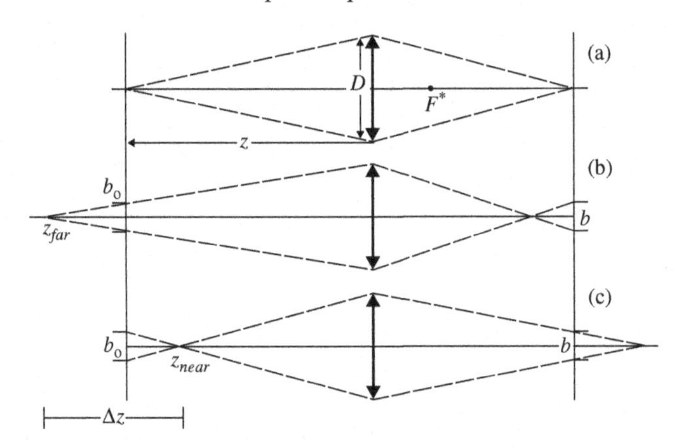

Figure 9.16 (a) In-focus object and image; (b) object at the far point for a blur circle of diameter, b; (c) object at the near point of the depth of field for a blur circle at diameter b.

The depth of focus for an object at infinity is given by Equation (9.6). The depth of focus for an object at a finite distance is given by Equation (9.7).

9.5 Hyperfocal distance

Objects located within the depth of field appear to be in equal focus. To obtain the ranges of object distance that are in focus, z_{near} to z_{far}, the Gaussian equation can be used to determine these limits. See Figures 9.15 and 9.16.

$$z_{\text{near}} = \frac{f^*(z' + \Delta z'(1/2))}{f^* - (z' + \Delta z'(1/2))}, \tag{9.9}$$

$$z_{\text{far}} = \frac{f^*(z' - \Delta z'(1/2))}{f^* - (z' - \Delta z'(1/2))}. \tag{9.10}$$

The resulting depth of field (Δz) is

$$\Delta z = z_{\text{far}} - z_{\text{near}} = \frac{4\Delta z'(f^*)^2}{(\Delta z')^2 - 4(f^* - z')^2}. \tag{9.11}$$

To maximize the depth of field, one can force the denominator to zero, which causes the object to be at infinity as would be expected!

You may ask, "how far is infinity?" or, "do objects really have to be at infinity to focus the rear focal point?" In fact, for optical systems, any object beyond the hyperfocal distance acts as if it were at infinity.

Referring to Figure 9.16, by similar triangles (where b_o is the blur in object space for a blur diameter of b on the image),

$$\frac{D}{z_{\text{near}}} = \frac{-b_o}{z - z_{\text{near}}}$$

or

$$z_{\text{near}} = \frac{zD}{D - b_o}. \tag{9.12}$$

With transverse magnification (M_t),

$$b = M_t b_o$$

$$D = \frac{f^*}{F/\#},$$

$$z_{\text{near}} = \frac{z(f^*/F/\#)}{\dfrac{f^*}{F/\#} - \dfrac{b}{M_t}}.$$

Rearranging,

$$z_{\text{near}} = \frac{M_t f^* z}{M_t f^* - b(F/\#)}.$$

Recalling

$$M_t = \frac{f^*}{z + f^*},$$

$$z_{\text{near}} = \frac{z(f^*)^2}{(f^*)^2 - b(F/\#)(z + f^*)}. \tag{9.13}$$

Similarly for z_{far}:

$$z_{\text{far}} = \frac{z(f^*)^2}{(f^*)^2 + b(F/\#)(z + f^*)}. \tag{9.14}$$

The hyperfocal distance that simulates infinity for an optical system is found by setting the far field point, z_{far}, to infinity and therefore the denominator to zero:

$$(f^*)^2 + b(F/\#)(z + f^*) = 0.$$

Solving for z:

$$z_{\mathrm{H}} = -f^*\left(\frac{f^*}{b(F/\#)} + 1\right). \tag{9.15}$$

Evaluating the quantity inside the brackets:

$$\frac{f^*}{b(F/\#)} = \frac{f^*}{b(f^*/D)} = \frac{D}{b} \gg 1.$$

We can drop the 1 in Equation (9.15), since the diameter of the optics will be much larger than the spot size:

$$z_{\mathrm{H}} = -\frac{(f^*)^2}{b(F/\#)}. \tag{9.16}$$

This is the hyperfocal distance, z_{H}. When the spot size, b, is diffraction limited:

$$z_{\mathrm{H}} = \frac{D^2}{2.44\lambda}. \tag{9.17}$$

The near field is

$$z_{\mathrm{near}} = \frac{z_{\mathrm{H}}}{2}. \tag{9.18}$$

Therefore, from one half the hyperfocal distance to infinity, all objects are at the focal plane and in focus.

Problems

9.1 A 5 mm diameter aperture is placed 2 cm behind a 10 diopter thin lens.
  (a) Where is the entrance pupil located and what is its size?
  (b) Where is the exit pupil and what is its size?

9.2 A thin lens with a 50 cm effective focal length and $(F/\#)_\infty$ equal to 12.5 is used to image an object 20 cm in front of it. A 2 cm (diameter) aperture is placed 3 cm to the left (in front of) the lens.
  (a) What is the aperture stop?
  (b) Where are the entrance and exit pupils?

9.3 An object is 25 cm in front of a 60 cm focal length, $(F/\#)_\infty = 6$ lens. An aperture 10 cm in diameter is 2 cm behind the lens.
  (a) Where is the entrance pupil and what is its size?
  (b) Where is the exit pupil and what is its size?

9.4 A 50 mm, $F/1$ lens is used as an imager. An aperture, 10 mm in diameter, is placed 10 mm behind the lens.
   (a) Locate the entrance pupil.
   (b) Locate the exit pupil.

9.5 Two thin lenses (L_1 and L_2) separated by 30 cm have effective focal lengths of 20 cm and 15 cm, and diameters of 12 cm and 10 cm, respectively. An aperture with a diameter of 8 cm is placed 10 cm in front of the 20 cm *efl* lens (L_1).
   (a) For an object 50 cm in front of lens 1, locate the entrance and exit pupils.
   (b) For an object 25 cm in front of lens 1, locate the entrance and exit pupils.

9.6 Two thin lenses with *efl* of 16 (L_1) and 4 cm (L_2) are separated by 8 cm. Each lens has a diameter of 4 cm.
   (a) What are the values for object distances that make lens 1 the aperture stop?
   (b) What are the values for object distances that make lens 2 the aperture stop?

9.7 Two thin lenses, 5 cm in diameter, with effective focal lengths of +10 cm and –6 cm, are placed 4 cm apart.
   (a) What is the optical power of the system?
   (b) Which element is the aperture stop for an object at infinity?

9.8 Two thin lenses, both 10 cm in diameter, with focal lengths of 20 cm and 12 cm, respectively, are placed 8 cm apart. A 3 cm aperture is placed halfway between them. The object is at infinity.
   (a) Where is the entrance pupil and what is its size?
   (b) Where is the exit pupil and what is its size.

9.9 A lens of 10 diopters is mounted 6 cm in front of a negative lens with a focal length of −7 cm. A 1 cm diameter aperture stop is placed halfway between the two lenses.
   (a) What are the location and diameter of the entrance pupil?
   (b) What are the location and diameter of the exit pupil?

9.10 Calculate the location and size of the entrance and exit pupils of the thin lens system below for the following object locations (all dimensions in millimeters):
   (a) object at infinity;
   (b) object located 200 mm in front of the first lens.

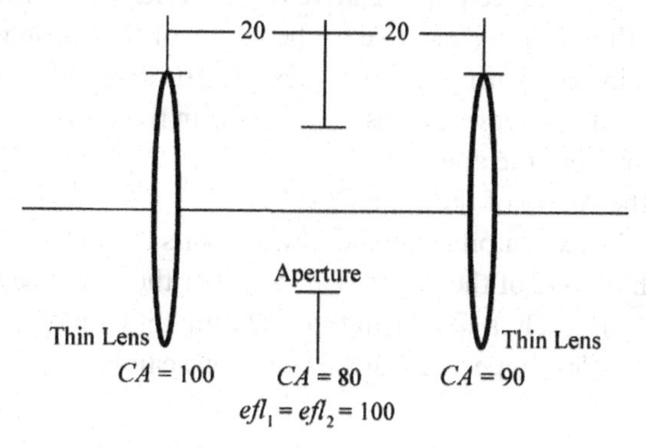

9.11 For a 35 mm slide (standard size format 24.5 mm × 36.5 mm), at the focal plane of a thin lens with a 100 mm effective focal length, what is:
(a) the field of view (FOV) in the horizontal direction?
(b) the FOV in the vertical direction?
(c) the FOV on the diagonal?

9.12 Two thin lenses, each 5 cm in diameter, with focal lengths of +12 cm and +5 cm are placed 3 cm apart. An aperture stop 1 cm in diameter is set halfway between the lenses. Find the locations and diameters of the entrance pupil and the exit pupil.

9.13 A thin lens of +10 cm focal length is mounted 7 cm in front of another lens with a −7 cm focal length. When a stop, 2 cm in diameter, is placed halfway between the two lenses, what are the locations and diameters of the entrance pupil and the exit pupil?

9.14 An object is 20 cm in front of a 6 cm diameter, $F/1$ lens, and an Astop is 3 cm behind the lens. If the Astop is 2 cm in diameter:
(a) What are the location and diameter of the entrance pupil?
(b) What are the location and diameter of the exit pupil?
(c) Where is the image of the object?

9.15 For a 500 mm diameter telescope, $F/3.5$, with a detector size of 7 μm, what is the hyperfocal distance? (Objects from ½ hyperfocal distance to infinity are in focus.)

9.16 The optical system shown below contains three thin lenses, of 5, −3, 7 diopters each, with 10 cm between each lens. If the second lens has a diameter of 5 cm and is the stop of the system:

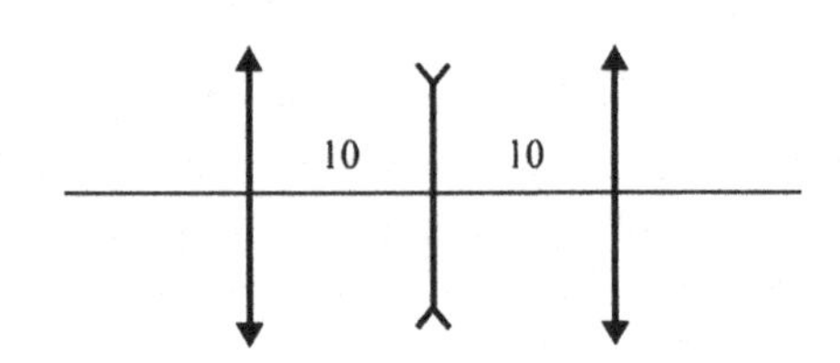

(a) Where is the entrance pupil relative to the vertex of the first lens?
(b) Where is the exit pupil relative to the vertex of the last lens?

9.17 Consider a thin lens with a 25 mm focal length and a diameter of 3 cm. An aperture with a diameter of 2 cm is placed 3 cm in front of the lens, and an object is located 8 cm from the aperture.
(a) What is the Astop of the system?
(b) Where is the exit pupil located and what is its diameter?

9.18 For the depth of field of the eye, we can consider the eye to be three cones with a diameter of 3 μm each, a focal length of 22 mm, and a pupil diameter of 4 mm. What is the smallest hyperfocal distance the eye can have?

9.19 Two thin lenses are separated by 20 cm and have focal lengths of 30 cm. If a 3 cm diameter stop is placed halfway between them, where are the entrance and exit pupils located, and what are their diameters?

9.20 A thin lens with an effective focal length of +100 mm has a clear aperture (CA) of 50 mm. A 2 cm high object is located 200 mm in front of the lens. The lens is located centrally between two apertures which are separated by 100 mm but have different hole diameters. The hole on the object side is 50 mm, and the hole on the image side is 25 mm, as shown below.

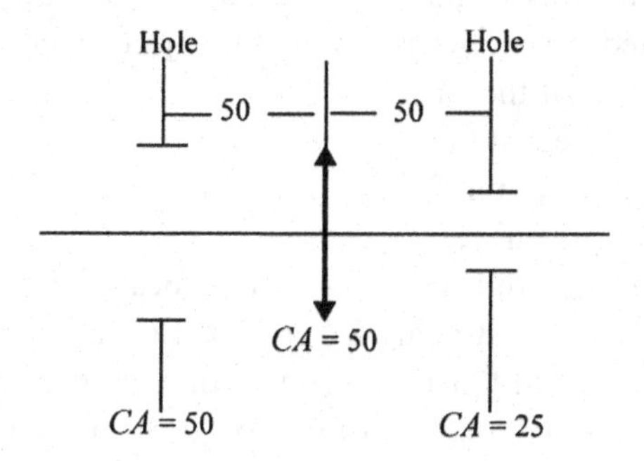

(a) Which aperture is the stop of the optical system?
(b) What are the size and location of the entrance pupil?
(c) What are the size and location of the exit pupil?
(d) What are the size, location and handedness of the image?

9.21 What is the hyperfocal distance for a 50 mm diameter, $F/2$ lens if a 25 μm detector is used to detect the image?

9.22 A thin lens with an aperture of 5 cm and a focal length of +3.5 cm has a 2.0 cm stop located 1.5 cm in front of it. An object 1.5 cm high is located with its lower end on the axis, 8.0 cm in front of the lens.
(a) Locate the position of the exit pupil.
(b) What is the size of the exit pupil?
(c) Locate the image of the object.

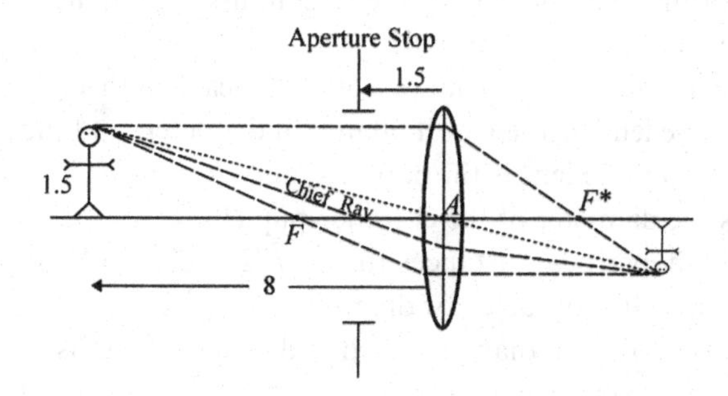

9.23 An astronomer on Kitt Peak uses a 100 μm diameter detector on an 84 inch diameter, $F/4$ telescope. What is the hyperfocal distance?

9.24 A thin lens with a focal length of +5.0 cm and a clear aperture of 6.0 cm has a 3.80 cm diameter stop located 1.60 cm behind it. An object 2.20 cm high is located on the axis 8.0 cm in front of the lens. Locate:
(a) the position of the entrance and exit pupils;
(b) the size of the entrance pupil;
(c) the image location and size.

9.25 A thin lens with a focal length of −6.0 cm and a clear aperture of 7.0 cm has a 3.0 cm stop located 3.0 cm in front of it. An object 2.0 cm high is located on the axis 10.0 cm in front of the lens. Find:
(a) the position of the exit pupil;
(b) the size of the exit pupil;
(c) the position of the image.

9.26 A 6.0 cm focal length thin lens has a stop located 1.0 cm to the right. The diameters of the lens and stop are 2.0 cm and 1.6 cm, respectively.
(a) For an object located 18.0 cm in front of the lens, determine the aperture stop and find the size and location of the entrance and exit pupils.
(b) For an object located 10.0 cm in front of the lens, determine the aperture stop and find the size and location of the entrance and exit pupils.

9.27 A telephoto lens consists of two thin lenses and an aperture. The first lens has a focal length of 4.00 cm and a diameter of 3.00 cm. The aperture is located 2.00 cm behind the first lens, and has a diameter of 1.125 cm. The second lens is 1.00 cm behind the aperture. The second lens has a focal length of −1.25 cm and a diameter of 0.75 cm. For an object at infinity:
(a) Determine the aperture stop.
(b) Find the size and location of the entrance and exit pupils.
(c) What is the $F/\#$ for this system?
(d) What is the effective focal length (*efl*)?

9.28 A thin lens of 50 mm focal length has a diameter of 4 cm. A stop 2 cm in diameter is placed 3 cm to the left of the lens, and an axial point object is located 20 cm to the left of the stop.
(a) Which of the two, the stop or the lens, limits the bundle?
(b) Where is the exit pupil located?

9.29 An object is placed 25 cm in front of a 60 mm focal length lens, and there is a stop 2 cm behind the lens. If the lens if 50 mm in diameter, and the stop 20 mm:
(a) How far from the lens is the entrance pupil?
(b) What is the diameter of the entrance pupil?

9.30 Calculate the hyperfocal distances for an $F/2$ and $F/16$ lens that has an *efl* of 100 mm, using a 100 μm detector diameter.

9.31 Two thin lenses, 5 cm in diameter each and of focal lengths +10 cm and +6 cm are placed 4 cm apart. An aperture stop 2 cm in diameter is set halfway between the lenses.

(a) Find the diameters of the entrance pupil and the exit pupil for the object at infinity.
(b) Find the diameter of the entrance pupil and the exit pupil if the object is 8 cm in front of the first lens.

9.32 An optical system with two lenses and three apertures is shown below. All distances are in centimeters. The focal lengths of lenses 1 and 2 are 10 cm and 7 cm, respectively. The diameters of lenses 1 and 2 are both 5 cm, and they are 8 cm apart. The diameters of apertures A1, A2, and A3 are 4.5, 3, and 4 cm respectively. An object 1 cm tall is located 9 cm in front of lens 1.

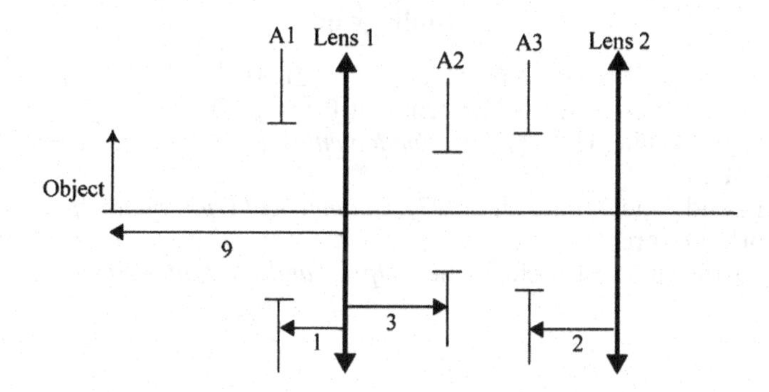

(a) Which lens or aperture is the aperture stop?
(b) What should the diameters of the three apertures and lens 2 be to make lens 1 the Astop without changing the size of lens 1?

9.33 A thin lens with an optical power of 5 diopters and a diameter of 5 cm has a 35 mm field stop diameter at F^*.
(a) What is the $(F/\#)_\infty$?
(b) What is the full field of view (in degrees)?
(c) What is the radius of the aperture stop (in centimeters)?

9.34 Two equal positive optical power thin lenses ($\phi = 5$ diopters) are separated by 15 cm. A CMOS chip sensor, 1 cm × 1 cm, is placed at the focal point (F^*).
(a) What is the effective focal length?
(b) What is the full field of view of the diagonal?

9.35 For a positive equi-convex thick lens with $|R_1| = |R_2| = 25$ cm, an axial thickness of 4 cm, a refractive index of 1.7 ($n = 1.7$), and a diameter of 5 cm:
(a) What is the optical power (in diopters)?
(b) What is the full field of view for a 35 mm diameter field stop?
(c) If an object is 50 cm from the front vertex, what is the working $F/\#$?

9.36 A CMOS array with a pixel size of 4 μm, 1024 × 1024 pixels (assume 100% fill factor) is used to record a visible image (400–700 nm wavelength). The objective lens is an $F/2$, 100 mm *efl*, used with a magnification of −0.5 ($M_t = -0.5$).
(a) What is the depth of field, or determine the near and far object distances?

(b) If used for distant objects, what is the hyperfocal distance ($M_t = 0$)?

(c) Where should the CMOS array be placed relative to the lens (*BFD*) for maximum depth of field ($M_t = 0$)?

9.37 Prove that numerical aperture and $F/\#$ are related by:

$$NA = \frac{n}{\sqrt{4(F/\#)^2 + 1}}.$$

Bibliography

Hopkins, R., Hanau, R., Osenberg, H. *et al.* (1962). *Military Standardized Handbook 141 (MIL HDBK-141)*. US Government Printing Office.

Jenkins, F. A. and White, H. E. (1976). *Fundamentals of Optics*, fourth edn. New York: McGraw-Hill.

Mouroulis, P. and MacDonald, J. (1997). *Geometrical Optics and Optical Design*, New York: Oxford.

Wallin, R. (2000). Sharpest Field Tables. *Optics and Photonics News*, **11**, 36.

10

Paraxial ray tracing

Paraxial ray tracing is a technique used in geometrical optics for predicting the paths light will take through an optical system. Its primary application is in the design of lens and mirror systems.

Rays may be viewed as comprising streams of photons emanating from a light source and propagating toward surfaces throughout the optical system. As discussed in Chapter 2, four things may happen as rays strike optical surfaces: they may be refracted (transmitted), reflected, scattered, or absorbed. These effects may occur singly or in combination, and all of the energy in the incident beam must be accounted for by these mechanisms.

To review, refracted light enters a transparent medium at an angle different from the incident angle. Some of the energy in the incident beam will also be reflected by the surface. The surface may scatter some of the incident light in one or more directions. Additionally, it may absorb some of the energy, resulting in a loss of intensity in the beam. Refracted and reflected rays may strike other surfaces, at which the same mechanisms will again take effect, and at which Snell's law may be applied.

Paraxial ray tracing is applied to systems in which diffraction and interference effects are insignificant. It is useful for modeling and optimizing the design of lens systems and instruments, (e.g. minimizing the effects of aberrations) before optical components are ordered or fabricated. The technique finds utility in many applications, a particular example being illumination engineering. In this application, it is used to locate light energy within the system, and to direct the energy properly toward the focal plane. It does not depend on the wavelength of light propagating through the system, although wavelength-dependent indices of refraction of the optical components must be taken into account.

Before computers were used in optical design, ray tracing calculations were performed using tables of logarithms. This process was tedious and required many hours of calculation. An approximation called the paraxial approximation is made at the beginning stages of a design and allows calculations to be performed without incorporating the use of sine and tangent functions, reducing processing time and simplifying the ray trace by avoiding trigonometric calculations of angle sines and tangents. The paraxial ray used in this approximation lies close to the optical axis throughout the system, at a small angle (θ) from the axis. Therefore, $\sin\theta \approx \tan\theta \approx \theta$ and $\cos\theta \approx 1$ (for θ in radians). The equations governing ray traces performed using this simplification (also referred to as first-order ray tracing) are linear.

There is a disadvantage to the paraxial method: it predicts "perfect" images, which never occur in optical systems due to diffraction effects and the non-ideal behavior of spherical surfaces. However, the method provides a necessary starting point for any optical design or evaluation because it is quick, simple, and effective in determining design feasibility.

10.1 Ray tracing worksheet

There are two approaches used in paraxial ray tracing to obtain the object–image relationships. Each approach requires that two rays be traced. Thus far we have been using the parallel ray method to find the image location. In this method, two unique rays, one parallel to the optical axis and another propagating through the front focal point, are traced as shown in Figure 10.1. In image space, the ray corresponding to the input ray parallel to the optical axis emerges through the rear focal point; the ray propagating through the front focal point emerges parallel to the optical axis, as shown. The parallel ray method locates the image and determines its size; it does not provide a means of determining image brightness.

To trace a paraxial ray through a lens system, the refraction and transfer equations developed in Chapter 5 are needed, and these are repeated below:

$$n'u' = nu - \phi y; \tag{10.1}$$

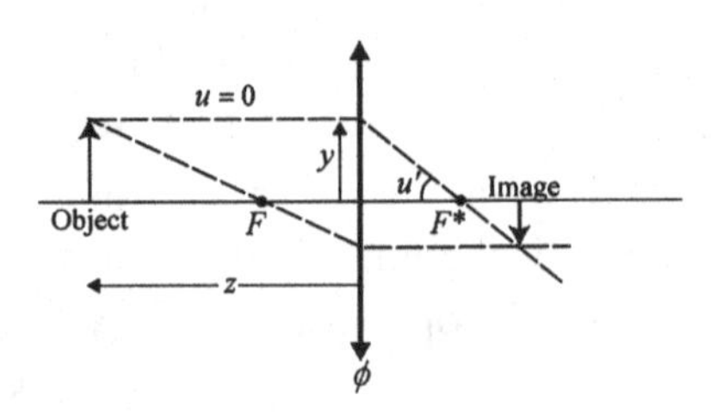

Figure 10.1 Parallel ray trace method.

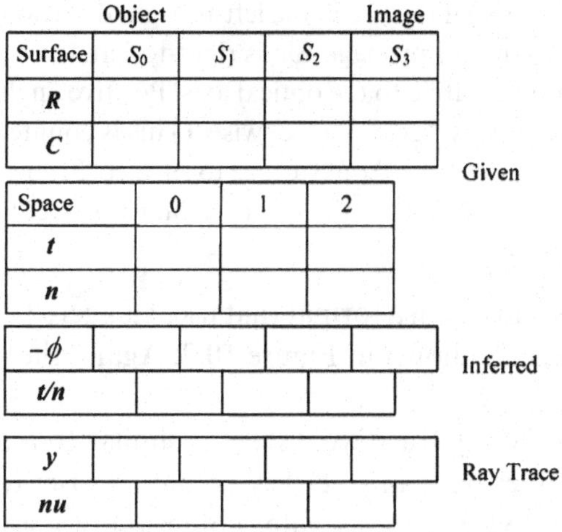

	Object			Image	
Surface	S_0	S_1	S_2	S_3	
R					
C					Given
Space	0	1	2		
t					
n					
$-\phi$					Inferred
t/n					
y					Ray Trace
nu					

Figure 10.2 Paraxial ray trace worksheet.

$$y_{+1} = y + n'u'\left(\frac{t'}{n'}\right). \tag{10.2}$$

These equations provide the basis for transferring a ray across a surface from one space to a second space, as well as for evaluating the ray height (y) along a ray in any given space. The ray is a straight line in both the physical and virtual regions of that homogeneous space.

To trace paraxial rays through an optical system, Equations (10.1) and (10.2) are applied to a coordinate system moving in the z direction from surface to surface. The results from these equations are placed into the ray trace worksheet shown in Figure 10.2. The worksheet provides a systematic method for mathematically tracing a propagating ray from object space to image space across surfaces and through homogeneous spaces. The ray trace sign convention is the same as that described in Table 5.1, and is repeated here for clarity.

(1) A right-hand coordinate system is used, with the z axis as the optical axis. The y–z plane is the meridional (or tangential) plane and the x–z plane is the sagittal plane.
(2) Light travels from left to right
(3) Refractive indices are positive for light traveling left to right, and negative for light traveling right to left.
(4) Surface curvatures are positive when the center of curvature lies to the right of the surface, and negative when the center of curvature lies to the left.
(5) Surfaces are numbered corresponding to the order in which light propagates through them, beginning with the object at the left. The object is considered to be at surface zero (0).

(6) Thicknesses are positive if they lie in the left-to-right direction of propagation and negative if the direction of propagation is right-to-left.
(7) Angles are measured relative to the optical axis. Positive angle values are counter-clockwise, and negative values are clockwise. (This is counterintuitive.)
(8) The coordinate system moves from surface to surface. Primes (′) on quantities are used to differentiate quantities in the space beyond a surface from quantities in the space before a surface.

The worksheet is laid out in a format that uses staggered columns for surface and space properties, as shown in Figure 10.2. Again, the object is at surface zero.

The structure of the ray trace worksheet is similar to a brick wall with the surface property values lying in a vertical column. Offset to the side by half a column width is the space property value column. For a given optical system, the ray trace worksheet has three sections:

(1) the "givens" that characterize the optical elements (radius, curvature, thickness, and indices);
(2) the "inferred" quantities derived from the givens (negative optical power $-\phi$, reduced thickness t/n); and
(3) the actual "ray trace" quantities determined by the initial starting points. Negative optical power is used in the worksheet to facilitate calculation of ray trace quantities.

Figure 10.3 shows the givens of the optical system as they would appear in the worksheet. The values for each surface are entered in one column of the worksheet in the positions of R_0, R_1, etc., and the corresponding space characteristics are entered in the intermediate columns directly below, as shown.

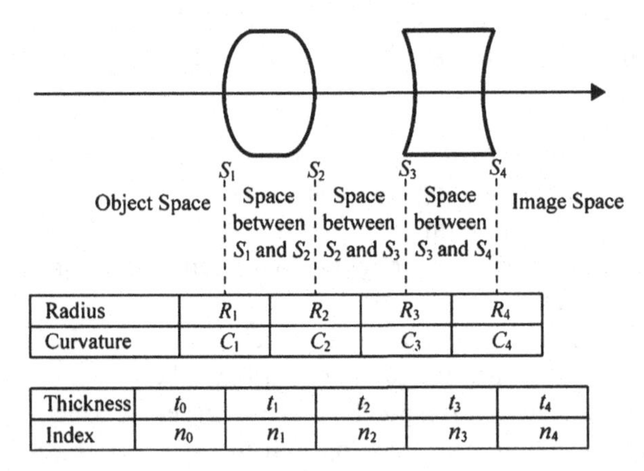

Radius	R_1	R_2	R_3	R_4
Curvature	C_1	C_2	C_3	C_4

Thickness	t_0	t_1	t_2	t_3	t_4
Index	n_0	n_1	n_2	n_3	n_4

Figure 10.3 Paraxial ray trace givens.

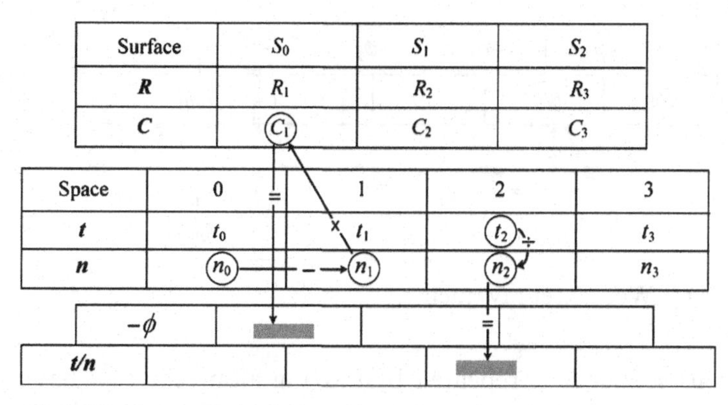

Figure 10.4 Worksheet calculations for inferred values.

The example expresses the design of a doublet with four optical surfaces and five optical spaces.

The two inferred values of the ray trace worksheet (negative optical powers and reduced thicknesses) are computed for each surface and space, respectively. Recall that the optical power (ϕ) is equal to $C(n' - n)$, so in the ray trace table we can determine the negative (−) optical power of a surface by subtracting the successive refractive index value from the current refractive index value and multiplying that difference by the curvature (C) of the intervening surface, which results in a negative optical power on the worksheet. This is shown schematically in Figure 10.4 by a triangle of numbers visually depicting the calculation $C_1(n_0 - n_1)$. The result of that calculation is the negative optical power ($-\phi$) for that surface. This value is entered in the row labeled $-\phi$ in the worksheet. Similarly, the reduced thickness (t/n) can be directly calculated and entered in the worksheet for all the spaces of the optical system.

A similar triangle multiplication takes place to calculate the next entry in the ray trace table using Equations (10.1) and (10.2) to propagate the ray through the optical system. A value is multiplied by the number directly above it on a surface or in a space column. Then, the resulting value is added to the previous number in the row indicated and entered into the ray trace worksheet as illustrated in Figure 10.5.

The value we have just traced for ray height y_1 on surface 1 is found from Equation (10.2) and illustrated by the triangle mathematics in Figure 10.5. The arrows indicate multiplication in the space column of $(n_0u_0) \times (t_0/n_0)$ and addition of y_0 to produce the value for the next entry in the y row, y_1. Similarly, the ray reduced angle n_3u_3 is found by multiplying on surface 3 $y_3 \times (-\phi_3)$ and adding n_2u_2 to give the ray reduced angle value of n_3u_3 in that row for optical

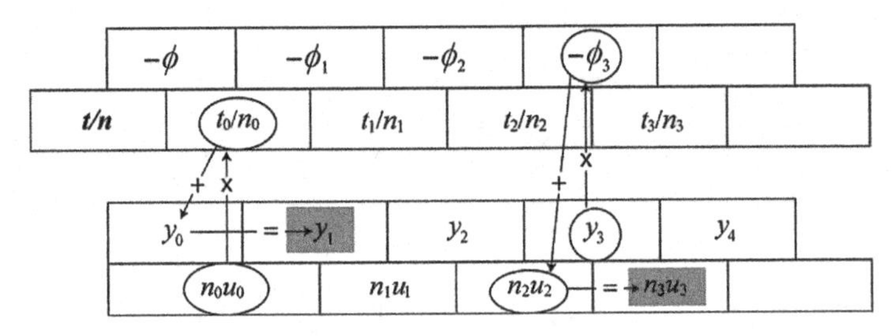

Figure 10.5 Worksheet ray tracing.

space 3. This sequence is repeated by moving along the row having the ray heights and the row containing the angles to fill in all the values. The ray trace worksheet can also be used in reverse order, so that values are filled in for regions to the left using Equations (10.1) and (10.2); however this approach is not as straightforward.

Example 10.1 illustrates a ray trace problem using, first, Equations (10.1) and (10.2), and second, the ray trace worksheet.

Example 10.1

Trace a paraxial ray through a lens and find the back focal distance (*BFD*.) This can only be done when an object is at infinity, so $u = 0$. (All dimensions in millimeters.)

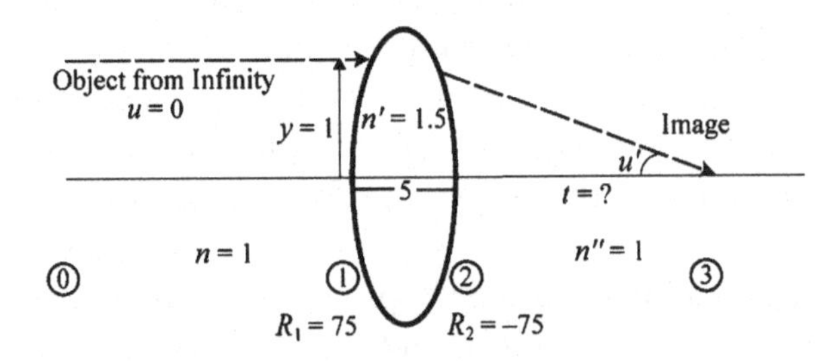

Using the refraction and transfer equations, the object is at the first surface (0): *refraction occurs at surface 1*:

$$n'u' = nu - \phi y = 0 - ((1.5 - 1)/75) = -0.006\,666$$

$$y_2 = 1 + (5/1.5)(-0.006\,666) = 0.977\,777$$

refraction at surface 2:

$$n'u' = -0.006\,666 - 0.006\,666(0.977\,777) = -0.013\,185$$

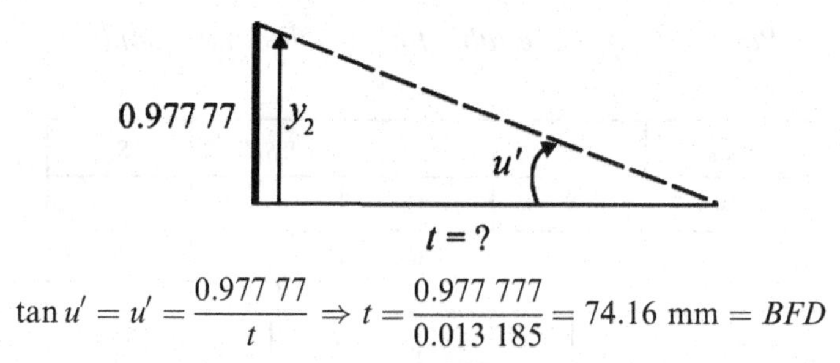

$$\tan u' = u' = \frac{0.977\,77}{t} \Rightarrow t = \frac{0.977\,777}{0.013\,185} = 74.16 \text{ mm} = BFD$$

Paraxial ray trace table for Example 10.1

Surface	S_0	S_1	S_2	S_3
R	∞	75	−75	∞
C	0	0.0133	−0.0133	0

Space	0	1	2
t	∞	5	?(*BFD*)
n	1	1.5	1

Given

$-\phi$	0	−0.00 667	−0.00 667	0
t/n	∞	3.3333	?/1	

Inferred

y	1	1	0.977 778	0
nu	0	−0.006 667	−0.013 185	

Ray Trace

At image $y = 0$:

$$y_{+1} = y + n'u'(t'/n') \Rightarrow 0 = 0.977777 + (-0.013185)t \Rightarrow t = 74.16$$

A second example illustrates the use of the method to ray trace an achromatic doublet.

Example 10.2

Ray trace an achromatic doublet for an object at infinity. Find (a) *BFD*, (b) the effective focal length, (c) *P**, and (d) *F*/#.

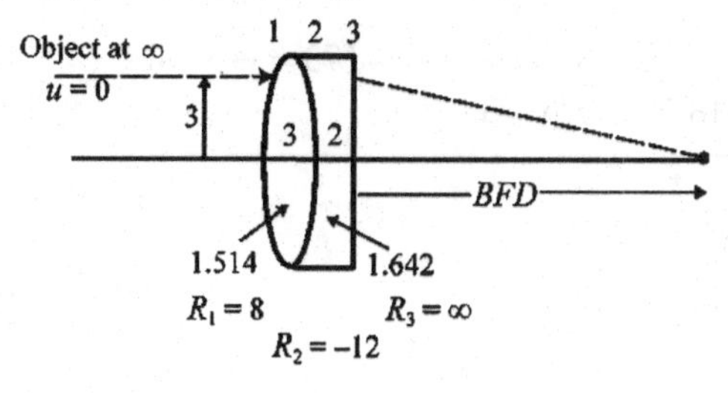

Paraxial ray trace table for an achromatic doublet

Given

Surface	S_0	S_1	S_2	S_3	S_4
R	-	8	−12	∞	-
C	-	0.125	−0.0833	0	-

Space	0	1	2	3
t	∞	3	2	? (*BFD*)
n	1	1.514	1.642	1

Inferred

Surface	S_0	S_1	S_2	S_3
$-\phi$	-	−0.064 29	0.010 648 3	0

Space	0	1	2	3
t/n	∞	1.981 087	1.217 952 6	?/1

Ray Trace

Surface	S_0	S_1	S_2	S_3	S_4
y	3	3	2.617 903	2.416 953	0

Space	0	1	2	3
nu	0	−0.192 87	−0.164 994	−0.164 994

To find the image location, which is where the ray crosses the optical axis, recall y at the image surface 4 is zero: $y_4 = 0$ by definition for the image, so we can find t/n, or *BFD* (vertex 3 to image):

$$y_4 = y_3 + \frac{t}{n}(nu)$$

$$0 = 2.416\,953 + (t/n)(-0.164\,994)$$

$$\therefore t/n = 14.64873 = ?/1.$$

Solution: Now we can answer parts (a)–(d) of the example:

(a) $t/n = t/1 = 14.648\,73$.

(b)

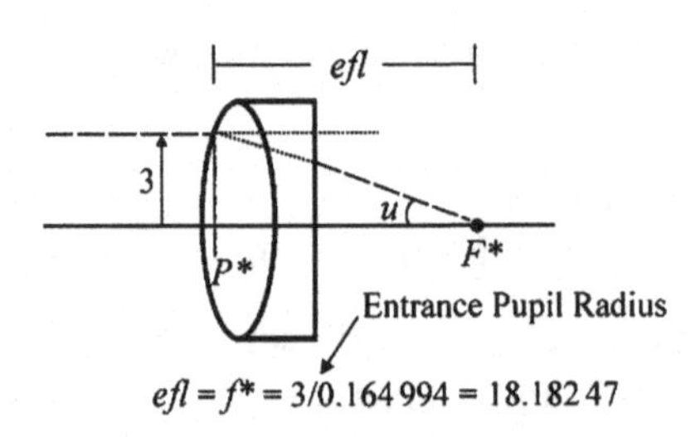

(c) P^* to surface 3: $-18.182\,47 + 14.648\,73 = -3.54$.

(d) $F/\# = efl/D = 18.182\,47/6 = 3.03$.

10.2 Chief and marginal rays

Rays that will be discussed in this section are confined to the meridional plane (y–z plane) and are called meridional rays. There are rays in the orthogonal plane called sagittal, which will be discussed in Chapter 11. So far we have been using the parallel ray trace method to locate the image and determine its size, but we have not considered the field of view and image brightness, as discussed in Chapter 9. Since any two rays in a linear system may be used to locate the image, we can choose rays associated with the aperture and field stops. The method utilized is called the throughput method. Figure 10.6 depicts the rays used in the two methods for ray tracing through a thin lens.

The throughput method uses two special rays: the chief ray and the marginal ray. As the name suggests, this method provides a powerful tool to obtain information that can be used to calculate radiant flux as well as to determine the image's location and size.

The marginal ray begins at the axial object point and proceeds to the edge of the entrance pupil. It defines pupil size and image location. The chief ray begins at the edge of the field of view and proceeds to the center of the entrance pupil. The chief ray defines pupil locations and image size. Both rays are meridional and lie in the tangential plane of an optical system. The chief ray's heights and angles are barred (bars over the top of the symbols) while the marginal ray heights and angles are unbarred. Their relationship is shown in Figure 10.7.

Wherever a marginal ray crosses the axis, an image is located and the radius of the image is determined by the chief ray height in that plane. Wherever a chief ray crosses the axis, a pupil is located and the radius of the pupil is determined by the marginal ray height in that plane. Tracing these rays is a very powerful method of finding the image and radiation throughput.

The marginal ray is shown in Figure 10.8 among all the rays emanating from the axial point. Note the marginal ray in object space continues in a straight line into virtual object space.

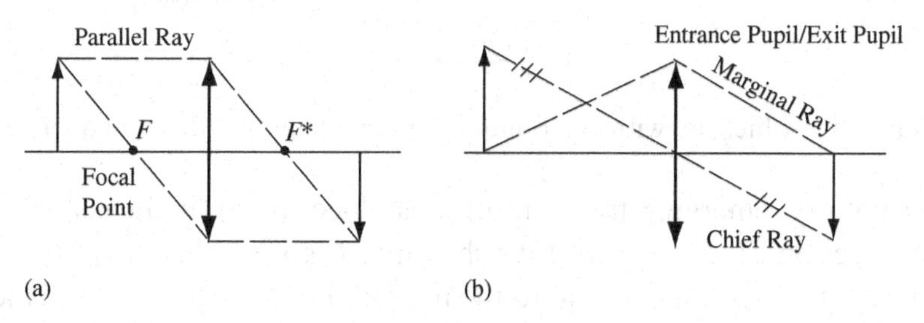

Figure 10.6 Rays used in (a) parallel ray and (b) throughput ray tracing.

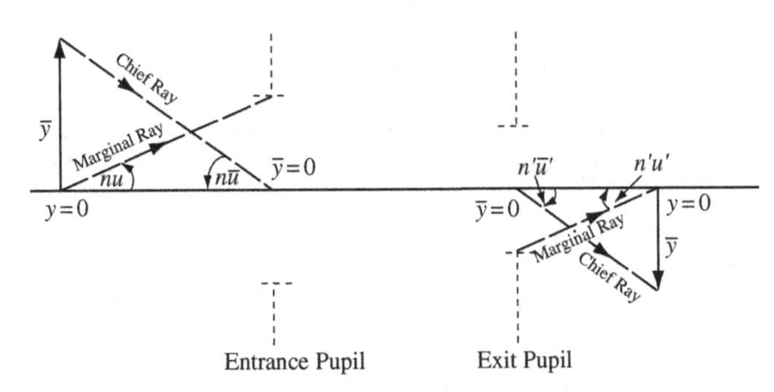

Figure 10.7 Marginal and chief ray definitions relating to entrance and exit pupils, objects, and images.

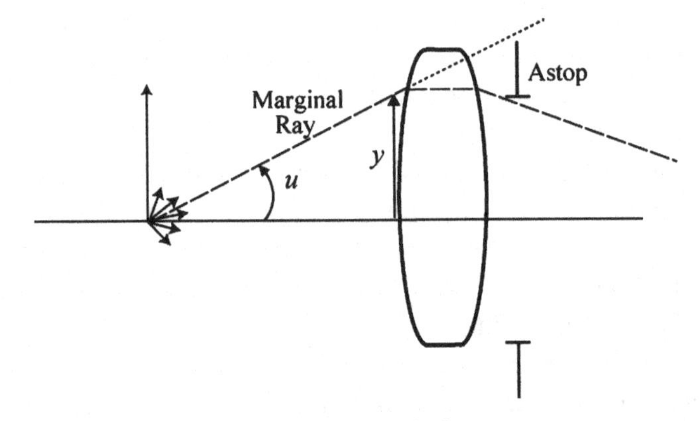

Figure 10.8 Marginal ray in the meridional plane.

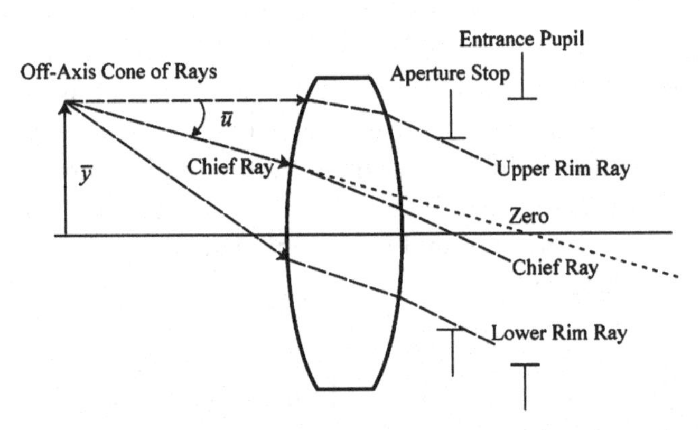

Figure 10.9 Chief ray with surrounding rays propagating through a lens.

The chief ray, emerging from an off-axis object point, is the central ray in the cross section of the cone of rays shown in Figure 10.9. The radius of the cone at any cross section is equal to the marginal ray height in that plane. The chief ray passes through the center of the aperture stop. The entrance pupil is located by its extension into virtual object space as shown in Figure 10.9 (dotted). The limiting ray from an off-axis point determines the boundary of the cone of rays.

The corresponding refraction and transfer equations for the marginal and chief rays are:

$$\left.\begin{aligned} n'u' &= nu - y\phi \\ y_{+1} &= y + \frac{t}{n}(n'u') \end{aligned}\right\}\text{marginal}$$

and

$$\left.\begin{aligned} n'\bar{u}' &= n\bar{u} - \bar{y}\phi \\ \bar{y}_{+1} &= \bar{y} + \frac{t}{n}(n'\bar{u}') \end{aligned}\right\}\text{chief.}$$

10.3 Optical invariants

The optical invariant is a constant value for rays traced through an optical system, and it takes on many names and values. For example, the Helmholtz invariant takes into account the chief and marginal ray values at image planes. The Smith–Helmholtz invariant refers to the entrance pupil location where the chief ray height is zero, and is calculated as the product of the field of view, or semi-field of view, times the entrance pupil radius. The Lagrange invariant is used for any plane or surface employing the chief and marginal ray values, while the more general term of optical invariant is used for any two rays that can pass through an optical system.

The Lagrange invariant will be used here and throughout the remainder of this book. It is a subset of the optical invariant, and reduces to the Helmholtz invariant at an image plane, and to the Smith–Helmholtz invariant at the system's entrance pupil. Using the standard notation for the chief and marginal rays, the Lagrange invariant is defined as

$$\text{Lagrange invariant} = n\bar{u}y - nu\bar{y}. \tag{10.3}$$

The Lagrange invariant takes on a simplified form $(-nu\bar{y})$ at object and image locations $(y = 0)$. At aperture stops and pupils $(\bar{y} = 0)$ the invariant is simplified to $n\bar{u}y$. The proof that this quantity is a constant is pursued below using the refraction and transfer equations for the chief and marginal rays shown in Figures 10.10 and 10.11, respectively.

The refraction equations for the chief and marginal rays at any optical surface are shown below and illustrated in Figures 10.10:

$$n'\bar{u}' = n\bar{u} - \bar{y}\phi; \tag{10.4}$$

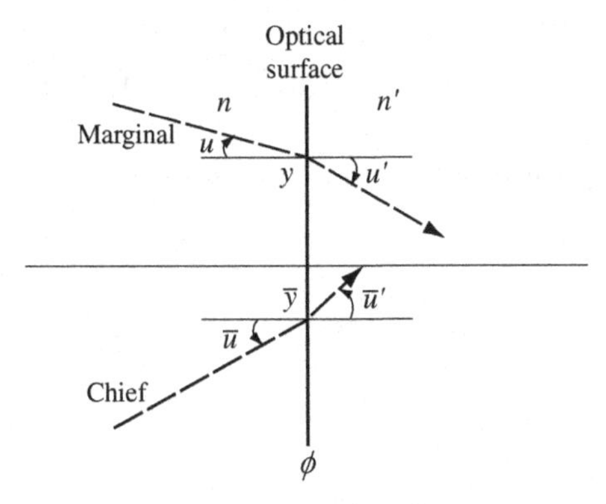

Figure 10.10 Chief and marginal ray refractions at a surface between two spaces.

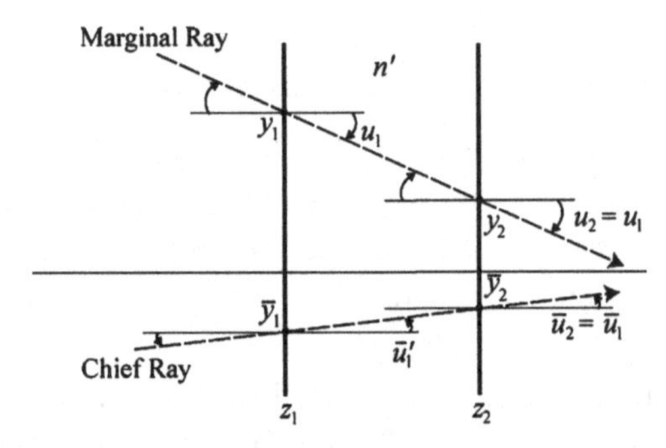

Figure 10.11 Ray values for generalized z locations within an optical space.

$$n'u' = nu - y\phi. \tag{10.5}$$

These equations hold true for any surface in an optical system constituting an interface between one space and an adjacent space, characterized by the refractive indices n and n'. Since the surface optical power is fixed for either the chief ray or the marginal ray passing through the surface, we can solve Equations (10.4) and (10.5) for surface optical power, ϕ:

$$\phi = \frac{n'\bar{u}' - n\bar{u}}{-\bar{y}}; \tag{10.6}$$

$$\phi = \frac{n'u' - nu}{-y}. \tag{10.7}$$

Therefore we can equate Equations (10.6) and (10.7):

$$\frac{nu - n'u'}{y} = \frac{n\bar{u} - n'\bar{u}'}{\bar{y}}. \tag{10.8}$$

Rearranging,

$$nu\bar{y} - n'u'\bar{y} = n\bar{u}y - n'\bar{u}'y. \tag{10.9}$$

Collecting all terms related to n space and all terms relating to n' space,

$$n\bar{u}y - nu\bar{y} = n'\bar{u}'y - n'u'\bar{y}, \tag{10.10}$$

a result which can also be expressed as:

$$nu\bar{y} - n\bar{u}y = n'u'\bar{y} - n'\bar{u}'y. \tag{10.11}$$

This quantity is invariant on refraction from one space to a second space across a boundary surface.

Below is the proof that this same invariant exists at any location within a space as shown in Figure 10.11. The transfer equations from axial location z_1 to z_2 in Figure 10.9 within n' space are

$$y_2 = y_1 + \frac{t}{n'}(n'u'), \tag{10.12}$$

$$\bar{y}_2 = \bar{y}_1 + \frac{t}{n'}(n'\bar{u}'). \tag{10.13}$$

Solving for the common quantity in both equations,

$$\frac{t}{n'} = \frac{y_2 - y_1}{n'u'}, \tag{10.14}$$

$$\frac{t}{n'} = \frac{\bar{y}_2 - \bar{y}_1}{n'\bar{u}'}. \tag{10.15}$$

Setting Equations (10.14) and (10.15) equal to each other and rearranging terms

$$n'\bar{u}'y_2 - n'\bar{u}'y_1 = n'u'\bar{y}_2 - n'u'\bar{y}_1. \tag{10.16}$$

Selecting quantities related to axial points z_1 and z_2 on opposite sides of Equation (10.16) and rearranging terms,

$$n'\bar{u}'y_2 - n'u'\bar{y}_2 = n'\bar{u}'y_1 - n'u'\bar{y}_1. \tag{10.17}$$

This quantity is invariant along the two rays in a given space, and it is the same quantity derived for refraction between two spaces, in Equation (10.11). Therefore, the Lagrange invariant is valid for any axial location along an optical system.

In some locations, such as optical surfaces, one must choose the side of the surface from which the chief and marginal rays will be picked, as these axial points have two sets of values associated with them, one for space n and another for space n'.

Changes in the Lagrange invariant affect axial (z axis) values only. Doubling the Lagrange invariant while retaining the entrance pupil value results in halving axial quantities and distances. The Cyrillic character, ж, ("Zhuh", the Zh pronounced as in "Zhivago") is used to denote the Lagrange invariant:

$$\text{ж} = n\bar{u}y - nu\bar{y}. \tag{10.18}$$

In the following example, calculation of the Lagrange invariant allows us to answer specific questions about the design of an optical system.

Example 10.3

Consider an optical system with a 500 mm diameter entrance pupil and 2° full field of view, used to detect a bright missile in the sky. What is the Lagrange invariant of this system? What is the smallest detector array one can use with this system? What is the largest possible nu in image space?

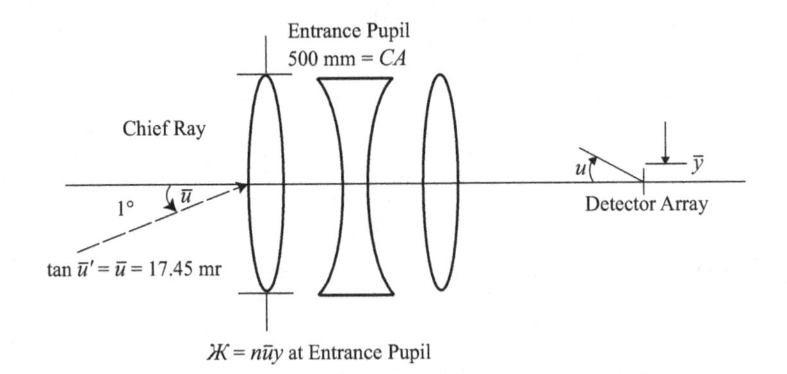

$$\text{ж} = 1(0.017\,45)(250\text{ mm}) = 4.364\text{ mm (constant)}.$$

Recall $nu < 1$ Why? Practically choose ($u = 90°$, $n = 1$):

$$nu \to n \sin u \leq 1$$

$$\therefore F/\# \geq \frac{1}{2n \sin u} \geq \frac{1}{2}$$

so $u < 90°$; choose $F/1$ optics, $u = 0.5$

$$4.364 \text{ mm} = -nu\bar{y} = -0.5 \rightarrow \bar{y} = -8.728 \text{ mm}.$$

If a smaller array is used, there is no possible marginal ray angle (u) (it cannot be greater than 90°) and we cannot build the optical system.

10.4 Marginal and chief ray trace table

In Section 10.1 we traced an arbitrary ray using a ray trace table. Now, two special rays will be traced simultaneously through the optical system using the ray trace table for the chief and marginal paraxial rays. The rays within a given optical space are straight lines, both in the physical space as well as in the virtual region of that space.

Numerical ray tracing is very important as a first-order evaluation of an optical system. There are four reasons to establish a paraxial design using this method:

(1) It provides a reference point for high-order real ray calculations.
(2) It provides an approximate optical system layout quickly.
(3) It establishes radiometric throughput; bigger $\varkappa$ values give brighter images.
(4) It checks the feasibility of a design for given system requirements.

The table used to systematically trace chief and marginal rays is shown in Figure 10.12.

This worksheet contains rows for two ray traces for each ray type. The first ray trace table, labeled y_{p} and nu_{p} for a "pseudo ray," is used to find the actual marginal ray. Recall that there is only one marginal ray (y_{m} and nu_{m}). The two ray traces with bars over the angles and heights represent rows for a pseudo chief ray ($\bar{y}_{\mathrm{e}}, n\bar{u}_{\mathrm{e}}$) and the actual chief ray, ($\bar{y}_{\mathrm{F}}, n\bar{u}_{\mathrm{F}}$). The Lagrange invariant can be found by multiplying the values of the marginal and chief rays at any surface. Figure 10.12 shows the selected values for obtaining the Lagrange invariant at surface 1, but any surface can be used to obtain this invariant.

Example 10.4

For the telephoto lens shown below with an aperture stop between the lens of CA = 1.2963 mm, calculate:

(a) *efl*;
(b) the optical power;
(c) *BFD*;
(d) $(F/\#)_{\infty}$;

(e) the exit pupil location and size;
(f) the entrance pupil location and size.

The marginal ray is coming from infinity at a height of 1 mm, and the chief ray has an angle of 3.712° in object space and intersects the first optical surface ① at −2.777 mm (all dimensions in millimeters).

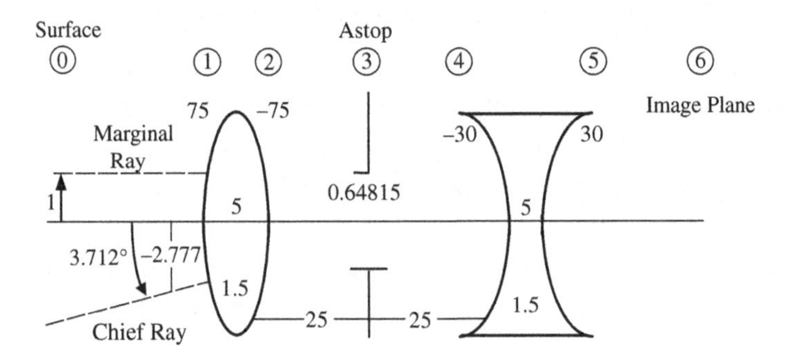

Use refraction and transfer equations to get the values in the worksheet:

Surface	S_0	S_1	S_2	S_3	S_4	S_5	S_6
R	-	75	−75	∞	−30	30	-
C	-	0.01333	−0.01333	0	−0.03333	0.03333	-

Space	0	1	2	3	4	5
t	∞	5	25	25	5	? (*BFD*)
n	1	1.5	1	1	1.5	1

$-\phi$	-	−0.00667	−0.00667	0	0.01667	0.01667
t/n	∞	3.333	25	25	3.333	?/1

y	1	1	0.97778	0.64815	0.318524	0.29225	0
nu	0	−0.0067	−0.013185	−0.013185	−0.00788	−0.00301	

$\bar{y}$		−2.777	−2.5	0	2.5	2.97245	
$n\bar{u}$	0.0648	0.083406	0.1	0.1	0.14175	0.19139	

The rays in image space provide information for *efl*, P^*; (*F*/#), the exit pupil, and *BFD*.

(a) $efl = \dfrac{y_1}{nu} = \dfrac{1}{0.00301} = 332.2\text{ mm} = 0.3322\text{ m};$

(b) $\phi = \dfrac{1}{efl} = \dfrac{1}{0.3322\,\text{m}} = 3.01\text{ diopters};$

(c) $BFD = \frac{y_5}{n' u'_5}$;

at image $y_6 = 0$, $y_6 = y_5 + \frac{t}{n} nu = 0.29225 + \frac{t}{n}(-0.00301) = 0$

$\frac{t}{n} = 97.1\ \text{mm} \equiv BFD$;

(d) $(F/\#)_\infty = \frac{1}{2n' u'_5} = \frac{1}{2(0.00301)} = 166.1$;

(e) The exit pupil location and size are found by projection of the chief and marginal ray, as shown below, from ray trace worksheet, as illustrated below:

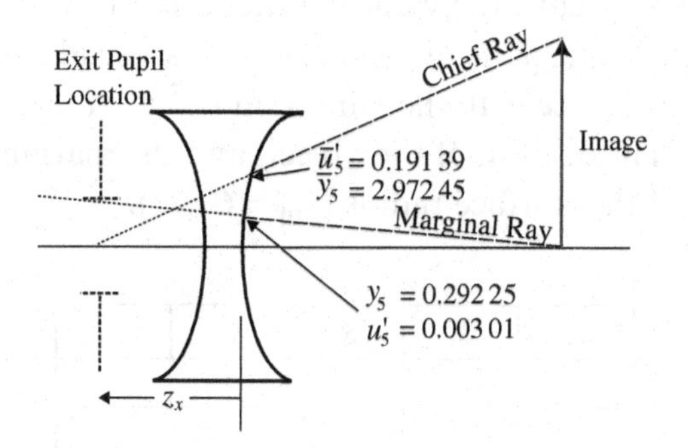

distance from last vertex:

$$Z_x = \frac{\bar{y}_5}{\bar{u}'_5} = \frac{-2.97245}{0.19139} = -15.5\ \text{mm}$$

diameter of exit pupil:

$$\frac{D_{ex}}{2} = y_5 + z_e u'_5 = 0.29225 + (-15.5)(-0.00301)$$

$$= 0.339\ \text{mm}$$

$$D_{ex} = 0.678\ \text{mm};$$

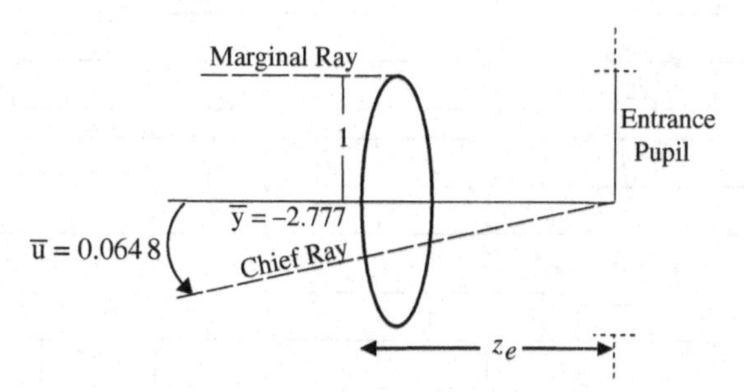

distance from surface:

$$\bar{y}_e = 0 = \bar{y}_1 + \frac{t}{n}(n\bar{u}_0) = -2.777 + z_e(0.0648)$$

$$z_e = 42.87\ \text{mm};$$

diameter of entrance pupil:
since marginal ray is parallel to axis

$$D_e = 2y_1 = 2(1) = 2 \text{ mm}.$$

10.4.1 Stops and pupils

To determine which surface acts as the aperture stop, trace a ray from the object's axial point to the image location at which this ray crosses the optical axis in image space. This is done in the rows labeled y_p and nu_p in the worksheet in Figure 10.12. Also record the values of the ratio of the clear aperture (CA) for each surface over the y height value for this pseudo marginal ray. The aperture stop is the surface with the minimum CA/y_p value. This surface is the aperture stop for that location of the object, and the marginal ray should just osculate the edge of this surface radius ($y_m = CA/2$).

Surface	S_0	S_1	S_2	S_3	S_4	S_5	S_6	S_7
R	-							
C	-							

Space	0	1	2	3	4	5	6
t							
n							

	S_0	S_1	S_2	S_3	S_4	S_5	S_6	S_7
$-\phi$	-							
t/n								
y_p								
nu_p								

$$\alpha_a = \frac{0.5\, CA_{p\,(Astop)}}{y_{p\,(Astop)}}$$

	S_0	S_1	S_2	S_3	S_4	S_5	S_6	S_7
y_m								
nu_m								
$\bar{y}_e$								
$n\bar{u}_e$								

$$\alpha_f = \frac{\bar{u}_{desired}}{\bar{u}_{existing}}$$

$= Ж$

	S_0	S_1	S_2	S_3	S_4	S_5	S_6	S_7
$\bar{y}_F$								
$n\bar{u}_F$								

	S_1	S_2	S_3	S_4	S_5	S_6	S_7
CA							
CA/y_p							
$CA/(\lvert y_m\rvert + \lvert \bar{y}_F\rvert)$							

Figure 10.12 Ray trace worksheet for chief and marginal rays.

The real marginal ray is then found by calculating the ratio $(CA/2)/y_p$ at that surface to get the aperture scale factor, α_a:

$$\alpha_a = \left(\frac{CA/2}{y}\right)_{min}. \tag{10.19}$$

This factor (α_a) is used to multiply the angles and heights in the worksheet for the pseudo ray to get the marginal ray trace values.

The chief ray height is known to be zero ($\bar{y} = 0$) at the aperture stop surface. If the parameters for the chief ray are given for object space, they can be transferred to the ray trace worksheet. More often, however, the required field of view is provided, and the following method can be employed to calculate the chief ray values.

First, choose an angle at the aperture stop surface (e.g. $n\bar{u} = 0.1$) for $\bar{y}_c = 0$, and then propagate this pseudo chief ray through the ray trace worksheet. Once the angle in object space is found for this pseudo chief ray, divide the desired field of view by the field of view of the pseudo chief ray. This is the field scale factor needed to obtain the real chief ray:

$$\alpha_f = \frac{FOV_{desired}}{FOV_{from\ pseudo\ ray}}. \tag{10.20}$$

Once the field scale factor is known, the real chief ray values can be obtained by multiplying by α_f. All the values of the pseudo chief ray ($\bar{y}_e$, $n\bar{u}_e$) in the worksheet of Figure 10.12 are multiplied by α_f to get the values of the real chief ray ($\bar{y}_F, n\bar{u}_F$).

At this point, the optical system's marginal and chief rays are known, so the system can be checked for vignetting.

10.4.2 Image location and size

The primary objective in a design is to find the location and size of the image relative to the system's last optical surface (fiducial surface.) We defined the image location by the intersection of the marginal ray and the optical axis, which locates the image base. We must then locate the image and calculate its size. The slope of the marginal ray in object space is the entrance pupil radius divided by the distance to the object. Therefore, for a finite distance, the marginal ray angle is typically positive. However, recall that we have a rotationally symmetric system. This means that if the object is at infinity, the marginal ray crosses the optical axis at the object. Therefore, the marginal ray trace angle for an object at infinity (a plane wavefront) is zero ($nu = 0$).

For an infinite object location, one can ray trace via the worksheet and eventually calculate the ray height and slope in image space at the last optical surface. These values, as shown in Figure 10.13, can be used to locate the image along the optical axis.

From Figure 10.13, the axial distance from the last surface to the image can be calculated as

$$\tan u_{\mathrm{m}} = \frac{y_{\mathrm{m}}}{t} \cong u_{\mathrm{m}}, \tag{10.21}$$

or as an angle given in radians for the paraxial approximation

$$t = y_{\mathrm{m}}/u_{\mathrm{m}}. \tag{10.22}$$

The size of the image is determined by the value of the chief ray at the image plane. The chief ray in object space was the semi-field of view (half the full field of view, FFOV). As shown in Figure 10.14, tracing this ray to image space gives the height y_{F} and the slope u_{F} at the last optical surface.

Projecting the chief ray to the image plane determines the size of the image:

$$\bar{y}_{\mathrm{I}} = \bar{y}_{\mathrm{F}} + \bar{u}_{\mathrm{F}} t, \tag{10.23}$$

where the quantity, $\bar{y}_{\mathrm{I}}$, is equal to half the image size.

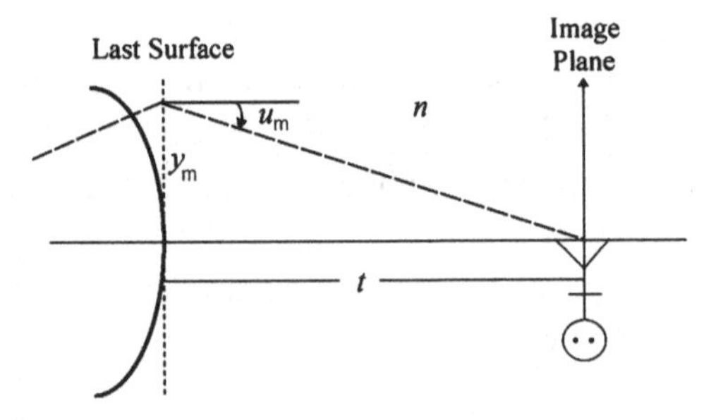

Figure 10.13 Location of an image for an object at infinity.

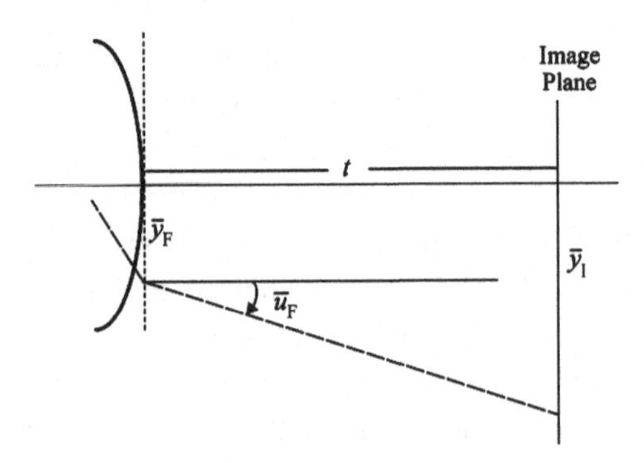

Figure 10.14 Chief ray location at the image plane.

10.4.3 Vignetting analysis

For a perfect optical system, no surface or aperture other than the stop limits the beam (frustums of rays in three dimensions) for a given field of view. For the system shown in Figure 10.15, the off-axis bundle (cone) of rays is rotationally symmetric, with the chief ray at the center. This cone of rays is not cut off by any of the optical elements, and it propagates to the edge of the field stop as shown. For the situation shown, the upper rim ray just barely passes through the first element, but passes through all others for this field of view (FOV). However, if a larger field stop was used to increase the field of view, the larger points on the object, which would reach the image plane and field stop at this higher chief ray angle, would have vignetting. In the case shown in Figure 10.15, the first lens would prevent some of the rays of the bundle from entering the optical system, i.e. the upper rim ray would not make it to the image, thus energy or image brightness is lost. Any element that results in a loss of rays or the blocking of rays causes vignetting, which is the reduction of image intensity at the periphery or outer edge of the image. For some fields of view, a surface cuts off a portion of the circular beam so light does not uniformly fill the exit pupil. Therefore, in a design, we must specify the degree of vignetting: unvignetted, 50% vignetted, etc.

To illustrate this effect, a vignetting diagram is used to show the bundle of rays in the x–y plane centered around the chief ray at a surface that vignettes or stops some of the ray propagation. A shear of two circles indicates the effect of vignetting, with their center displaced by the chief ray height at that surface, as shown in Figure 10.16. The shaded area represents the fraction of light collected and transferred to an image point for an off-axis field point. The exit pupil illuminated area is the overlap of two unequal circles with a common chord.

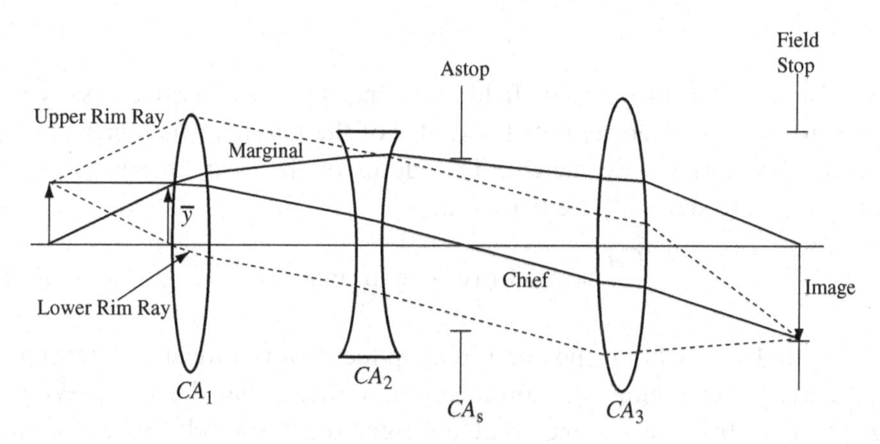

Figure 10.15 Perfect unvignetted system.

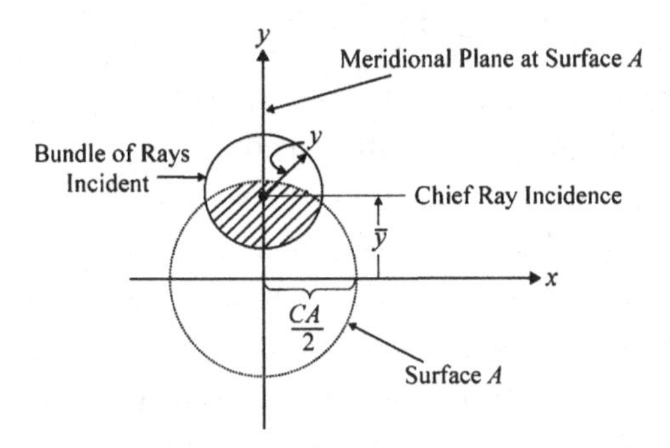

Figure 10.16 Vignetting diagram.

In the meridional plane, the height of the upper rim ray (urr) is:

$$\bar{y} + y$$

and that of the lower rim ray (lrr) is:

$$\bar{y} - y.$$

To prevent vignetting by any element or structure within the optical system, these rays must lie within the clear aperture of a surface. The vignetting diagram determines where rays pass through off-axis points within the optical surfaces. The bundles of rays at a given field of view have a maximum height equal to that of the chief ray and marginal ray heights at an element. For a symmetrical system, the sum of the chief and marginal ray heights must be referred to in absolute terms; half the clear aperture ($CA/2$) of any element must be greater than or equal to the absolute values of the sum of chief and marginal ray heights:

$$\frac{CA}{2} \geq |\bar{y}| + |y|. \tag{10.24}$$

The bundle of rays for some field angle incident on a surface is shown in Figure 10.17. The chief ray is at the center of the bundle, which has a radius equal to the marginal ray height. The radius of the clear aperture must be greater than the sum of those two radii:

$$\frac{CA}{2} > |\text{chief ray}| + |\text{marginal ray}|. \tag{10.25}$$

It should be noted that the aperture stop location is critical in determining vignetting. On changing the stop location, surfaces that did not previously vignette may become surfaces that do vignette. It should also be pointed out that sometimes vignetting surfaces are used to increase the resolution

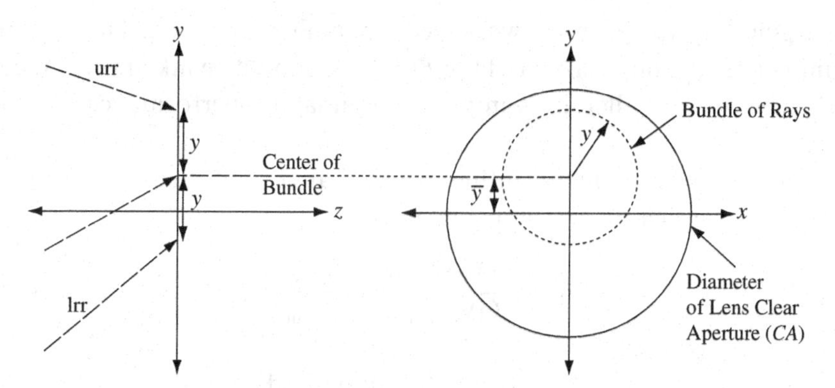

Figure 10.17 Radius of the surface element must be greater than the sum of chief and marginal ray heights.

performance of the system, so vignetting within a system is not always a negative characteristic.

To determine which surfaces vignette using our ray trace worksheet, Figure 10.12, simply add the magnitudes of y_{m} and $\bar{y}_{\mathrm{F}}$ and divide the result into the CA value in the worksheet on the last row. If the value is less than 2, the surface vignettes. If two or more surfaces within a system vignette, the maximum vignetting will occur with the smallest $CA/(|y_{\mathrm{m}}| + |\bar{y}_{\mathrm{F}}|)$ value.

10.5 Scaling of chief and marginal rays

The paraxial ray trace equations are linear, and therefore the marginal and chief rays can be scaled to any desired value by a constant factor. For example, if a fixed FOV or a specific $F/\#$ is required for an application, the existing ray trace design can be scaled by applying an appropriate scale factor. This scale factor provides a new ray through the optical system by simple multiplication of the factor with the existing chief or marginal ray values. Ray scaling factors differ, depending upon the ray chosen. The aperture scaling factor, α_{a}, and the field angle scale factor, α_{f}, are valid within an existing optical design with the marginal and chief rays already traced. More specifically, if we want a $\pm 20°$ FOV, but the system was designed for a $\pm 10°$ FOV, we can scale the chief ray by a factor of 2 and then evaluate this new design's vignetting characteristics.

The aperture scale factor, α_{a}, is used when the entrance pupil is not large enough to collect a sufficient amount of light from a given object, and should be made larger, resulting in a larger Lagrange invariant. It is defined as

$$\alpha_{\mathrm{a}} = \frac{\text{aperture radius desired}}{y \text{ height at existing aperture}} = \frac{0.5CA_{\mathrm{p\ (Astop)}}}{y_{\mathrm{p\ (at\ Astop)}}}. \tag{10.26}$$

All marginal ray values in the worksheet are multiplied by α_a. This computation influences the pupil sizes and the $F/\#$. Care should be taken when increasing $F/\#$, because smaller $F/\#$s improve the imaging performance of optical systems.

The field angle scale factor influences field angle and image size. It can be determined by either of the following means:

$$\alpha_f = \frac{\mathrm{FOV}_{\mathrm{desired}}}{\mathrm{FOV}_{\mathrm{existing}}} = \frac{\bar{u}_{\mathrm{desired}}}{\bar{u}_{\mathrm{existing}}}; \tag{10.27}$$

$$\alpha_f = \frac{\text{Image height desired}}{\bar{y} \text{ at image}}. \tag{10.28}$$

The existing chief ray value is then multiplied by this scale factor (α_f). Remember, the system must be rechecked for vignetting. The following example provides an illustration of the use of scale factors.

Example 10.5

An existing system has the ray trace worksheet as shown below.

Surface	S_0	S_1	S_2	S_3	S_4	S_5	S_6	S_7
R	-							
C	-							
$-\phi$								
y		0.08031	0.2046	0.7098	0.8056	0.8739	1.1216	
$\bar{y}$		−0.18382	−0.1415	0	0.02	0.03407	0.0878	
CA		0.75	0.685	1.0	1.75	2.0	2.75	
CA/y_m		9.3	3.3	1.4	2.17	2.2	2.45	

Space	0	1	2	3	4	5	6
t							
n							
t/n							
nu	−0.3515	−0.3361	−0.3339	−0.47885	−0.482	−0.4566	−0.2944
$n\bar{u}$	0.0793	0.1145	0.1	0.1	0.1002	0.099 05	0.0862

It is desirable to have an $F/2.4$ system and a 20° FFOV. What are the scale factors to get these new rays?

$$F/\#_{\text{existing}} = \frac{1}{2(0.2944)} = 1.698,$$

$$\alpha_{\text{a}} = \frac{1/2.4}{1/1.698} = 0.708, \quad \alpha_{\text{f}} = \frac{10}{0.0793(180/\pi)} = 2.2.$$

The new marginal and chief rays are

y		0.05686	0.14485	0.5	0.5703	0.6187	0.7941	
nu	0.2488	0.2380	0.2364	0.3390	0.3413	0.3233	0.2084	

$\bar{y}$		−0.404	−0.311	0	0.044	0.0749	0.193	
$n\bar{u}$	0.174	0.252	0.22	0.22	0.2204	0.217	0.1896	

It is necessary to check for vignetting. Which surfaces vignette?

$$\frac{CA}{2} = |\bar{y}| + |y|.$$

The surfaces that vignette are 1 and 2.

10.6 Whole system scaling

One of the more important tasks that an optical designer needs to do is scale an existing design to a new system for a different application. There are many designs that have been optimized for a particular spectrum and focal length that can be scaled to the new requirements of the researcher. Therefore, if a design exists that has been used successfully for a particular effective focal length, why reinvent the wheel? By using a whole system scaling factor (α_{w}), a new optical system can be developed very easily, and is a good starting point for a new design. The whole system scale factor is the same in all directions, and the result of uniform scaling is proportional (in the geometric sense) to the original. For example, if you have a required focal length, and there is an existing optical design, you can calculate the whole system scaling factor by

$$\alpha_{\text{w}} = \frac{\textit{efl}\text{ you want}}{\textit{efl}\text{ you have}}. \tag{10.29}$$

Once the scale factor is determined, it is used as a multiplier for all the dimensions in the optical components, i.e. radius, thicknesses, clear apertures.

Example 10.6

We have a design for the lens shown below. Find a new lens design that has a focal length of 200, and sketch the design.

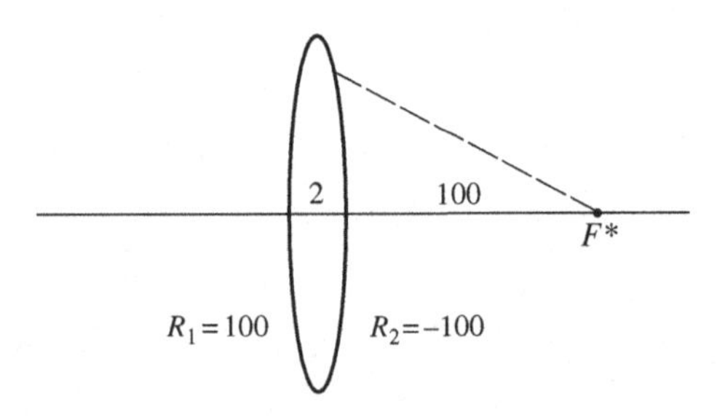

To design a lens with a focal length of 200, calculate the system scaling factor, and multiply all existing lens dimensions by that factor:

$$\alpha_W = \frac{\textit{efl} \text{ you want}}{\textit{efl} \text{ you have}} = \frac{200}{100} = 2.$$

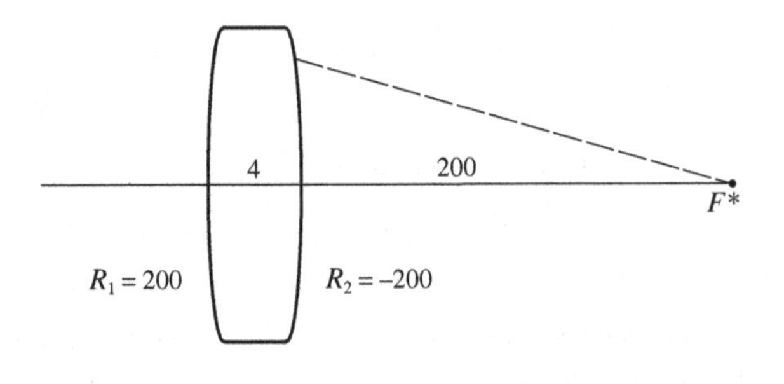

Let us do another example problem, that of designing an equi-convex singlet lens ($F/10$) for an *efl* of 100 mm in air with an index of 1.5 (glass). Start with the thin lens equation ($t = 0$):

$$\phi = (n-1)\left(\frac{1}{R_1} - \frac{1}{R_2}\right) \rightarrow \frac{1}{100} = (1.5-1)\left(\frac{1}{R} + \frac{1}{R}\right).$$

Solve for the radius:

$$R = 100 \text{ mm}.$$

Since a lens with a negative edge thickness, as shown in Figure 10.18, is not feasible, we must next find a value for the lens thickness.

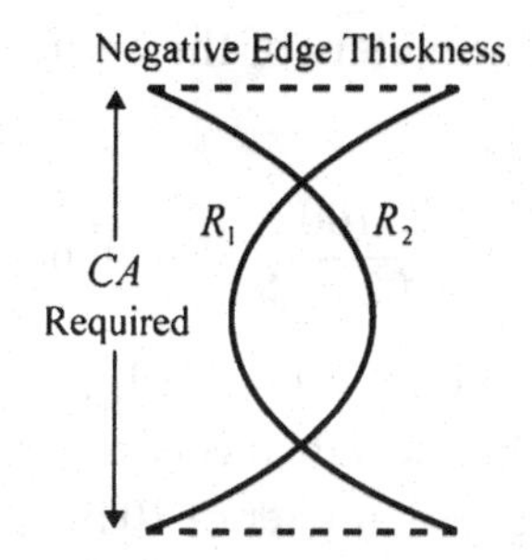

Figure 10.18 Equi-convex lens with negative edge thickness.

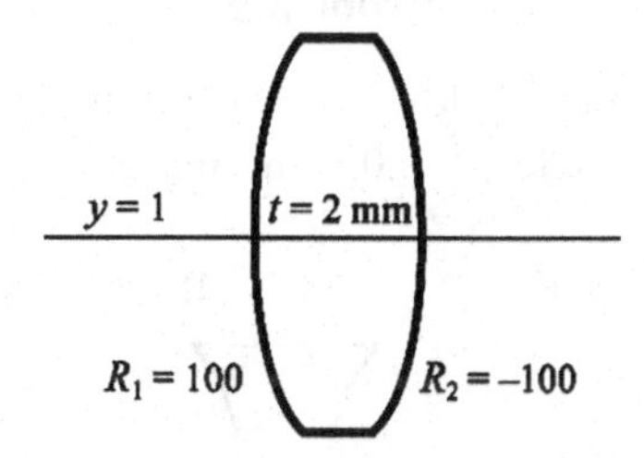

Figure 10.19 Layout for the example equi-convex lens.

Surface	S_0		S_1		S_2		S_3
R			100		−100		
C			0.01		−0.01		
t		∞		2			
n		1		1.5		1	
$-\phi$	0		−0.005		−0.005		0
t/n		∞		1.3333		99.66	
y	1		1		0.993333		0
nu		0		−0.005		−0.009 966 66	

Figure 10.20 Ray trace for the example lens.

By a trial and error approach, we choose an appropriate thickness of 1/5 of the diameter of the lens, or about 2 mm ($t = 2$ mm). Calculating the sag for each surface (Equation (5.11)) shows that the lens has a positive edge thickness. The final layout of the lens is shown in Figure 10.19.

A paraxial ray trace can now be done on the lens for an object at infinity, as shown in Figure 10.20. This gives an *efl* of 100.3344, not the required 100 specified initially.

So we need to scale the parameters of the entire system by a system scaling factor of

$$\alpha_W = \frac{100}{100.3344} = 0.99667.$$

This provides our final first-order design, with $R_1 = 99.667$, $R_2 = -99.667$ and a thickness of 1.99 mm. This design meets the required specifications of a 100 mm focal length lens with a diameter of 10 mm from the $F/10$ requirement.

Problems

10.1 For the Cooke triplet shown below, with an aperture stop diameter of 2 cm at the first lens, the chief ray angle is −0.05, and marginal ray angle is 0.

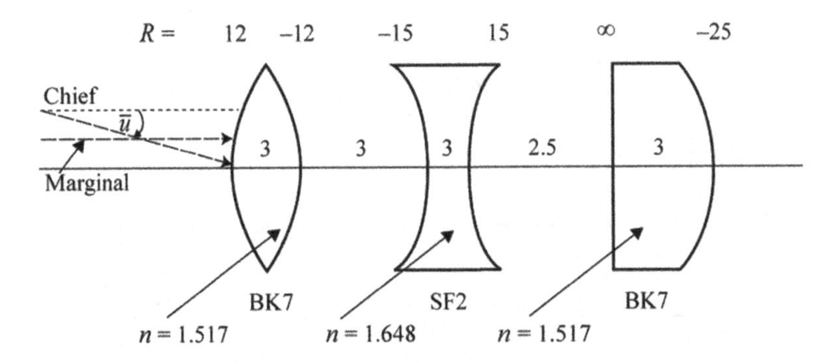

For the layout, ray trace the marginal ray to find:

(a) *BFD*;
(b) *efl*;
(c) P^* location;
(d) $(F/\#)_\infty$.

10.2 Paraxial ray trace the marginal and chief rays for the lens system below:

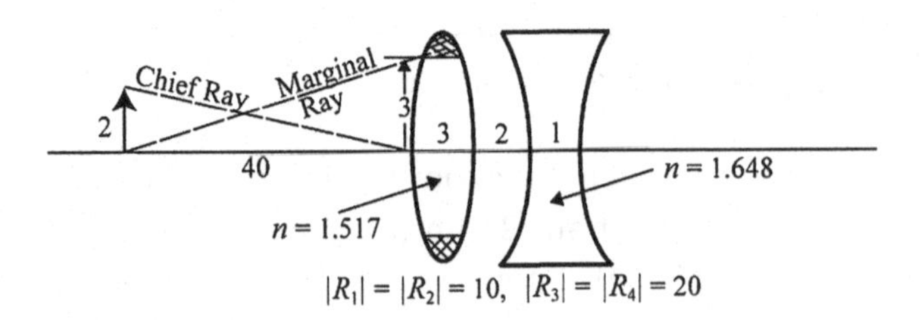

(a) What is the Lagrange invariant?
(b) What is the back focal distance (*BFD*)?

Table 10.1.

Surface	Radius	*t* to next surface	Material	*CA*
Obj (0)		Infinity	Air	
1	24.84	6	SSKN5(1.658)	30
2	–671.1491	4	Air	30
3	–123.0032	2	SF6(1.805)	5
4	58.17	27.5	Air	5
5	–14.29	1.5	FK5(1.487)	10
6	–39.56	1.0	Air	10
7	500	5	BaFN10(1.670)	30
8	–41.35341	???	Air	30
Image				

10.3 What is the Lagrange invariant for a 50 mm diameter, $F/2.5$ thin lens used to image a person 6 ft tall at a distance of 60 ft? What is the size of the field stop (*CA*)?

10.4 Cheesehead Optical Design Co. is in a real bind and has hired you as a consultant. Apparently they have lost all the information pertaining to the telephoto lens system shown below with the exception of the data provided in Table 10.1.

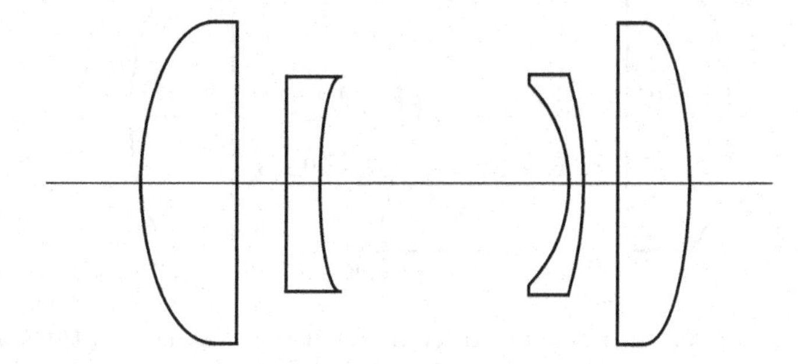

Your job is to do a paraxial ray trace (marginal and chief ray) for an object at infinity. The FOV is $\pm 5^\circ$. The marginal ray height is 13 mm at surface 1.

(a) What is the effective focal length (*efl*) of the system?
(b) What is the back focal distance (*BFD*) of the system? Where is the second principal plane for the system, P^*?
(c) Which surface is the aperture stop for the system?
(d) Knowing that the intended field of view is 10°, what is the size of the image at the image plane?
(e) Where is the entrance pupil (relative to the first surface) and what is its size?
(f) Where is the exit pupil (relative to the last surface) and what is its size?
(g) What is the $F/\#$ for the system?

10.5 Set up a paraxial ray trace for the following lens for a ray of height 1 ($y = 1$) coming from infinity.

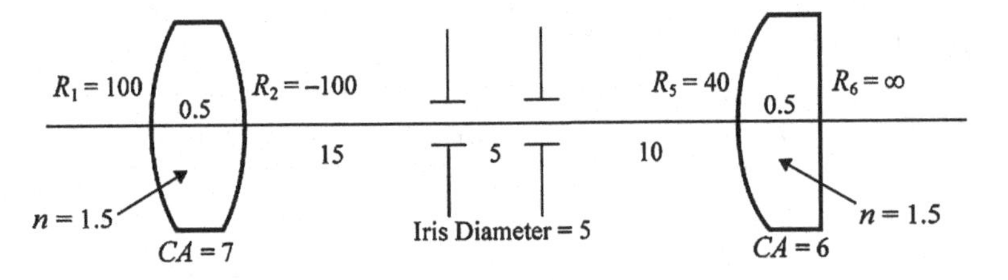

(a) Find the *efl*.
(b) Find the back focal distance (*BFD*).
(c) Find the aperture stop.

10.6 For the triplet below (all dimensions in centimeters) fill in the ray trace table and find the following:
(a) image location (*BFD*)
(b) *efl*
(c) image height
(d) entrance pupil relative to first surface of lens 1
The marginal ray is from infinity and has a height of 1 cm. The chief ray is zero in object space at the first surface and is at an angle of +0.02.

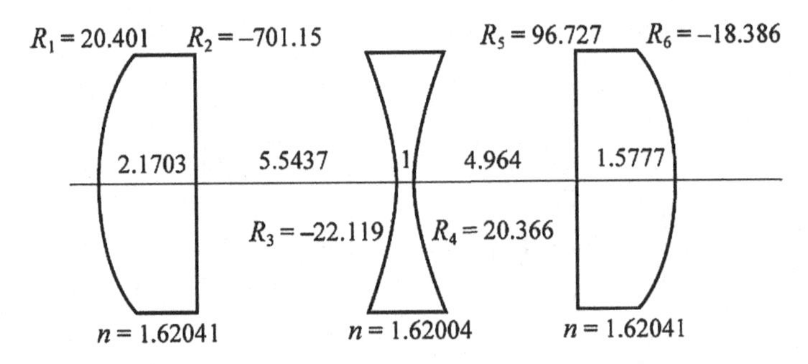

10.7 Using a ray from infinity, ray trace this marginal ray ($y = 1$) through the lens system below. Determine:

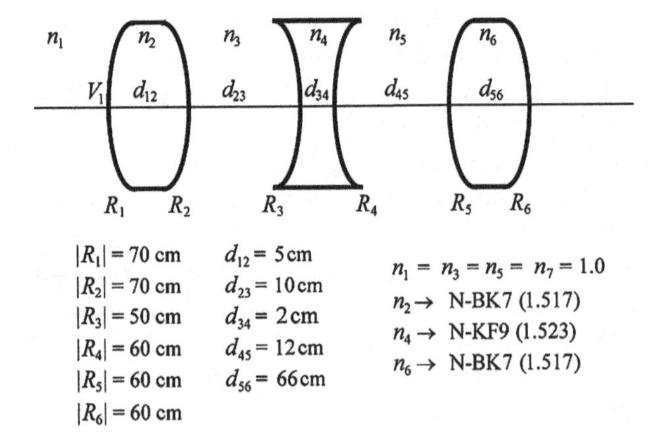

(a) the back focal distance (*BFD*);
(b) the effective focal length (*efl*);
(c) the location of the principal planes (P and P^*) relative to V_1.

10.8 For the system shown below fill in the ray trace table for the marginal ray from infinity at height 3 and the chief ray angle of 0.1 at the left side of the stop surface.

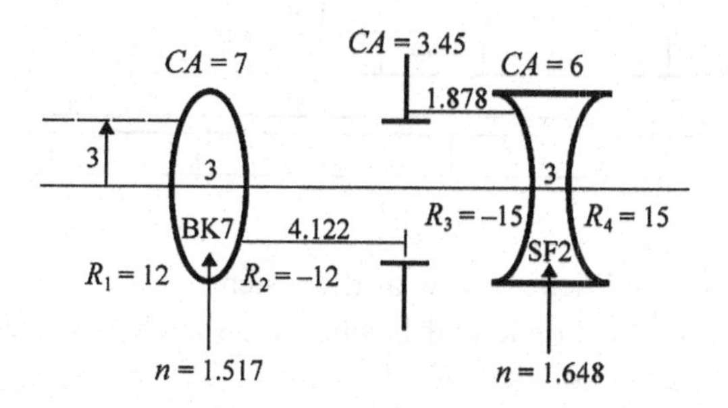

(a) What is the effective focal length?
(b) What is the back focal distance (*BFD*)?
(c) What is the full field of view (in degrees)?
(d) Where is the Astop?

10.9 You are given an optical system with the ray trace worksheet shown below.

$-\phi$																	
t/n																	
y			2.5		2.378 32		2.167 59		2.134 48		2.086 00		2.053 68		1.985 53		
nu		0		−0.0369		−0.0421		−0.0291		−0.0162		−0.0162		−0.0207		−0.049 99	
$\bar{y}$	∞		−0.3864		−0.2525		−0.0466		0		0.122 71		0.204 51		0.337 85		
$n\bar{u}$		0.0349		0.040 63		0.0412		0.040 90		0.040 90		0.0409		0.0405		0.354 62	
CA			10		10		9		8		8		10		10		
CA/y																	

(a) The given marginal ray is not correct (it is a pseudo ray), because it does not touch the edge of the stop. Which of the six surfaces is the Astop of the system?
(b) What are the correct ray trace values for the marginal ray for this Astop?
(c) Where is the image relative to the vertex of the last surface?
(d) What is the *FFOV* (full field of view) for this system (in degrees)?

10.10 Using the ray trace worksheet below:

$-\phi$															
t/n															
y	2.150 44		2.150 44		2.0406		1.5		1.4508		1.1142		0.9362		0
nu		0		−0.027 74		−0.054 064		−0.054 064		−0.0306		−0.044 97		−0.057 04	
$\bar{y}$			−1.682 36		−1.251 07		0		0.113 872		1.510 34		1.935 69		3.2898
$n\bar{u}$		0.087 266		0.108 968		0.125 108		0.125 108		0.126 952		0.107 468		0.0825	
CA			5		5		3		3		5		5		
CA/y							2								

(a) What is the full field of view of the system?
(b) Give the values (angle and heights) for the chief ray when changing the system to a field of view of $\pm 12.5°$.
(c) For the $\pm 12.5°$ field of view, which surfaces vignette?
(d) Did the Astop change its position when changing to the $\pm 12.5°$ field of view?

10.11 Set up the paraxial ray trace for the lens shown below. Keep five significant figures in your results. All dimensions are in centimeters. Assume $CA = 10$ for all lenses.

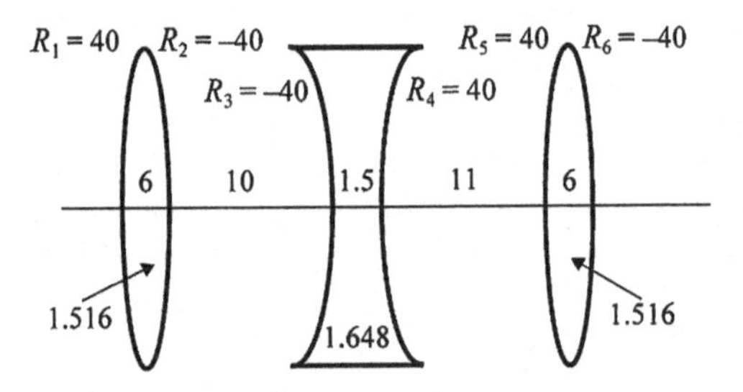

(a) Trace a pseudo marginal ray through the system for an object at infinity at a height of 1.0 centimeter. Which surface shown is the Astop? Trace a pseudo chief ray through the system with an angle (nu) of 0.1 radians through the aperture stop (Astop).
(b) What aperture scaling factor, α_a, puts the marginal ray through the edge of the aperture stop?
(c) Find the field scaling factor, α_F, which gives a full field of view equal to 10°.
(d) Set up a ray trace table for these new scaled rays from parts (b) and (c).

Use the table from part (d) for the remainder of the problem:

(e) Find $(F/\#)_\infty$.
(f) Find the effective focal length (*efl*) and the back focal distance (*BFD*).
(g) Find the Lagrange invariant value;

(h) Find the entrance pupil location relative to the first surface and its diameter.
(i) Find the exit pupil location relative to the last surface and its diameter.
(j) Which surfaces vignette? Specify whether the upper or lower rim ray is doing the vignetting.

10.12 For a Cooke triplet (designed by Taylor) with the radii and thicknesses shown in the table below, ray trace the marginal and chief rays to answer the following questions. The chief ray angle before the stop is 0.05 radians.

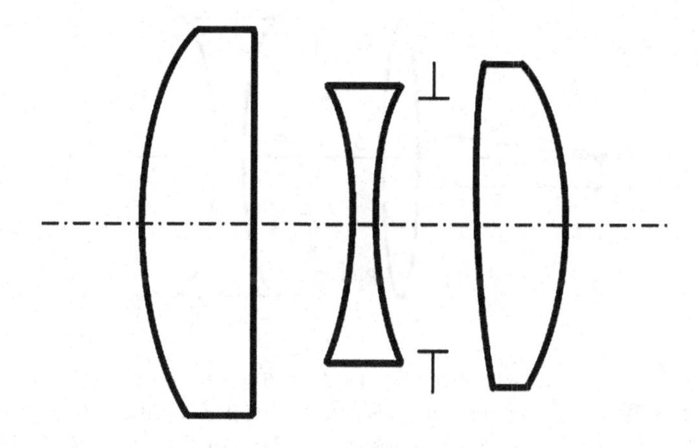

Surface Number	Radius (mm)	Thickness (mm)	Glass	*CA*
1	23.713	4.831	LaK9(1.691)	10
2	7331.288	5.86		10
3	–24.456	0.975	SF5(1.673)	7
4	21.896	2.822		7
5 (Aperture)	?	2		8
6	86.759	3.127	LaK9(1.691)	8
7	–20.4942			8

(a) Find:
- the back focal distance *BFD*;
- the effective focal length (*efl*).

(b) Find:
- the Lagrange invariant;
- *F*/#.

(c) Find:
- the distance from the front vertex to the front principal plane (*P*);
- which surface is the Astop.

(d) Find:
- which surfaces vignette;
- what the dimensions of the field stop are.

(e) Find:
- the full field of view of the triplet;
- Where the entrance and exit pupils are relative to the first vertex and last vertex respectively.

10.13 For the system shown below:

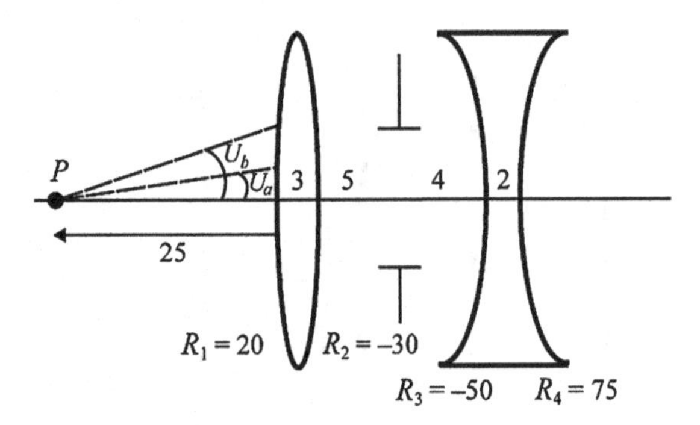

Paraxial ray trace two arbitrary rays (a and b) from the axial point ($P=$ surface #0) at two different angles (U_a and U_b): $U_a = 0.05$, $U_b = 0.10$.

(a) Paraxial ray trace two new rays: $U_c = 3U_a = 0.15$, $U_d = 4U_b = 0.20$.

(b) At surface 4, what is the ratio of ${}^dY_4$ to ${}^bY_4$?

(c) At surface 4, what is the ratio of ${}^cU_4$ to ${}^aU_4$?

(d) At surface 4, what is the ratio of ${}^dU_4$ to ${}^bU_4$?

10.14 The rays traced below are pseudo marginal and pseudo chief rays for a particular system (i.e. the actual marginal and chief rays can be obtained by scaling them appropriately).

(a) What are the correct values for the marginal ray?

(b) For a field of view of $\pm 10°$, determine the chief ray trace table.

(c) For the $\pm 10°$ field of view, which surfaces vignette?

$-\phi$															
t/n															
y			3		2.744		1.260		0.9087		0.7545		0.4508		
nu		0		−0.1297		−0.247 39		−0.192 98		−0.1537		−0.1537		−0.1630	
$\bar{y}$			0		0.1978		0.9719		1.137		1.465		1.84		
$n\bar{u}$		0.1		0.1		0.1237		0.1657		0.1657		0.1354			
CA			8		9		9		7		7		7		
CA/y															

10.15 Calculate the Lagrange invariant for a photographic objective $F/2.8$, $efl = 100$ mm, and a field of view of $\pm 2.5°$.

10.16 If a camera lens of 30 mm diameter is $F/4$ and is used to focus an object 15 cm away, what is the working $F/\#$ if the object is 5 mm in height?

10.17 Ray trace a paraxial ray through the two lenses shown below.

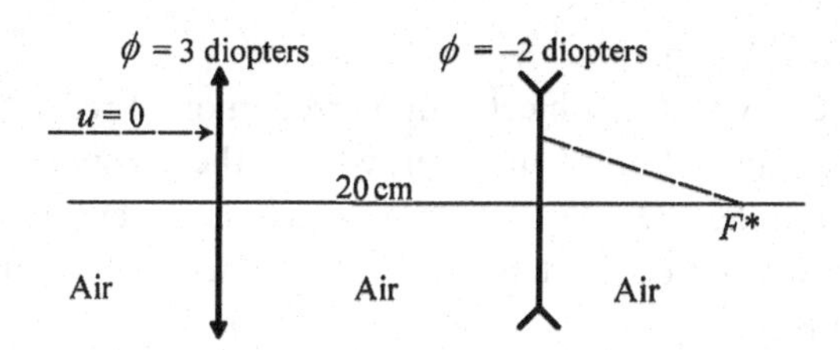

Determine the *BFD* distance (F^* relative to the negative lens ($\phi = -2$)).

10.18 Using a ray from infinity, ray trace this marginal ray at $y = 2$ through the lens below, and determine:

(a) the location of the principal planes (P^*) relative to V_6;
(b) the effective focal length (*efl*);
(c) the back focal distance (*BFD*);
(d) $(F/\#)_\infty$.

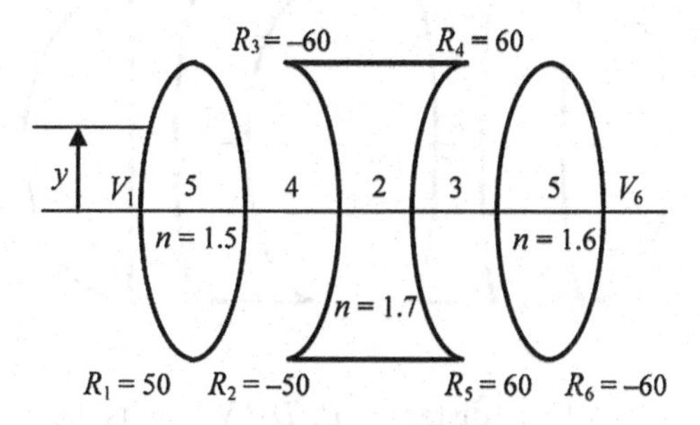

10.19 Using a ray trace table, find the following for the system below:

(a) the image location (*BFD*);
(b) *efl*;
(c) the image height;
(d) the entrance pupil relative to 1st surface of lens 1;
(e) the exit pupil relative to last surface of lens 3.

The marginal ray is from infinity with a height of 1 cm. The chief ray is at an angle of $+0.02°$ crossing the first lens surface at -3 cm. The triplet is shown below (all dimensions in centimeters).

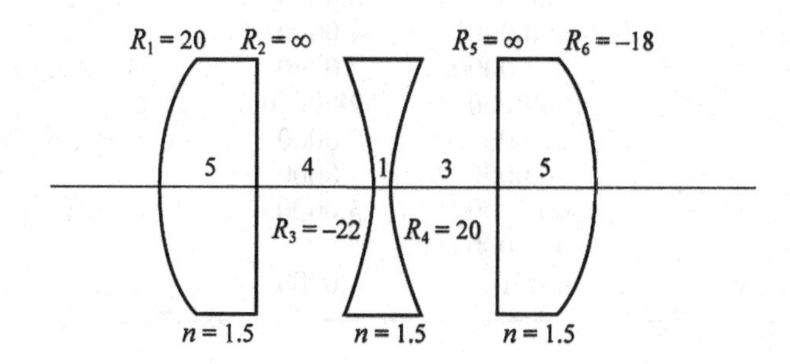

10.20 For an optical system that has an F number of 2 ($F/2$) from the object to the entrance pupil:
(a) What is the NA in air?
(b) What is the NA if the object is immersed in oil of refractive index 1.63?

10.21 For a lens with an NA of 0.5 in water, what is the marginal ray angle?

10.22 If a positive thin lens ($F/1$) of 50 mm effective focal length is used to focus an object 30 cm in front of this lens into image space (behind the lens, filled with water), what is:
(a) the NA in object space;
(b) the NA in image space?

10.23 Infrared Industries has just designed a system with the lens and dimensions shown in Table 10.2. Your job is to paraxial ray trace the marginal and chief angles for an object at infinity. The chief ray at the first surface is 10° ($\bar{u} = 10°$), and is incident at $\bar{y} = -4.2$ mm. The marginal ray is 13 mm at the first surface (all dimensions in mm).

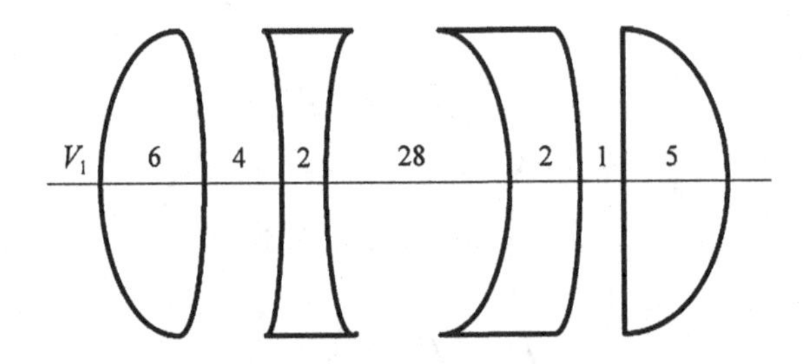

(a) What is the back focal distance (BFD)? Where is the second principal plane for the system, P^*?
(b) Knowing that the intended field of view is 10°, what is the size of the image at the image plane?

Table 10.2.

Surface #	y radius	Thickness	Glass
Object	Infinity	Infinity	
1	25.0000	6.0000	S-BSL7
2	–400.0000	4.0000	
3	–125.0000	2.0000	S-LAH60
4	100.0000	28.0000	
5	–25.0000	2.0000	S-BAM12
6	–40.0000	1.0000	
7	500.0000	5.0000	S-FPL53
8	–41.0000		
Image	Infinity	0.0000	

(c) Where is the entrance pupil (relative to the first surface) and what is its size?
(d) Where is the exit pupil (relative to the last surface) and what is its size?
(e) What is the *F*/# for the system?

Bibliography

Ditteon, R. (1997). *Modern Geometrical Optics.* New York: Wiley.
Geary, J. M. (2002). *Introduction to Lens Design: With Practical Zemax Examples.* Richmond, VA: Willman-Bell.
Glassner, A. (Ed.). (1989). *An Introduction to Ray Tracing.* New York: Academic Press.
Hopkins, R., Hanau, R., Osenberg, H., *et al.* (1962). *Military Standardized Handbook 141 (MIL HDBK-141).* US Government Printing Office.
Shannon, R. R. (1997). *The Art and Science of Optical Design.* Cambridge: Cambridge University Press.

11
Aberrations in optical systems

Our tour through paraxial optics has only considered perfect images of scenes in which a point source in object space was mapped to a point in image space using paraxial rays. Gaussian optics also produced perfect images outside the paraxial region. Paraxial optics, however, is only a first-order approximation to a real optical system. Realizable optical systems do not produce perfect point images from point sources (represented mathematically as a delta function). In real optical systems, there is some blur or spreading of the point image.

11.1 Diffraction

The complex propagation of light passing through an aperture stop of a lens system will form a less than perfect image (for a detailed explanation of Huygens' wavefronts and propagation see Mahajan (2001)). In fact, the best one can do is to make the system "diffraction-limited." Diffraction occurs when a wavefront (radiant beam) impinges upon the edge of an opaque screen or aperture. Light appears outside the perfect geometrical shadow because the light has been diffracted by the edge of the aperture. The effect this has on our simple rotationally symmetric optical systems is that a point does not map to a point, but is blurred or smeared. You may have observed the effect of the diffraction of light from a portal where there is light beyond what would be defined as the geometrical shadow boundary. If a wavefront, as shown in Figure 11.1, passes through a circular aperture, it does not continue as a circular disc. The edges are diffracted away from the geometrical shadow, depending on the wavelength of light, aperture size, and distance from screen. The actual analysis of the energy distribution in the wavefront after passage through the circular aperture is very difficult and involves Rayleigh–Sommerfeld diffraction equations from Huygens' wavelets (Gaskill, 1978). Far-field effects (in what are

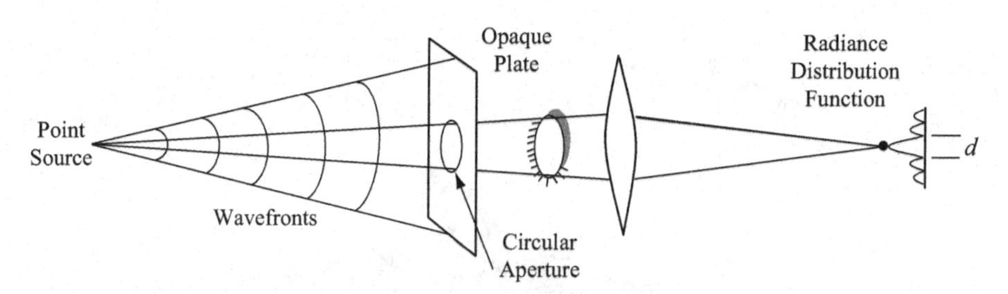

Figure 11.1 Aperture edges causing diffraction.

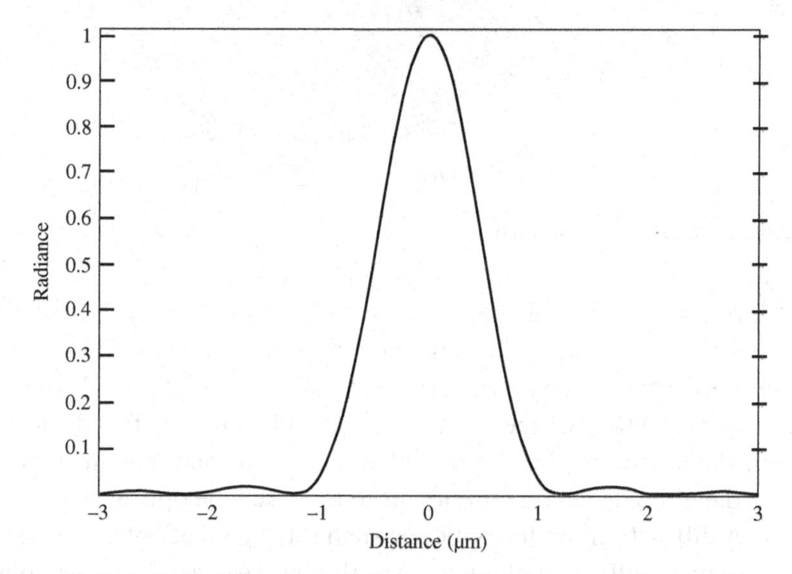

Figure 11.2 A Besinc function of the zeroth order squared describing the radiance of a point source for $\lambda F/\# = 1$ μm.

known as Fraunhofer diffraction regions), where the pattern observed is a long distance from the aperture, are special cases. The patterns caused by diffraction can be analyzed via Fourier optics of a circular aperture for rotationally symmetric optical systems. The results are in the form of a Bessel function. Clearly, this mathematical analysis is beyond the scope of this geometrical optics book. The results, however, are significant and important enough to be summarized here. The blur diameter, as determined from the diffraction analysis, can be expressed as

$$d = 2.44\,\lambda\,(F/\#). \tag{11.1}$$

The constant 2.44 is used because it corresponds to the diameter of the central lobe of the Besinc function $J_1(\pi r)/\pi r$ for a circular aperture, as shown in Figure 11.2. Therefore, the actual radiant energy distribution in the image of a point differs from the point because of diffraction. This diffraction spot is called the "Airy disc," a three-dimensional representation of which is given in Figure 11.3. The diameter

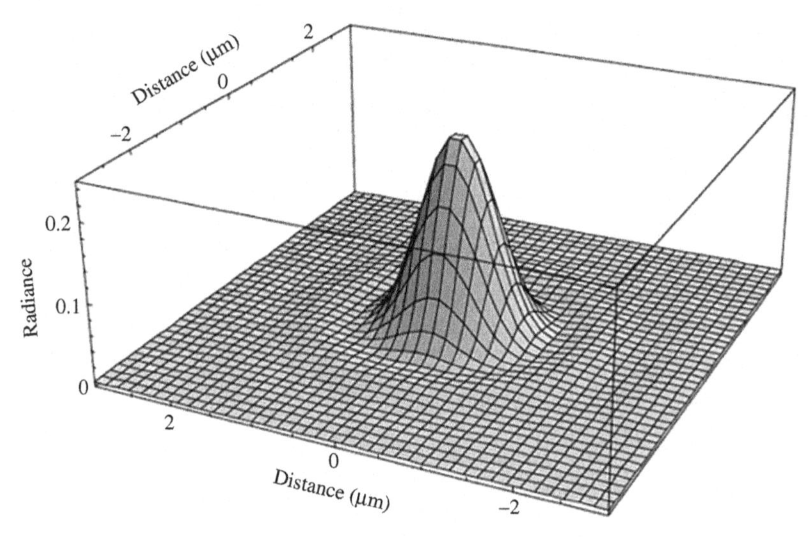

Figure 11.3 Airy disc pattern.

of the "Airy disc" is given as the first zero shown in Figure 11.2, by definition or accepted practice. About 84% of the radiant power is contained in this central disc. The relative maximum for each ring peak shown in Figure 11.2 is 0.0175 for the second ring, 0.0042 for the third ring, and 0.0016 for the fourth ring (Born and Wolf, 1959). From a design point of view, the diffraction limit is the "Holy Grail," or the ultimate design criterion as to how well the optical system creates an image. A diffraction-limited optical system is the goal of optical designers. A perfect system is diffraction-limited. An alternative means of describing the diffraction limit is the angular resolution, α, which can be expressed as

$$\alpha = 2.44\frac{\lambda}{D}, \tag{11.2}$$

where D is the diameter of the aperture. The angle α (in radians) is interpreted as an angle that can be used for evaluating the maximum resolution of an optical system in image or object space.

The angular diffraction blur can be interpreted as the best angular resolution a telescope can achieve for two point sources. The Rayleigh criterion α_r is half the diffraction blur, which is when the maximum of one point source lies at the zero of a second point source, or

$$\alpha_r = 1.22\frac{\lambda}{D}.$$

Angular diffraction is often used to describe the performance of optical systems while taking into account atmospheric "seeing" effects. The quality of atmospheric "seeing" is geographically dependent, but typical "good" seeing is around 1–2 arcseconds and "poor" seeing is usually considered greater than 8 arcseconds.

11.2 Diffraction and aberrations

The combination of aberrations and diffraction effects is even more involved; however, in the present discussion they will be considered separately. In the absence of aberration, one can imagine that the spherical wavefront in the exit pupil is perfect except for the region extending beyond its geometrical shadow. When aberrations are present, this spatial radiant energy distribution no longer holds true. Aberrations cause the spherical wavefront at the exit pupil to become deformed or warped. The amount of change in the wavefront determines the severity of the aberration. The Rayleigh criterion states that if the real wavefront differs from the spherical wavefront at the exit pupil by a quarter wavelength ($\lambda/4$), the system is considered good.

If the aberrations are kept small ($< \lambda/4$), the radiant energy in the outside rings is about 16%, and the central region, or Airy disc, is about the size predicted by the diffraction equation (Equation (11.1)).

11.3 Monochromatic lens aberrations

The object–image relationship developed thus far is based on paraxial optics. Paraxial optics uses small angle approximations for angles near the optical axis, and a constant index of refraction for all media. The small angle approximation comes from the Maclaurin series expansion of the sine function:

$$\sin I = I - \frac{I^3}{3!} + \frac{I^5}{5!} - \frac{I^7}{7!} \ldots, \tag{11.3}$$

where all terms but the first are dropped for the paraxial condition. This paraxial approximation leads to errors when the optical system has a finite aperture and large fields of view. Typically, the rays far from the optical axis do not cross within the diffraction blur. In addition, since the index of refraction is a function of wavelength, different colors focus at different positions along the axis. For example, blue light focuses nearer to the lens than red light because the refractive index is higher in most glasses for shorter wavelengths, and thus the same lens produces greater optical power for blue light.

How do we define a good optical system? What is an aberration-free lens system? If the aberrations are restricted to less than the size of the diffraction-limited blur, we have a well-corrected system (a perfect optical system, practically speaking) or a diffraction-limited system.

Aberrations are image defects that arise from characteristics of the spherical surfaces used to create optical power. These image imperfections are caused by the spherical surfaces themselves, which produce the reflection or refraction of light, and are not results of poor fabrication techniques, material properties, or mounting techniques. There are two general types or categories of aberrations: chromatic and monochromatic. The former category occurs in light of every color, whereas the latter can exist even in light of only one wavelength. Chromatic aberration is caused by the optical power of a given element being different because of the different indices of refraction for each wavelength ($n_F > n_C$). Monochromatic aberrations occur for a given wavelength due to the way a spherical surface is used or how the surface passes the radiation from the object. Monochromatic aberrations of the third order only will be considered. These occur when the sin I in Snell's law is approximated by the first two terms in the McLaurin series expansion shown in Equation (11.3).

11.3.1 Spherical aberration

The most fundamental monochromatic aberration (independent of wavelength) is spherical aberration. It is present across the entire field of view and, most interestingly, exists on-axis. No other third order monochromatic aberration exists on the optical axis. Rays traced at various heights (zones which are rotationally symmetric) within the entrance pupil focus at different heights in the image space, as illustrated in Figure 11.4. The paraxial location is the region furthest from the refracting surface, shown in Figure 11.4, where the rays close to the optical axis cross the optical axis. In that figure, one sees that the rays which are at the edge of the entrance pupil are focused closest to the refracting surface.

Each zone or ring of rays of radius y in the entrance pupil produces a circle at the paraxial image plane with its radius proportional to the radius in the entrance pupil (you can think of the entrance pupil as a summation of tiny donuts, each donut being a ray zone or ring). This is illustrated in Figure 11.4 in cross section for a finite object distance. As seen in Figure 11.4, there is no real focus. There is a saddle or minimum blur location where all the rays are contained within what is called the "circle of least confusion." It should be noted that spherical aberration occurs because the optical surface is a spheroid. If we consider two rays from an object at infinity impinging on a spherical surface, as shown in Figure 11.5, one ray at paraxial height (1 unit), the other far off axis (8 units), where are they focused? The paraxial ray ($y = 1$) for a spherical surface of radius equal to 10, with a refractive index of 1.5, is focused

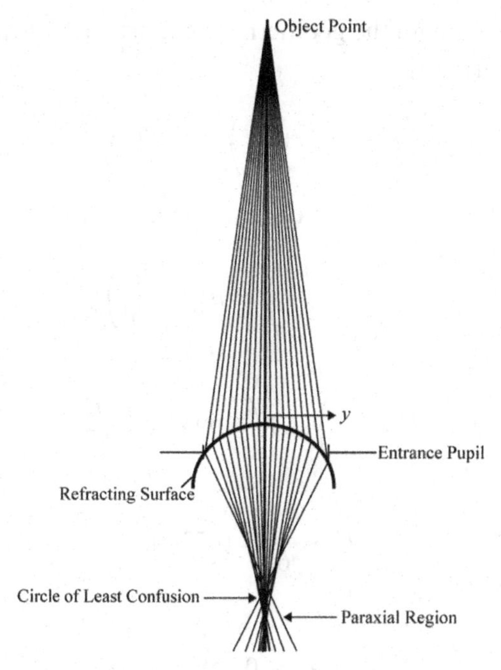

Figure 11.4 Rays traced from a point object to the paraxial image region to produce a blur, showing focus as a function of ray height in the entrance pupil.

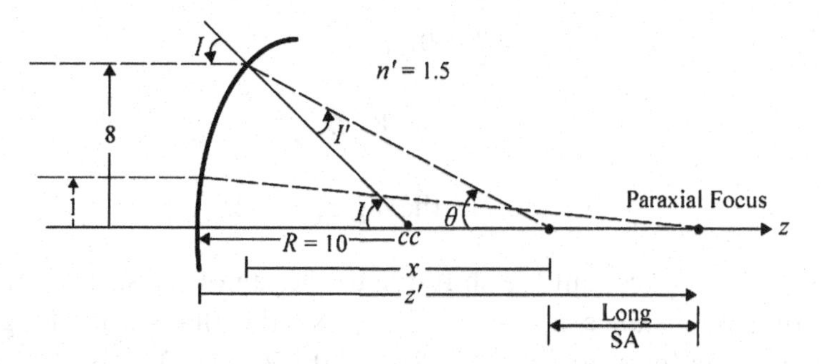

Figure 11.5 Longitudinal spherical aberration (Long SA) illustrated by the distance between paraxial focus and a ray imaged far off the axis.

at 20 units from the surface, or at the paraxial focus. This was calculated using the refraction and transfer equations from paraxial optics. Calculating where the ray of height 8 focuses takes a little more effort. In this case, we use Snell's law to determine where an exact ray would cross the axis. Instead of doing an exact ray trace, we can calculate where the outer ray crosses the axis if only the first two terms of the sin I expansion in Equation (11.3) are used. Employing only up to the third-order term, one can locate the change (shift) in axial point for the marginal ray. From the geometry of Figure 11.5, the angle of incidence on the refracting surface is

$$I = \sin^{-1}\left(\frac{8}{10}\right) = 0.9273.$$

By Snell's law, with the approximation at the first two terms,

$$n'\left(I' - \frac{I'^3}{3!}\right) = n\left(I - \frac{I^3}{3!}\right);$$

$$1.5\left(I' - \frac{I'^3}{6}\right) = 1\left(0.9273 - \frac{0.9273^3}{6}\right)$$

$$I' - \frac{I'^3}{6} = 0.5296.$$

Solving,

$$I' = 0.559,$$

$$\theta = I - I' = 0.9273 - 0.559$$

$$\theta = 0.3686,$$

$$\tan\theta = 8/x,$$

$$x = 20.71;\ \text{the sag } z_s = 3.2.$$

One must take into account the shift due to the sag of the surface; therefore, the longitudinal distance is $x + z_s = 23.91$, so the third-order longitudinal special aberration for a ray height of 8 is $= 20 - 23.91 = 3.91$.

Large spherical aberration causes the image to be smeared or blurred beyond the diffraction Airy disc. Longitudinal spherical aberration (LSA) can also be related to transverse spherical aberration (TSA) via the $F/\#$ of the optical system. It is related to the height below the optical axis where a marginal ray intersects the paraxial focal plane (see Figure 11.6). Transverse spherical aberration is more useful, since it is related to diffraction blur or the Airy disc, which is perpendicular to the optical axis. The longitudinal and transverse spherical aberrations are related by

$$TSA = 2(LSA)(\tan u). \tag{11.4}$$

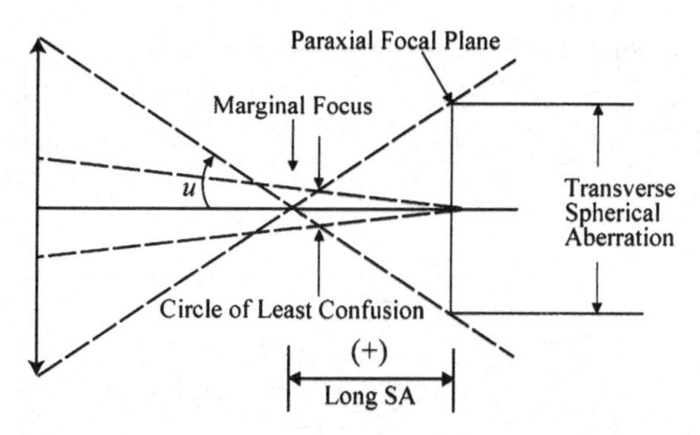

Figure 11.6 Spherical aberration showing the relationship between longitudinal aberration and transverse spherical aberration.

The correction of spherical aberration cannot be accomplished with a single lens with spherical surfaces. It can be minimized by lens shape changes, or bending of the lens, to reduce the spherical aberration. The glass chosen should have a high refractive index, so the radius used to produce a given optical power is lower, thus causing the spherical aberration to be lower. The "best shape" is the bending that gives the least amount of spherical aberration. Recall that the shape factor ($\mathcal{S}$) from Chapter 6 is

$$\mathcal{S} = \frac{C_1 + C_2}{C_1 - C_2}, \tag{11.5}$$

where C_1 and C_2 are curvatures of lens surfaces for thin lenses. The longitudinal spherical aberration can be expressed for a refractive index (n) in terms of shape factor as (Jenkins and White, 1976):

$$LSA = \frac{r_c^2}{8f}\left[\frac{1}{n(n-1)}\right]^2 \left[\frac{n+2}{n-1}\mathcal{S}^2 + 4(n+1)\mathcal{S}\not{p} + (3n+2)(n-1)^2\not{p}^2 + \frac{n^3}{n-1}\right]. \tag{11.6}$$

The object location is often described by a position factor $\not{p}$ relating the object and image distances:

$$\not{p} = \frac{z' + z}{z' - z}. \tag{11.7}$$

Plots for spherical aberration for various indices of refraction versus shape factor are shown in Figure 11.7. To determine the best shape for

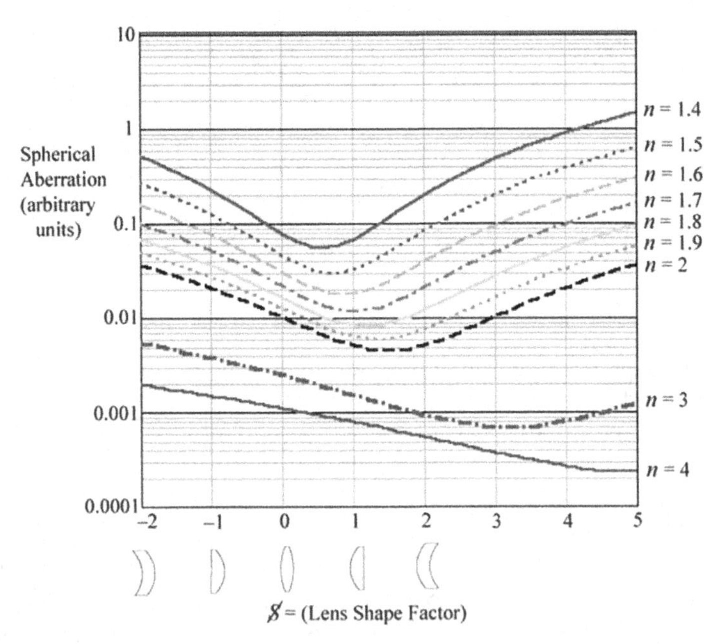

Figure 11.7 Spherical aberration as a function of shape factor for an object at infinity.

minimum longitudinal aberration, take the derivative with respect to the shape factor. Subsequently, for the minimum spherical aberration, the shape factor should be

$$\mathcal{S} = -\frac{2(n^2-1)}{n+2}\not{p}. \tag{11.8}$$

This is the shape factor that gives the minimum spherical aberration for a singlet for any optical power as long as the object is at infinity. The minima shown in Figure 11.7 for various indices of refraction demonstrate that a meniscus configuration is favored as the index of refraction increases.

Example 11.1

Find the shape and radii for a thin singlet (refractive index = 1.5) having a 50 mm focal length, for minimum spherical aberration with the object at infinity.

$$\not{p} = \frac{z'+z}{z'-z} = -1,$$

$$\mathcal{S} = -\frac{2(1.5^2-1)}{1.5+2}(-1) = 0.714,$$

$$R_1 = \frac{2f(n-1)}{\mathcal{S}+1} = \frac{2(50)(1.5-1)}{0.714+1} = 29.2\,\text{mm},$$

$$R_2 = \frac{2f(n-1)}{\mathcal{S}-1} = \frac{2(50)(1.5-1)}{0.714-1} = -174.8\,\text{mm}.$$

Complete elimination of spherical aberration can be accomplished by using aspheric surfaces on the lens. Consider a plano-convex lens, as shown in Figure 11.8. What is the surface profile needed to eliminate spherical aberration? A method to eliminate aberration is to aspherize the optical surfaces. Recall that we have already found in Chapter 9 that using a mirror with a parabolic curve instead of a spherical one produced perfect geometrical imaging on-axis. For a refracting lens, the surface is a hyperboloid. To prove it is a hyperboloid, we will force the optical path length (*OPL*) to be equal for all the rays. Using Fermat's principle, the *OPL* for a ray of any height, y, must be equal to the *OPL* of a ray on the axis, as shown in Figure 11.8.

To eliminate spherical aberration, we choose two general ray heights and force their *OPL*s to be equal. This results in a conic surface, instead of a sphere:

$$OPL = nt + 1, \tag{11.9}$$

where t is the lens axial thickness. The *OPL* for any ray in a zone height, y, above the axis is

$$OPL = n(t+z) + \sqrt{y^2 + (1-z)^2}. \tag{11.10}$$

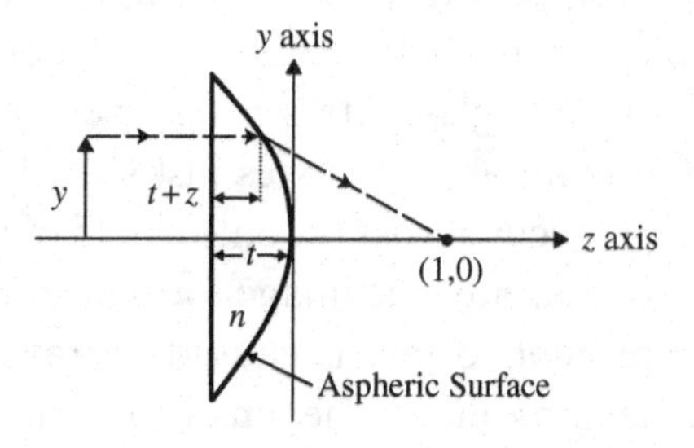

Figure 11.8 Using a spheric surface on a refracting lens to eliminate spherical aberration.

Setting the OPL for both rays equal to each other and solving,

$$nt + 1 = nt + nz + \sqrt{y^2 + (1 - z)^2}$$
$$(1 - nz)^2 = y^2 + (1 - z)^2$$
$$1 - 2nz + n^2z^2 = y^2 + 1 - 2z + z^2$$
$$z^2(n^2 - 1) - 2z(n - 1) - y^2 = 0$$
$$z^2(n - 1)(n + 1) - 2(n - 1)z - y^2 = 0. \quad (11.11)$$

Multiplying the equation by $n + 1/n - 1$ and completing the square,

$$z^2(n + 1) - 2(n + 1)z + 1 - y^2\left(\frac{n + 1}{n - 1}\right) = 1$$
$$(z(n + 1) - 1)^2 - y^2\left(\frac{n + 1}{n - 1}\right) = 1. \quad (11.12)$$

The equation is now starting to take the form of a hyperboloid:

$$\frac{[z - 1/(n + 1)]^2}{1/(n + 1)^2} - \frac{y^2}{[(n - 1)/(n + 1)]} = 1. \quad (11.13)$$

This is the equation of a hyperboloid centered at:

$$z = \frac{1}{n + 1}, \qquad y = 0. \quad (11.14)$$

For the lens surface, this is a surface of revolution about the z axis. This plano-convex lens with a hyperboloid convex surface produces no aberrations on axis, or no spherical aberration.

11.3.2 Coma

"Coma" is the comet-like appearance that a point off-axis at some field angle forms in the image plane. Coma aberration is the most pronounced just off the optical axis (at small field angles). Its strong aperture height dependence causes it to become dominant as one moves just slightly off axis. Recall that only spherical aberrations occur across the entire field of view, including on the optical axis. Figure 11.9 illustrates the image formed from a point source with only coma aberration present. Coma is caused by rays from a point object passing through a particular zone of the lens being imaged to a circle at the paraxial image plane, rather than a point. However, unlike spherical aberration, the center of the circle of rays is displaced from the central ray located on

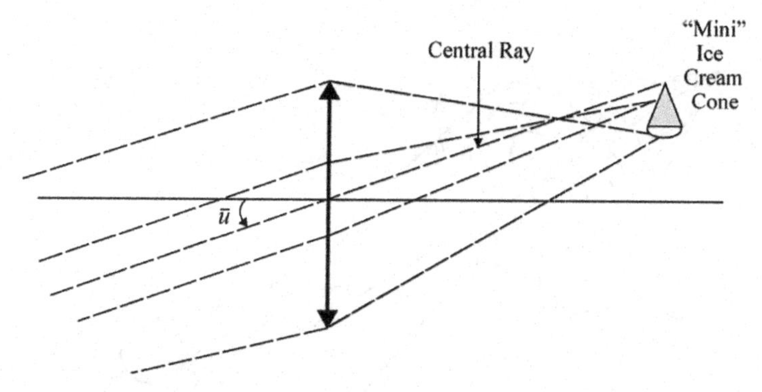

Figure 11.9 Coma illustrated with the apex pointing away from the optical axis for two zones in the exit pupil.

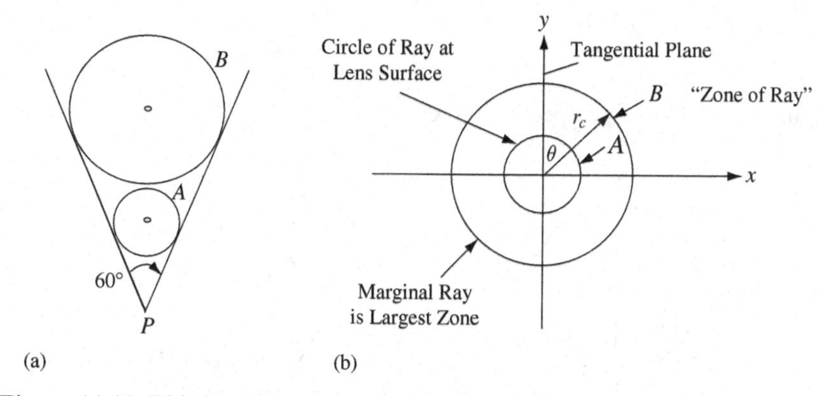

Figure 11.10 Third-order coma: (a) shown in the formation of a 60° wedge pattern; and (b) defining the rays in the exit pupil corresponding to rays at the image and a 60° coma pattern.

the paraxial image plane. These image circles are larger for larger zone radii, and are centered further from the reference point formed by the central ray. Therefore, the circles not only become larger, but shift down toward the axis. This is shown in Figure 11.9. Coma occurs due to a variation in magnification, dependent on the aperture zone in the exit pupil of the lens.

In Figure 11.10(a), the apex angle for third order coma is always 60°. Each zone of rays (A and B) is shifted, and the circles increase to produce a 60° apex angle. A perspective drawing is shown in Figure 11.11. Each zone of rays focuses to a circle whose size is proportional to the height (y) in the aperture, so magnification varies with aperture zone. The size of the coma pattern is linearly proportional to the field angle ($\bar{u}$). The rays causing coma for some off-axis field of view angles are illustrated in Figure 11.12. By numbering the rays in two zones at a given field angle, rays passing through these zones are indicated as prime and unprimed numbers (shown in Figure 11.12(a)). These rays in a zone of the

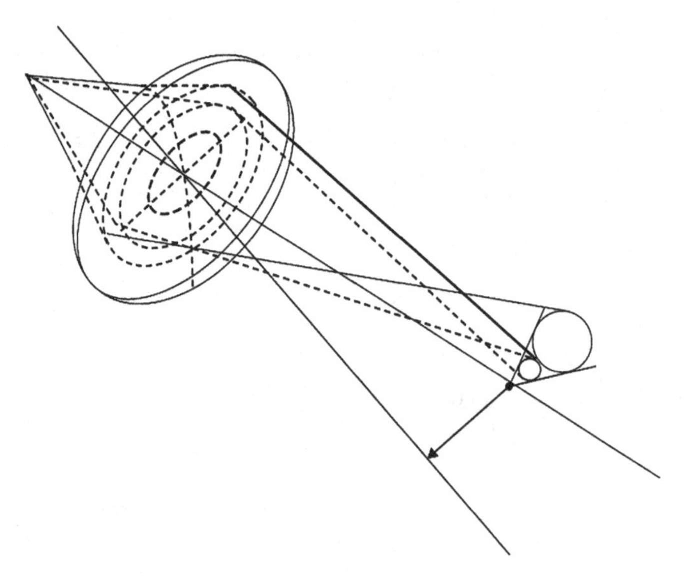

Figure 11.11 Coma perspective sketch.

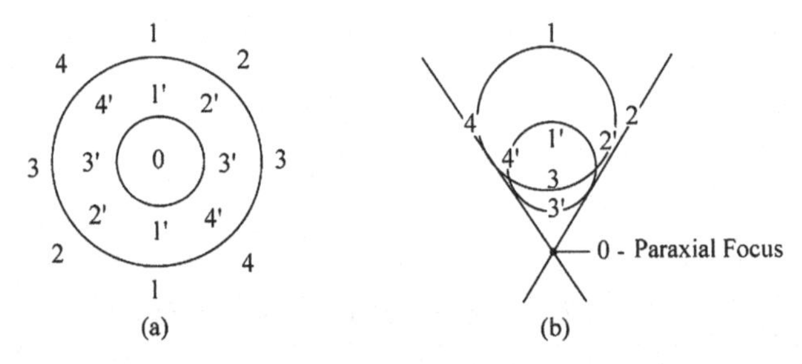

Figure 11.12 Unique rays in an aperture stop mapped to paraxial image plane circles: (a) ray zones in the exit pupil of an optical system; (b) ray crossing paraxial plane from the zones in (a).

pupil form a doubly degenerate circle, as shown in Figure 11.12(b), where the rays on the image are rotated by twice the angle they are rotated in the pupil. Rays that lie in the sagittal plane are indicated as 3 and 3′, and in the tangential plane as 1 and 1′. Note that the rays in two locations of the primed value zone (3 and 3′) in Figure 11.12(a) are converging to the same location, 3′, in Figure 11.12(b).

Correspondingly, rays in the tangential plane indicated by 1′ in Figure 11.12(a) are both focusing at the point 1′ in Figure 11.12(b). The other numbered rays arrive as illustrated, such that the image is a double passed circle. A single rotation in the exit pupil zone, Figure 11.12(a), maps out a double circle in the image plane, Figure 11.12(b). Similarly, for the larger unprimed zone, the image circle is larger and shifted upward from the paraxial focus as shown in

Figure 11.13 Positive coma pattern over full field of view in the paraxial image plane.

Figure 11.12(b). The image formed by all the zones superimposed has a comet-like appearance. The apex of the coma points away from the optical axis as shown in Figure 11.9 for a positive coma. The pattern that one would observe in the focal plane if pure positive coma were present (with no other aberration) is shown in Figure 11.13. By geometry, the dimensions of the coma pattern for sagittal and tangential directions due to the 60° apex angle, as shown in Figure 11.12(b), have a length ratio of 1:3, as expected for a third-order coma.

Bending the lens to the correct shape factor can control coma. Equation (11.15) shows how the sagittal coma (CMA_3) is related to the object position factor and the shape factor for the edge of the field of view:

$$CMA_3 = \left[\frac{6n+3}{4n}\mathscr{P} + \frac{3n+3}{4n^2-4n}\mathscr{S}\right]\frac{\bar{u}r_c^2}{3f}. \tag{11.15}$$

Therefore, by setting CMA_3 to zero, one can find the shape factor that eliminates coma in a single lens:

$$\mathscr{S} = -\frac{(2n^2-n-1)}{n+1}\mathscr{P}. \tag{11.16}$$

Interestingly, coma can be eliminated using a shape factor ($\mathscr{S} = 0.8$) that is near the shape factor needed to minimize spherical aberration ($\mathscr{S} = 0.714$) for a refractive index of 1.5 and an object at infinity.

11.3.3 Astigmatism

Astigmatism is a word often used to describe a defect of the human eye. However, in the formal optical sense, astigmatism refers to the producing of two images at two spatial locations, one each for the sagittal and tangential directions. Thus there are two different optical powers in the two orthogonal directions. Looking at a circular exit pupil from the image plane and from an off-axis point, the exit pupil projection appears as an ellipse, but the sags in each orthogonal direction are equal in magnitude. Therefore, the higher optical power in the tangential plane ($\phi_T > \phi_S$) causes that focus to be closer to the lens, and therefore the radii of curvature are different ($R_T < R_S$). Due to the different radii of curvature, a fan of rays in the tangential plane focuses at a nearer point than a fan of rays in the sagittal plane. Rays in the y–z plane (tangential) focus at a different length than rays in the x–z plane (sagittal). All rays from an object point are traced through optical systems at some angle. They are found to pass through a line perpendicular to the tangential plane (tangential focus) initially, then through another line at the sagittal plane (sagittal focus) as illustrated in Figure 11.14.

Between these two foci the rays form an elliptical cross section. At one spatial location, however, the ellipse turns into a circle. Since astigmatism is quadratically dependent on field angle ($\bar{u}^{-2}$), the separations of the tangential, sagittal, and paraxial foci increase drastically as the field of view becomes larger. Both the tangential and sagittal foci move toward the lens as the object moves further off-axis. If the object is a plane, perpendicular to the optical axis, the tangential image surface and a sagittal image surface are formed, as shown in Figure 11.15. Both of these surfaces are paraboloid in formation and intersect at the optical axis at the paraxial image focus. The objective in

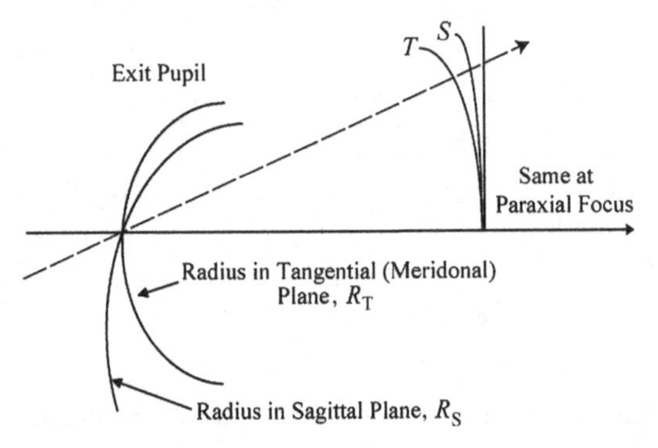

Figure 11.14 Astigmatism is due to optical power being different in tangential and sagittal planes.

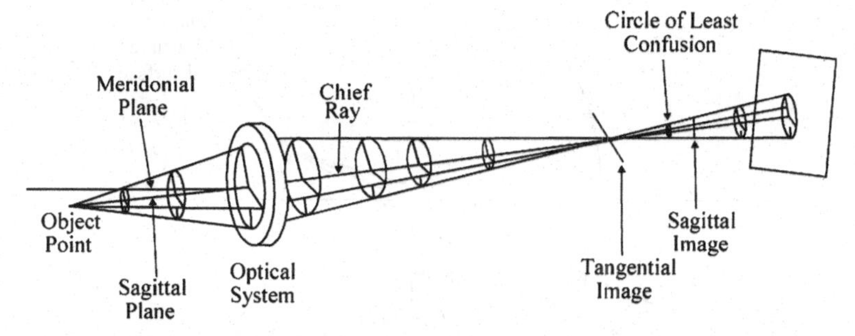

Figure 11.15 Formation of the sagittal and tangential images (lines) of an object point, off-axis, at some field angle.

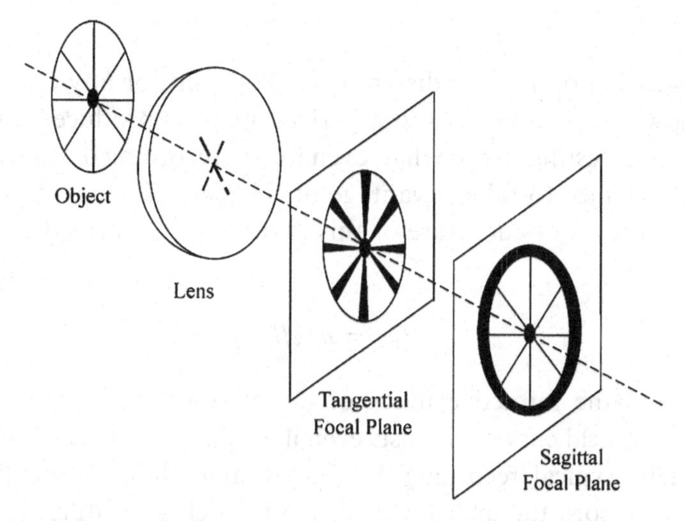

Figure 11.16 Images in the tangential and sagittal focal planes.

correcting astigmatism is to make the sagittal and tangential surfaces, as shown in Figure 11.15, coincide. With astigmatism, circles on the object concentric with the optical axis appear to be sharply imaged in the tangential image surface, and radial lines appear to be in focus in the sagittal image surface. This is illustrated in Figure 11.16.

11.3.4 Field curvature (Petzval) aberration

Every optical system has field curvature aberration, since a perfect image is formed at the paraxial focus as we rotate about the exit pupil, as shown in Figure 11.17. Thus perfect images are possible only on curved spherical surfaces.

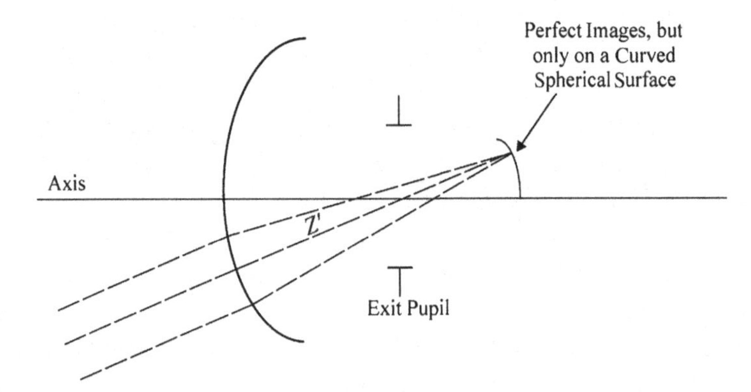

Figure 11.17 Petzval curvature (field curvature).

From Gaussian optics, the distance z' is constant for any off-angle position, as shown in Figure 11.7. So a perfect image is produced on a curved surface. It is interesting to note that, even today, recording devices may be set on a curved surface to take advantage of this perfect imagery on a curved surface. The radius of curvature of this perfect image formed on a Petzval surface is:

$$R_{\mathrm{FC}} = n \cdot efl. \tag{11.17}$$

Note that n is the refractive index of the glass of the lens. It should be mentioned that field curvature exists even if astigmatism has been eliminated. Unfortunately, typical recording film, solid state chips, and enlargers use flat planes to record the image. Correcting for field curvature is referred to as "field flattening," which is an unfortunate choice of words in optics, since solid state cameras use the same term to indicate uniform irradiance across the focal plane. Positive lenses introduce inward (uncorrected) curvature of the Petzval surface, and negative lenses introduce outward (overcorrected) curvature. Therefore, by combining a positive and a negative lens of equal but opposite optical powers, one can produce a flat field across the image plane.

11.3.5 Distortion

Distortion aberration is the easiest to visualize. There are two types: pincushion and barrel. As shown in Figure 11.18, in pincushion distortion the object is stretched by its corners, while in barrel distortion it is compressed at its corners. The distortion is due to the variation of magnification with field

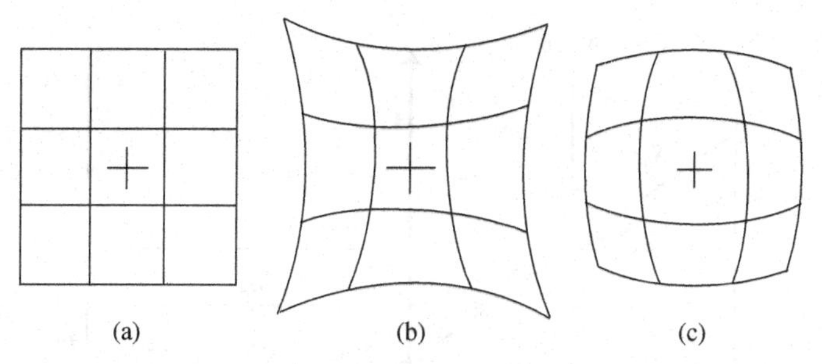

Figure 11.18 Distortion: (a) object; (b) pincushion distortion; (c) barrel distortion.

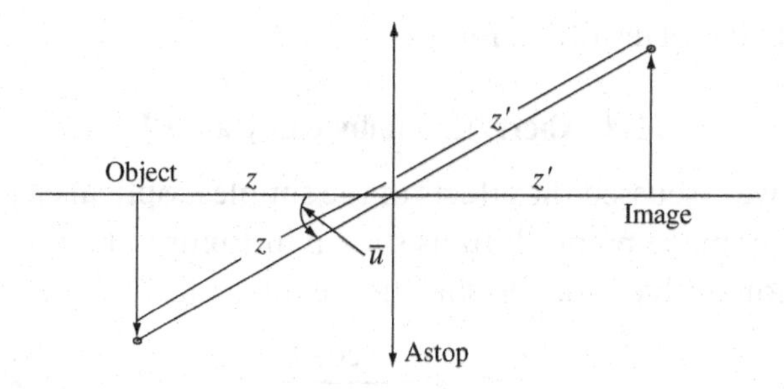

Figure 11.19 A thin lens with the aperture stop at the lens forces the magnification (z'/z) to be the same at all field of view angles, so no distortion is present.

angle: magnification varies as to the third power of field angle. As the field of view becomes larger, the magnification increases much faster. Note that a straight line through the axis of the object is imaged as a straight line, but the image of any straight line not through the optical axis is a curve.

Negative distortion leads to a pincushion image (Figure 11.18(b)), and positive distortion results in a barrel image (Figure 11.18(c)). If the stop is at the lens, as shown in Figure 11.19, there is no distortion. The reason for this is that the ratio of the object distance (z) to the image distance (z') is a constant for all field angles. Therefore, no distortion is present.

For a stop in front of the lens, as shown in Figure 11.20, the magnification changes for rays at different angles. The transverse magnification for ray A is greater than that for ray B; therefore, barrel distortion occurs.

Pincushion distortion occurs if the stop is after the lens in image space. An aperture stop located between two positive lenses cancels the two types of distortion, thus minimizing the overall distortion. In designing optical systems for recording media, it is best to develop a system symmetrical about the aperture stop in order to minimize distortion.

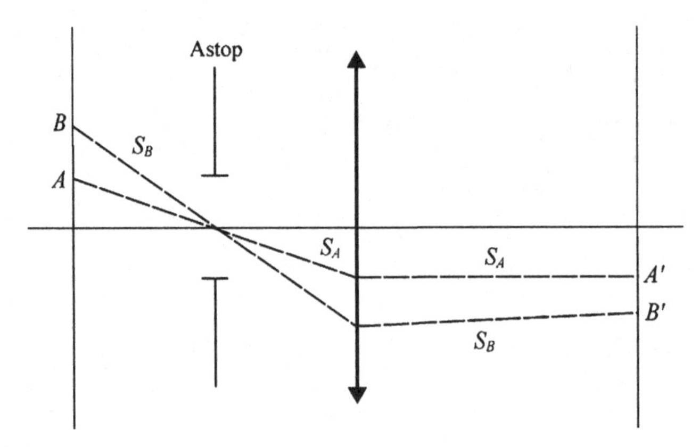

Figure 11.20 Barrel distortion formed.

11.4 Aberration induced by a PPP

In Section 2.6, we discussed the effects of the simplest optical element, the PPP, which caused an image plane shift, as shown in Figure 11.21.

The expression derived for the shift in the image location, $\Delta z'$, was:

$$\Delta z' = t - \frac{t \cos U}{n \cos U'}. \tag{11.18}$$

However third-order spherical aberration is also introduced. The third-order spherical aberration wavefront aberration is (Wyant and Creath, 1992):

$$W_{040} = -\frac{tu^4(n^2 - 1)}{8n^3}, \tag{11.19}$$

or in terms of $F/\#$:

$$W_{040} = -\frac{t}{(F/\#)^4}\left[\frac{(n^2 - 1)}{128n^3}\right]. \tag{11.20}$$

Example 11.2

For a PPP of thickness 10 mm, made of N-BK7 (517642), placed in an $F/5$ lens system, how much spherical aberration is introduced (in micrometers and with a wavelength of 632.8 nm)?

For this problem, $t = 10$ mm, $n = 1.517$, $F/5$, and $\lambda = 632.8$ nm. To find the spherical aberration of a PPP, we use the following equation:

$$W_{040} = -\frac{t}{(F/\#)^4}\left[\frac{(n^2 - 1)}{128n^3}\right].$$

Plugging in the values gives $W_{040} = -0.0466\ \mu\text{m} = -0.0736$ waves.

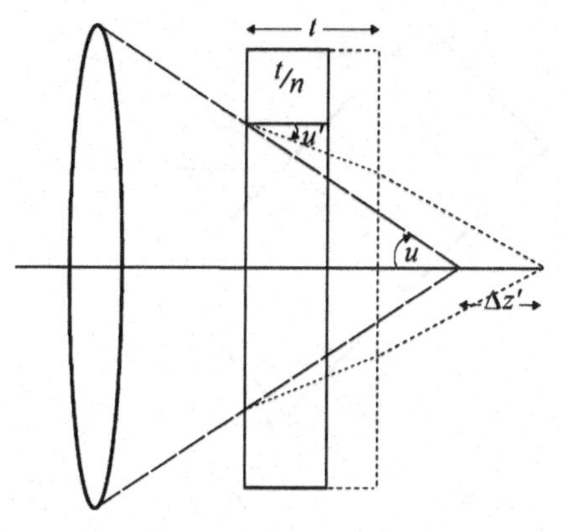

Figure 11.21 PPP located in a converging beam.

Often, a window is introduced in the image space of the detector array to isolate the array from the surroundings via a vacuum. This window must be thick enough not to implode, as discussed in Chapter 4, but not too thick, since Equation (11.19) shows that spherical aberration increases with thickness (t).

If the thickness and $F/\#$ are fixed for a particular system, the spherical aberration increases for the index range $n = 1.0$ to $n = \sqrt{3}$: it is maximum at $n = \sqrt{3}$. As n increases beyond $\sqrt{3}$, the amount of third-order spherical aberration decreases. For a tilted PPP, as shown in Figure 11.22, we get a displacement laterally as well as longitudinally. This change of image in the lateral direction (d) was derived in Equation (2.26) (where ψ replaces θ in that equation)

$$d \approx \frac{t\psi(n-1)}{n}. \tag{11.21}$$

This tilted PPP introduces both coma and astigmatism. The sagittal third-order coma that a tilted PPP at angle ψ introduces is

$$W_{131} = -\frac{tu^3\psi(n^2-1)}{2n^3}\cos\theta, \tag{11.22}$$

where θ is the angle in the pupil plane, as described in Figure 11.10 (Wyant and Creath, 1992). For sagittal coma, $\theta = 0$, so $\cos\theta$ is maximum in terms of $F/\#$:

$$W_{131} = -\frac{t\psi}{(F/\#)^3}\left[\frac{(n^2-1)}{16n^3}\right]\cos\theta. \tag{11.23}$$

The third-order wavefront astigmatism introduced is given by

$$W_{222} = -\frac{tu^2\psi^2(n^2-1)}{2n^3}\cos^2\theta, \tag{11.24}$$

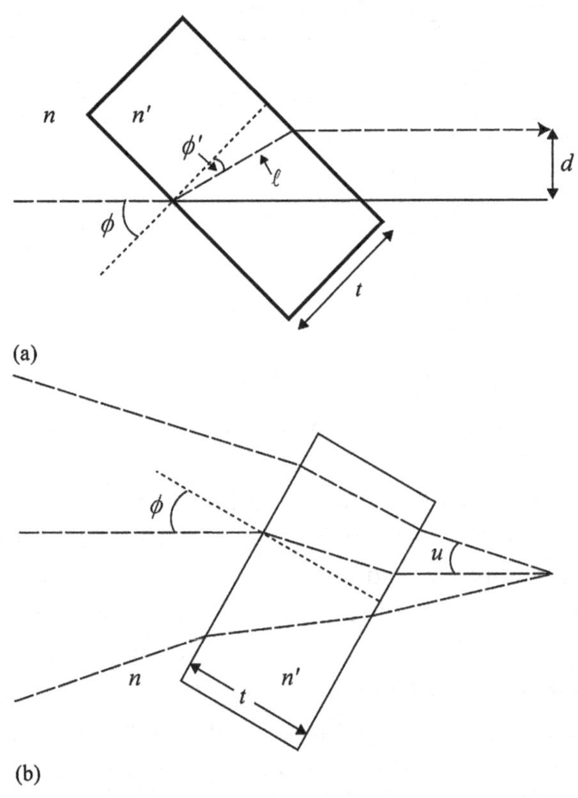

Figure 11.22 Tilted PPP: (a) lateral displacement of the beam; (b) $F/\#$ does not change.

or

$$W_{222} = \frac{t\psi^2}{(F/\#)^2}\left[\frac{n^2-1}{8n^3}\right]\cos^2\theta. \tag{11.25}$$

It should be noted that this PPP introduces chromatic aberration.

Example 11.3

For a 10 mm thick PPP in an $F/5$ optical system, what is the maximum tilt angle allowable to have a wavefront error of $\lambda/4$ at 632.8 nm using N-BK7(517642) for spherical aberration, coma and astigmatism?

Spherical aberration is unaffected by tilt angle, but there is a maximum thickness of the PPP that allows a $\lambda/4$ wavefront error that can be calculated:

$$W_{040} = \frac{-t}{(F/\#)^4}\left[\frac{n^2-1}{128n^3}\right] = \frac{-632.8\text{ nm}}{4}$$

$$\rightarrow t_{\max} = \frac{(-158.2\ \text{nm})(5^4)(128)(1.517^3)}{1.517^2 - 1} = 33.95\,\text{mm}.$$

The maximum tilt angle for coma using $t = 10\,\text{mm}$:

$$W_{131} = \frac{-t\psi}{(F/\#)^3}\left[\frac{n^2-1}{16n^3}\right]\cos\theta = \frac{-632.8\ \text{nm}}{4}$$

$$\rightarrow \psi_{\max} = \frac{(-158.2\,\text{nm})(5^3)(16)(1.517^3)}{(-10\,\text{mm})(1.517^2 - 1)(\cos 0)} = 0.0848828\ \text{rad} = 4.86642°.$$

Note that θ, the angle in the pupil plane, is zero for sagittal coma.

The maximum tilt angle for astigmatism using $t = 10\,\text{mm}$ and $\theta = 0$:

$$W_{222} = \frac{-t\psi^2}{(F/\#)^2}\left[\frac{n^2-1}{8n^3}\right]\cos^2\theta = \frac{-632.8\ \text{nm}}{4}$$

$$\rightarrow \psi_{\max} = \sqrt{\frac{(-158.2\ \text{nm})(5^2)(8)(1.517^3)}{(-10\ \text{mm})(1.517^2 - 1)(\cos^2 0)}} = 0.092\ \text{rad} = 5.28°.$$

11.5 Chromatic aberration

A lens will not focus all the colors (wavelengths) of light to exactly the same place, because the focal length depends on the index of refraction. As demonstrated in Figure 4.20, which shows the dispersion curve for standard glasses, the blue wavelength (*F*, 486.1 mm) has a higher index of refraction than the red wavelength (*C*, 656.3 nm). The amount of chromatic (color) blur depends on the difference in refractive index of *F* and *C* light. For a positive lens, the chromatic aberration is shown in Figure 11.23.

While the Seidel aberrations are monochromatic (for a single color), chromatic aberrations only occur for polychromatic light. These aberrations are not only on

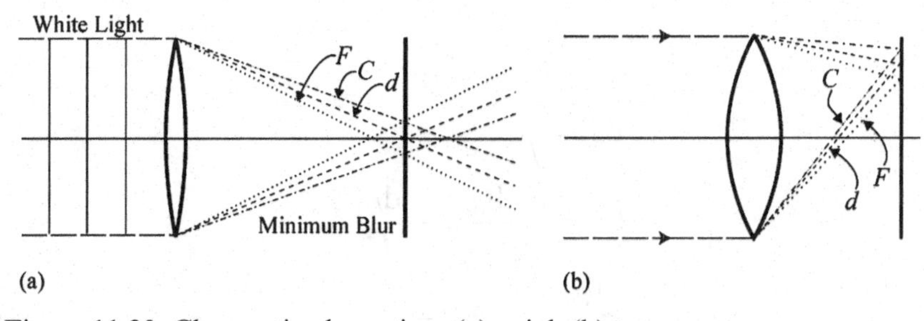

Figure 11.23 Chromatic aberration: (a) axial; (b) transverse.

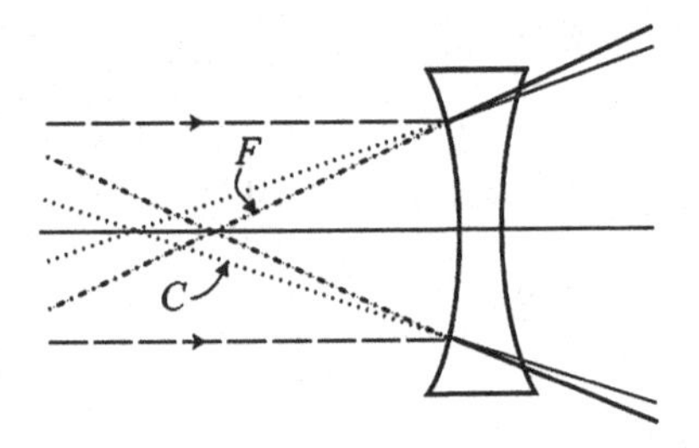

Figure 11.24 Negative lens axial chromatic aberration.

axis, as shown in Figure 11.23(a), but exists off axis as a function of field angle, as shown in Figure 11.23(b). These types of chromatic aberrations are referred to as longitudinal chromatic aberration (Figure 11.23(a)) and transverse (or lateral) chromatic aberration (Figure 11.23(b)). The amount of chromatic aberration is dependent on the glass chosen. Small Abbe number (higher dispersion) glasses have relatively larger chromatic blur. Exotic glass materials containing fluorites have been developed to produce low dispersion for singlets. A negative lens provides the opposite effect on the color dispersion, as shown in Figure 11.24.

The optical power difference between the F and C wavelengths for a thin lens in air can be expressed as (assuming all cases are in air):

$$\Delta\phi_{FC} = \phi_F - \phi_C = (n_F - 1)\left(\frac{1}{R_1} - \frac{1}{R_2}\right) - (n_C - 1)\left(\frac{1}{R_1} - \frac{1}{R_2}\right), \quad (11.26)$$

$$\Delta\phi_{FC} = (n_F - n_C)\left(\frac{1}{R_1} - \frac{1}{R_2}\right). \quad (11.27)$$

If we multiply the numerator and denominator by $(n_d - 1)$:

$$\Delta\phi_{FC} = \left(\frac{n_F - n_C}{n_d - 1}\right)(n_d - 1)\left(\frac{1}{R_1} - \frac{1}{R_2}\right). \quad (11.28)$$

Equation (11.28) has terms which were defined previously for Abbe numbers ($V^{\#}$) (Equation (4.33)) and the optical power of a thin lens (Equation (6.14)), so:

$$\Delta\phi_{FC} = \frac{\phi_d}{V^{\#}}. \quad (11.29)$$

Since

$$\phi_d = \frac{1}{f^*_{\mathrm{d}}},$$

$$\mathrm{d}\phi_d = -\frac{\mathrm{d}f_d}{f^{*2}_{\mathrm{d}}} = -\frac{\mathrm{d}f^*_{\mathrm{d}}}{f^*_{\mathrm{d}}}\phi_d,$$

$$\frac{\mathrm{d}\phi_d}{\phi_d} = -\frac{\mathrm{d}f^*_{\mathrm{d}}}{f^*_{\mathrm{d}}} = \frac{1}{V^{\#}}. \quad (11.30)$$

The approach in going from Equation (11.29) to Equation (11.30) is to make the assumption that the difference in optical power for F and C light ($\phi_F - \phi_C$) is equal to the differential $\mathrm{d}\phi_d$ ($\mathrm{d}\phi_d = \phi_F - \phi_C = \Delta\phi_{FC}$). The important interpretation of Equation (11.30) is that the spread, for a single lens in the z direction, is as shown in Figure 11.23(a), for colors from blue to red.

The results for a thin lens, see Equation (11.30), show that the longitudinal chromatic aberration is the focal length for d light divided by the Abbe number of the glass of the lens. For an N-BK7 glass lens, this value is about 1/64.2 or 1.6% of the effective focal length, and therefore for a 100 mm effective focal length, $F/4$ lens, the longitudinal chromatic aberration is 1.6 mm and the blur off-axis is about 0.4 mm.

There exists a focal plane, as shown in Figure 11.23(a), that has the minimum blur, which is called the circle of least confusion. However, the blur diameter can be greatly reduced by combining two lenses to form an achromat. Recall that a negative lens has the chromatic aberration in reverse order. Therefore, the combination of a strong positive lens made from a low dispersion glass (crown) cascaded with a weaker high dispersion flint glass, as shown in Figure 11.25, can correct the chromatic aberration for two wavelengths. So F and C light have the same focal length, while d light is a different one. This improves the blur considerably.

Two thin lenses, however, in the general case, can cancel the chromatic aberration without being of opposite sign. Consider taking the difference of the power of two thin lenses from Equation (6.74):

$$\mathrm{d}\phi = \mathrm{d}\phi_1 + \mathrm{d}\phi_2 - (\phi_1 \mathrm{d}\phi_2 + \mathrm{d}\phi_1 \phi_2)t,$$

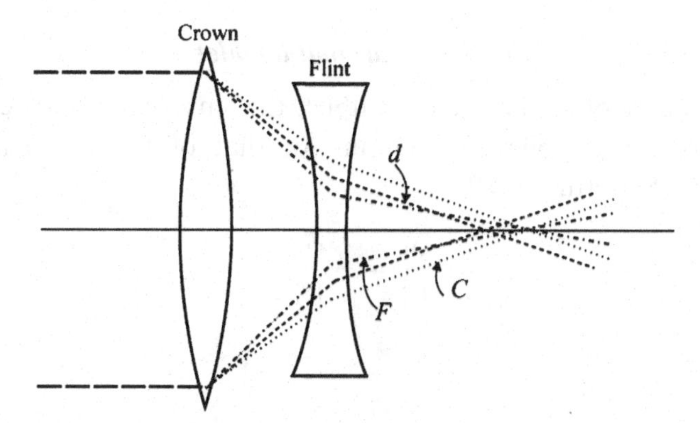

Figure 11.25 Positive crown and negative flint glass lenses canceling the chromatic aberration for F and C wavelengths.

where we interpret the differential power $d\phi$ as $\phi_F - \phi_C$ being the power spread around d light. Rearranging terms,

$$d\phi = (1 - t\phi_2)d\phi_1 + (1 - t\phi_1)d\phi_2. \tag{11.31}$$

From Equation (11.30) for lens 1 and lens 2:

$$d\phi_1 = \phi_1 / V_1^\#, \tag{11.32}$$

$$d\phi_2 = \phi_2 / V_2^\#. \tag{11.33}$$

Substituting into Equation (11.31)

$$d\phi = (1 - t\phi_2)\frac{\phi_1}{V_1^\#} + (1 - t\phi_1)\frac{\phi_2}{V_2^\#}. \tag{11.34}$$

To make the red (F light) and blue (C light) focus at the same point, the difference in the optical powers ϕ_F and ϕ_C must be zero ($d\phi = 0$). So from Equation (11.34):

$$\frac{\phi_1}{V_1^\#} + \frac{\phi_2}{V_2^\#} = t\left[\frac{\phi_1\phi_2}{V_2} + \frac{\phi_1\phi_2}{V_1}\right]$$

$$t = \frac{V_2^\#\phi_1 + V_1^\#\phi_2}{\left[V_1^\# + V_2^\#\right]\phi_1\phi_2} \tag{11.35}$$

This is the separation distance (t) needed between two lenses to make an achromat. There are two special cases:

(1) $t = 0$; an achromat doublet;
(2) the same glass is used for each lens ($V_1^\# = V_2^\#$).

11.5.1 Achromat doublet

In the special case of an achromat doublet, the lenses are cemented together, so the total optical power is the sum of that of the two lenses. From Equation (11.35), setting $t = 0$,

$$\frac{\phi_1}{V_1^\#} = -\frac{\phi_2}{V_2^\#}, \tag{11.36}$$

or

$$\phi_2 = -\frac{V_2^\#}{V_1^\#}\phi_1. \tag{11.37}$$

Since lens 1 is a positive lens, and the Abbe numbers are positive, the second lens has to have a negative power. For this achromat doublet, its optical power is the sum of the powers of the individual lenses:

$$\phi = \phi_1 + \phi_2 = \phi_1 - \frac{V_2^\#}{V_1^\#}\phi_1, \tag{11.38}$$

$$\phi = \phi_1 \left[\frac{V_1^\# - V_2^\#}{V_1^\#}\right]. \tag{11.39}$$

Rearranging terms,

$$\phi_1 = \phi \left[\frac{V_1^\#}{V_1^\# - V_2^\#}\right]; \tag{11.40}$$

similarly,

$$\phi_2 = -\phi \left[\frac{V_2^\#}{V_1^\# - V_2^\#}\right]. \tag{11.41}$$

Such doublets with optical power ϕ are cemented together to reduce the blur due to the refractive index variation for the visible spectrum. The achromat requires a positive and a negative lens with different Abbe numbers. The resulting focal length of the doublet is

$$\frac{1}{f} = \frac{1}{f_1} + \frac{1}{f_2}.$$

11.5.2 Air spaced achromat

For the second special case, in which the same glass material ($V_1^\# = V_2^\#$) is used for each lens, from Equation (11.35) the separation becomes

$$t = \frac{\phi_1 + \phi_2}{2(\phi_1\phi_2)} = \frac{f_2 + f_1}{2}. \tag{11.42}$$

An example of this case is the Ramsden eyepiece used on telescopes. The resulting optical power is

$$\phi = \frac{1}{f_1} + \frac{1}{f_2} - \left(\frac{f_2 + f_1}{2}\right)\left(\frac{1}{f_1}\right)\left(\frac{1}{f_2}\right)$$

$$\phi = \frac{1}{2f_1} + \frac{1}{2f_2} = \frac{f_2 + f_1}{2f_1 f_2}. \tag{11.43}$$

and the focal length is

$$f = \frac{2f_1f_2}{f_1 + f_2}. \tag{11.44}$$

Therefore, the effective focal length is determined by Equation (11.44) for the achromat. The lenses can be positive or negative, as long as the separation is set by Equation (11.42).

11.5.3 Secondary chromatic aberration

The achromat corrects the focal lengths for F and C wavelengths; however, as shown in Figure 11.26, there is still a residual chromatic aberration, since d light has a different focal length.

The residual chromatic aberration, defined as the difference in focal lengths for the d wavelength and the C wavelength (which is the same as that for the F wavelength), can be plotted as shown in Figure 11.27. Note: the d focus may not be the maximum focal shift from F, C.

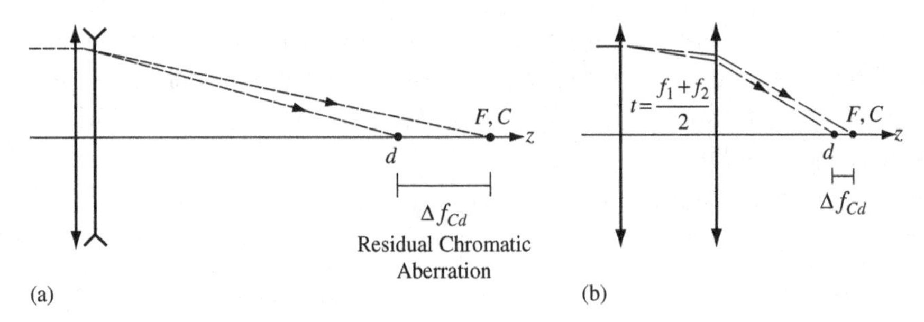

Figure 11.26 Residual chromatic aberration for: (a) an achromat doublet; (b) an air spaced achromat.

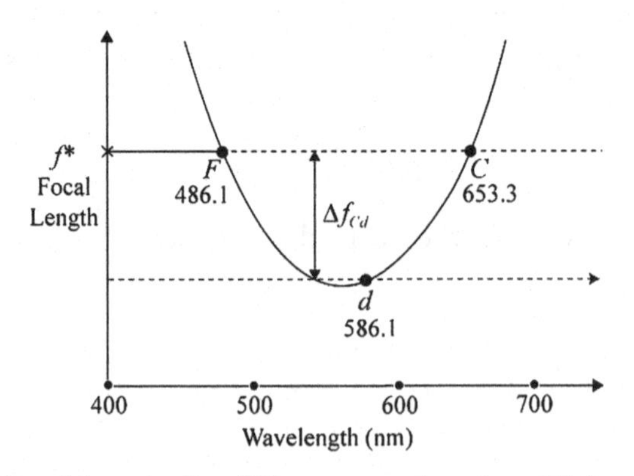

Figure 11.27 Focal lengths for different wavelengths of light.

The residual difference in focal lengths can be expressed in terms of optical power for the two lenses as

$$\Delta\phi_{dC} = \Delta\phi_{dC}^1 + \Delta\phi_{dC}^2, \tag{11.45}$$

where

$$\begin{aligned}\Delta\phi_{dC}^1 &= (n_d - 1)\left(\frac{1}{R_1} - \frac{1}{R_2}\right) - (n_C - 1)\left(\frac{1}{R_1} - \frac{1}{R_2}\right)\\ &= (n_d - n_C)\left(\frac{1}{R_1} - \frac{1}{R_2}\right)\\ &= (n_d - n_C)\left(\frac{n_F - n_C}{n_F - n_C}\right)\left(\frac{1}{R_1} - \frac{1}{R_2}\right).\end{aligned} \tag{11.46}$$

By using Equation (4.45) for the partial dispersion definition and Equation (11.27) for change in power,

$$\Delta\phi_{dC}^1 = p_1\Delta\phi_{FC}^1. \tag{11.47}$$

Similarly, for the second lens:

$$\Delta\phi_{dC}^2 = p_2\Delta\phi_{FC}^2. \tag{11.48}$$

Substituting into Equation (11.45),

$$\Delta\phi_{dC} = p_1\Delta\phi_{FC}^1 + p_2\Delta\phi_{FC}^2. \tag{11.49}$$

Using Equation (11.29) for $\Delta\phi_{FC}$ for each lens,

$$\Delta\phi_{dC} = \frac{p_1\phi_d^1}{V_1^\#} + \frac{p_2\phi_d^2}{V_2^\#}. \tag{11.50}$$

From Equation (11.36) for an achromatic doublet, rewriting in this form:

$$\frac{\phi_d^1}{V_1^\#} = -\frac{\phi_d^2}{V_2^\#}, \tag{11.51}$$

which leaves Equation (11.50) as

$$\Delta\phi_{dC} = (p_1 - p_2)\frac{\phi_d^1}{V_1^\#}. \tag{11.52}$$

In order to get the change in optical power of the two lenses for d and C wavelengths we substitute Equation (11.40) for ϕ_d^1 in Equation (11.52):

$$\Delta\phi_{dC} = \frac{p_1 - p_2}{V_1^\# - V_2^\#}\phi = \frac{\Delta p}{\Delta V^\#}\phi, \tag{11.53}$$

$$\frac{\Delta\phi_{dC}}{\phi} = \frac{\Delta f_{Cd}}{f} = \frac{p_1 - p_2}{V_1 - V_2} = \frac{\Delta p}{\Delta V^{\#}}, \tag{11.54}$$

where

$$\Delta\phi_{dC} = \phi_d - \phi_C; \quad \Delta f_{Cd} = f_C - f_d.$$

This longitudinal shift in the location of the focus for d to F, C light is also called secondary color aberration, as shown in Figure 11.26 as residual chromatic aberration.

Equation (11.54) relates the residual secondary color (Δf_{Cd}) to the change in partial dispersion and change in Abbe number for the glasses selected for the achromat doublet. From Figure 4.29, the slope between the two glass choices for these lenses on the partial dispersion versus Abbe number plot is $(p_1 - p_2)/(V_1^* - V_2^*)$. For recommended typical glasses, as shown in Figure 4.29, the slope is approximately

$$\frac{\Delta p}{\Delta V^{\#}} = \frac{1}{2400}. \tag{11.55}$$

Example 11.4

For an achromat thin lens doublet with an *efl* of 100 mm, made of N-BK7 and F2:

(a) What is the optical power of each lens?
(b) What is the residual secondary longitudinal color aberration?

$$\text{N-BK7:} \quad V^{\#} = 64.17, \quad p = 0.3075$$

$$\text{F2:} \quad V^{\#} = 36.37, \quad p = 0.2437$$

$$\phi = \frac{1}{efl} = 10^{-2}\ \text{mm},$$

$$\phi_1 = \phi\left[\frac{V_1^{\#}}{V_1^{\#} - V_2^{\#}}\right] = 10^{-2}\left[\frac{64.17}{64.17 - 36.37}\right] = 2.3(10^{-2})\ \text{mm}^{-1},$$

$$\phi_2 = -\phi\left[\frac{V_2^{\#}}{V_1^{\#} - V_2^{\#}}\right] = -10^{-2}\left[\frac{36.37}{64.17 - 36.37}\right] = -1.31(10^{-2})\ \text{mm}^{-1}.$$

$$\Delta f_{Cd} = f\left(\frac{\Delta P}{\Delta V^{\#}}\right) = 100\left(\frac{0.3075 - 0.2937}{64.17 - 36.37}\right) = 100\left(\frac{1}{2014}\right)$$

$$\Delta f_{Cd} = 0.05\ \text{mm}.$$

If one picks unique glasses such that the slope ($\Delta p/\Delta V^{\#}$) is smaller than the average, the secondary aberration goes down, but the lenses tend to become "fat."

11.5.4 Apochromat

Better color correction can be achieved than from the achromat by using three lenses. Using three lenses, three wavelengths (F, d and C) can have the same focal length.

$$f_F = f_d = f_C. \tag{11.56}$$

When the chromatic aberration for three wavelengths is eliminated, an apochromat is built. These apochromats became viable with the invention of exotic anomalous dispersion glasses. The glasses developed contained fluorite. Fluorite glasses display unusual behavior with regard to partial dispersion, and do not follow the general characteristics given in Figure 4.29. They are also very expensive and difficult to process.

In order to make an apochromat, all three glasses for the triplet cannot lie on the same line of the partial dispersion versus Abbe number plot. As shown in Figure 11.28, the three glasses in an apochromat must be chosen so that plots of their partial dispersion versus Abbe number form the sides of a triangle. The area of the triangle should be maximized by the proper choice of glasses, a, b, and c (Kingslake, 1978, p. 85).

New glass materials were developed for lens designers to reduce the secondary spectrum such that the image blur was no longer limited by secondary color. These fluorite glasses allow three colors (F, d, C) to focus in the same focal plane (three zero focus shifts) instead of two colors (F, C). In many astronomical systems with long focal lengths, the image quality is adversely affected by

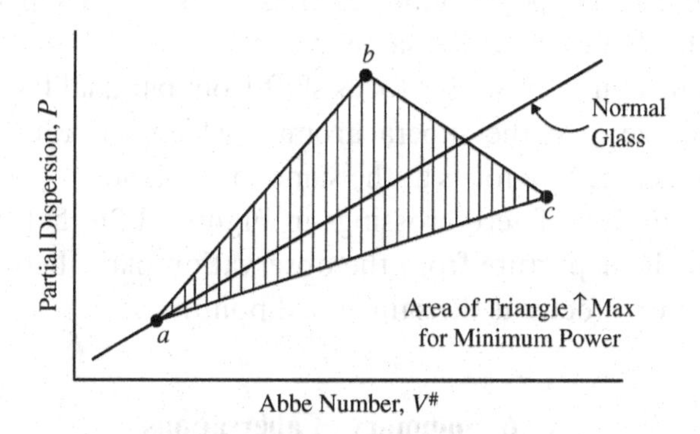

Figure 11.28 Glass choices for an apochromat, as shown in the figure using partial dispersion versus Abbe number – a, b, and c must be such that the area of the triangle formed by them is maximized.

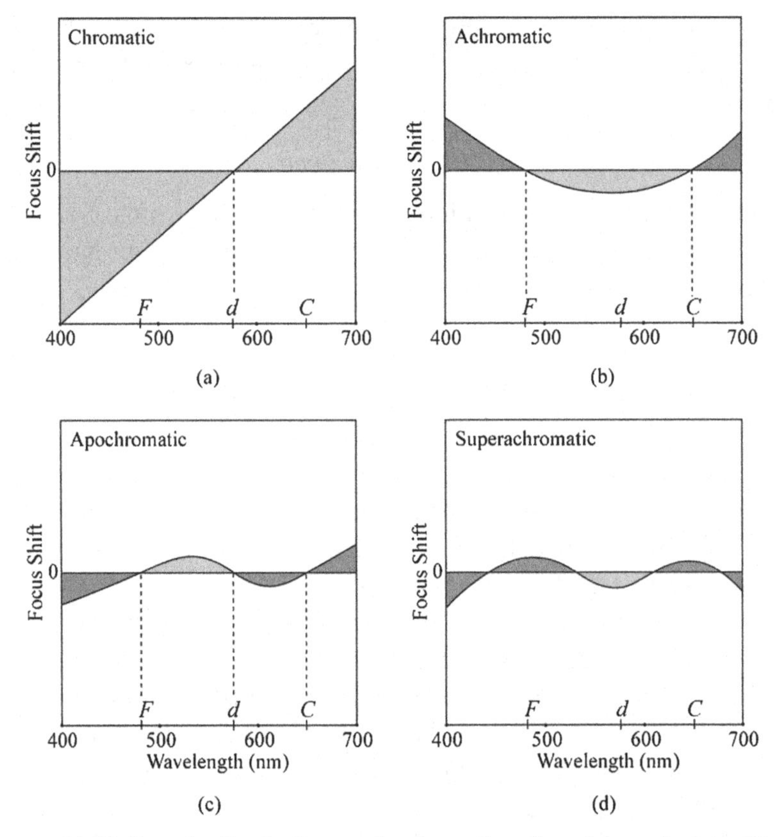

Figure 11.29 Longitudinal chromatic aberration for: (a) a singlet; (b) an achromat doublet with F and C achromatized; (c) an apochromat with three wavelengths (F, d, C) having the same focal length; (d) a superachromat with four wavelengths focusing to the same point.

secondary color, so an apochromatic design is necessary. A superachromat has four wavelengths that focus at the same focal plane, or four zeros, as shown in Figure 11.29. The figure shows the focus shift from paraxial focus for a singlet having only one zero to the superachromat, which has four zeros or four wavelengths that come to a focus at the same focal plane.

It is not the number of zero crossings in Figure 11.29 that determines the best image, but the departure from the observation plane between the wavelengths that come to focus at the same focal point.

11.6 Summary of aberrations

For ideal optical systems, ignoring diffraction, all rays of light from a point on an object would form the same point in the image plane. Aberrations are what

Table 11.1. *Aberration characteristics and possible corrective actions*

Aberration	Characteristic	Corrective action
(1) Spherical aberration	Monochromatic, on- and off-axis, image blur	Lens bending, high refractive index, aspherics gradient index, doublet
(2) Coma	Monochromatic, off-axis only, blur linear in field	Bending, spaced doublet with central stop
(3) Astigmatism	Monochromatic, off-axis blur is quadratic with field	Spaced doublet with stop
(4) Curvature of field	Monochromatic, off-axis blur, quadratic with field	Spaced doublet
(5) Distortion	Monochromatic, off-axis cubic field dependence	Symmetry about stop
(6) Chromatic aberration	Chromatic, on- and off-axis blur	Contact doublet, spaced doublet

cause these rays from a point on the object not to converge to a point on the image. Table 11.1 lists the types of aberrations and the corrective actions needed to minimize their effect on the image.

Problems

11.1 A lens made of a flint of $n = 1.72$ has a focal length of $+5$ m. For parallel incident light, determine the position and shape factors and the two radii of curvature necessary for the lens to have minimum spherical aberration.

11.2 If a lens of 25 mm diameter has a focal length of 50 mm, with a longitudinal spherical aberration of 2 mm, what is:
(a) the transverse spherical aberration;
(b) *F*/# of the lens;
(c) the longitudinal chromatic aberration.

11.3 A thin lens made of SF2 has a focal length of 2 m. For an object at infinity, and for the lens to have minimum spherical aberration:
(a) What shape factor should be used?
(b) What must be the radius of each surface of the thin lens?
(c) What must be the position factor?

11.4 What is the shape factor for minimum spherical aberration for an object at infinity using a glass of refractive index equal to 4?

11.5 What is the shape factor in order for a lens made of glass #780200.365 to have zero coma with an object at infinity?

11.6 A 5.3 diopter N-BK7 thin lens is to be combined with a flint thin lens of SF58 to make a contact doublet achromat for *F* and *C* wavelengths. What is the optical power of the flint lens?

11.7 An achromatic doublet consists of two positive lenses separated by 15 cm. The first lens has a power of 20 diopters. What is the optical power of the second lens? What is the total optical power of the system?

11.8 What is the longitudinal chromatic aberration for a singlet made of glass #565656 if its optical power is 10 diopters? If this were an achromatic doublet, what would the approximate longitudinal aberration value be?

11.9 We want a 10 diopter cemented achromat thin lens designed from glasses of Abbe numbers 30 and 60. What is the optical power of each lens?

11.10 For glass #755511, what shape factor should be used to minimize coma for an object at infinity? What is the value of the transverse coma?

11.11 For the following system, find:

(a) longitudinal chromatic aberration (F to C);

(b) transverse chromatic aberration at f_d* for both C and F light.

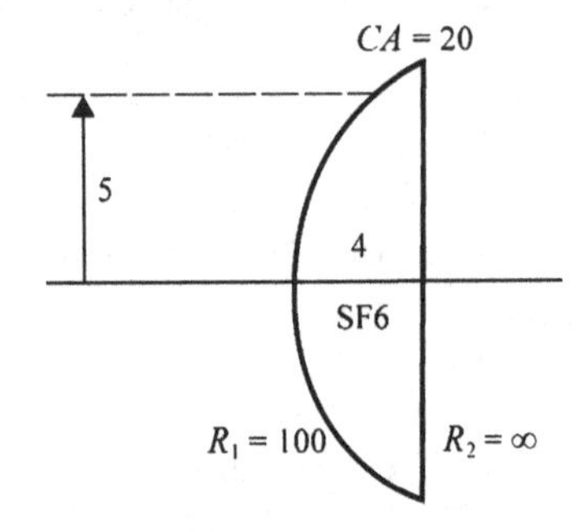

11.12 Design a lens to cement to the lens given in Problem 11.11 to make an achromatic doublet. What is your doublet's partial dispersion? (Specify the materials you used, all radii, thicknesses, and the optical power of your achromat.)

11.13 Given the system shown below, trace a paraxial ray ($y = 1$ mm) from an object at infinity. Also, trace a ray of $y = 70$ mm from an object at infinity.

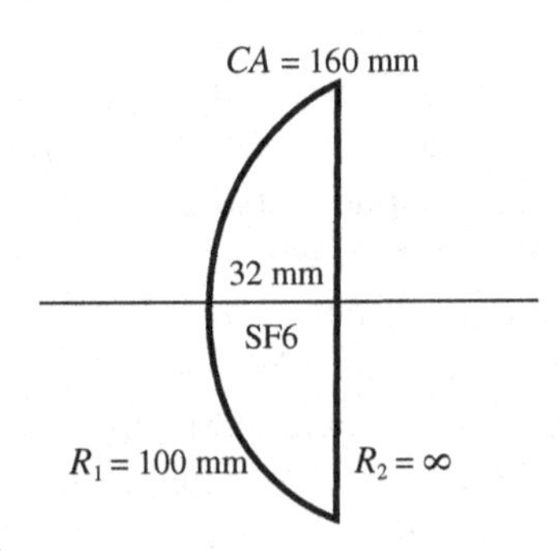

(a) Find the longitudinal spherical aberration from these rays.

(b) Find the transverse spherical aberration at paraxial focus.

11.14 Given the following system:

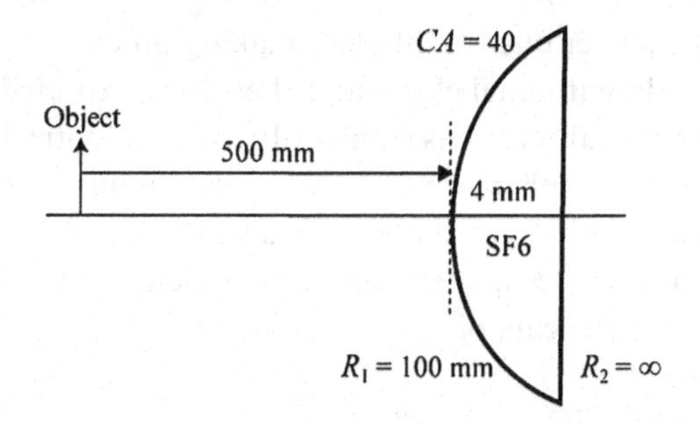

(a) Find the shape factor for minimum coma.
(b) This lens shape will be minimized for coma with what index of refraction?

11.15 Calculate the longitudinal spherical aberration (paraxial ray intercept to real ray intercept) for an input height of 2 cm for a spherical mirror with radius of curvature equal to 5 cm and a diameter (CA) of 4 cm.

11.16 For a thin lens (singlet) of 50 mm diameter (at the *d* wavelength), $F/\# = 4$, made of SF5:
(a) What is the effective focal length (*efl*) for the *F* wavelength?
(b) What is the *efl* for the *C* wavelength?
(c) What is the longitudinal chromatic aberration?
(d) What is the transverse chromatic aberration?

11.17 How much longitudinal aberration is present between paraxial focus (ray height of 1) and a second ray at a height 8 for a parabolic mirror with a focal length 5 ($y^2 = -20z$).

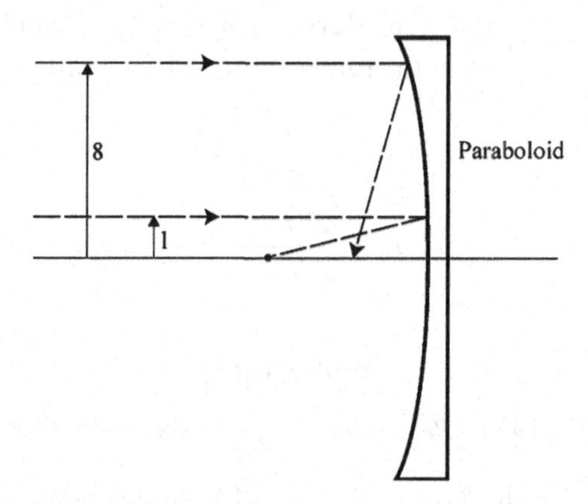

11.18 What is the shape factor for minimum spherical aberration for a lens of 5 diopters of 517642 glass used to image an object at infinity? Also sketch the expected lens shape.

11.19 A thin lens achromatic doublet has 4 diopters of optical power. If the Abbe numbers are $V_1^{\#} = 75$ and $V_2^{\#} = 25$:
(a) What is the power of each thin lens making up the doublet?
(b) What is the longitudinal chromatic aberration expected for this doublet?

11.20 List the third-order aberrations commonly found in optical systems.

11.21 What do you think a sphero-chromatic aberration might be?

11.22 Two thin lenses with 3 cm diameters have effective focal lengths of 5 cm and −25 cm, and are cemented together. A field stop of 5 cm (diameter) is placed at their back focal point.
(a) What is the $(F/\#)_\infty$?
(b) What is the full field of view?
(c) What is the diameter of the diffraction Airy disc at the field stop (for an on-axis point source at $-\infty$)?
(d) What is the angular diffraction blur?

11.23 Design a thin lens $F/2$, 5 cm achromat using 500800 and 750400 glasses. What would the longitudinal chromatic aberration be if this lens were a singlet?

11.24 For a point source ($\cdot$) and cross ($\times$) at infinity, what appears at the paraxial plane for each of the third-order aberrations? Consider each aberration separately.

11.25 Design a thin lens, $F/2$, 5 cm in diameter, cemented achromat using 500800 and 750400 glasses. Also, force $R_1 = -R_2$. The lens must be equi-convex.
(a) What is the longitudinal chromatic aberration?
(b) What is the transverse chromatic aberration?
(c) What would the longitudinal chromatic aberration be if this lens were a singlet?

11.26 Your boss wants an achromat of *efl* equal to 100 mm, but requires you to use only a single glass, KzFS4, to make it. Design a thin lens for this achromat and determine the optical power of each lens and separation.

11.27 Show that the expression for the radii of curvature (R_1 and R_2) for a thin lens with effective focal length, f, made of a glass of refractive index, n, has minimum spherical aberration when the following radii of curvature are used:

$$R_1 = \frac{2f(n-1)}{\mathcal{S}+1},$$

$$R_2 = \frac{2f(n-1)}{\mathcal{S}-1}.$$

Bibliography

Born, M. and Wolf, E. (1959). *Principles of Optics*, sixth edn. Cambridge: Cambridge University Press.

Hecht, E. (1998). *Optics*, third edn. Reading, MA: Addison-Wesley.

Jenkins, F. A. and White, H. E. (1976). *Fundamentals of Optics*, fourth edn. New York: McGraw-Hill.

Shannon, R. R. and Wyant, J. C. (1965). *Applied Optics and Optical Engineering*, Vol. III. New York: Academic Press.

Kingslake, R. (1978). *Lens Design Fundamentals*, New York: Academic Press.
Ray, S. F. (2002). *Applied Photographic Optics*, third edn. Oxford: Focal Press.
Smith, W. J. (2000). *Modern Optical Engineering*, third edn. New York: McGraw-Hill.
Gaskill, J. (1978). *Linear Systems, Fourier Transforms, and Optics*. New York: Wiley.
Mahajan, V. N. (2001). *Optical Imaging and Aberrations*. Bellingham: SPIE Optical Engineering Press.
Wyant, J. C. and Creath, K. (1992). Basic wavefront aberration theory for optical metrology. *Applied Optics and Optical Engineering* **11**, 1.

12

Real ray tracing

12.1 Approach

The act of image formation in our present understanding consists of reformatting diverging wavefronts from a source (object) to converging spherical wavefronts moving toward image points in the image plane. The transfer of wavefronts through an optical system can be done most easily, as has been accomplished so far, by the use of ray tracing. The tracing of rays through an optical system is determined purely by geometrical considerations and trigonometry. The assumptions made in ray tracing through an optical system are:

(1) Rays travel at a constant velocity in homogeneous media.
(2) Rays travel in straight lines.
(3) Rays follow Snell's law at the interface between media.
(4) At an interface, the reflected and refracted rays lie in the plane of incidence.
(5) Object and image surfaces are opaque.

Ray tracing through an optical system is best accomplished by a moving coordinate system using simple geometrical considerations and trigonometric functions, totally ignoring diffraction effects.

Thus far, only paraxial rays have been used to find the image location, size and brightness. The small angle approximation describes the optical system to first order; however, for object points at large distances from the optical axis, corresponding image points are clearly aberrated and not correctly predicted by paraxial ray tracing. Real ray tracing uses vectors starting from a point with direction cosines for the ray in each space (segment) as it is traced from the object point to the image point.

The sign convention for real ray tracing follows that of Chapter 10 and the previous chapters (see Table 5.1).

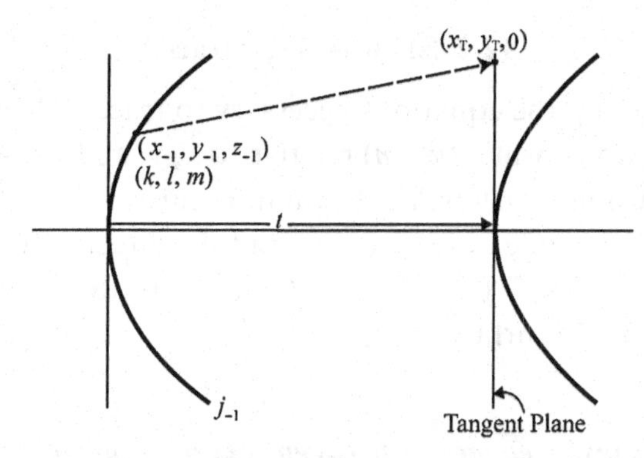

Figure 12.1 Real ray trace approach from the surface to the tangent plane.

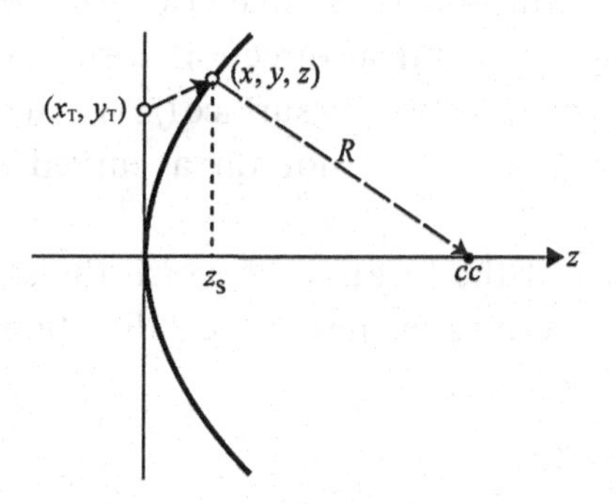

Figure 12.2 Real ray trace from the tangent plane to the tangent plane surface.

The various rays that can be traced through an optical system are:

(1) general ray – any ray transferring from an object to image point (may be scattered);
(2) meridional ray – a ray that lies in a plane which contains the optical axis;
(3) paraxial ray – a meridional ray that lies close to the optical axis;
(4) skew ray – any non-meridional ray in the system.

For real rays, the actual surfaces instead of just the vertex planes (tangent planes) are used, resulting in a three-dimensional situation. The approach used in this chapter will follow five general steps:

(1) Start at a point on one surface (j_{-1}), and transfer the ray to a tangent plane for the next surface (j). The direction cosines of the initial ray are known, as shown in Figure 12.1, at a surface j_{-1}.
(2) Transfer the ray from the tangent plane to the actual surface (Figure 12.2). Find the intersection of the real ray with the surface.
(3) Find the surface normal at the intersection point.
(4) Apply Snell's law of refraction to determine the new direction cosines in the plane of incidence after the surface.
(5) Transfer to the next surface. Repeat the sequence.

12.2 Skew real ray trace

This section will set up the equations necessary to trace a skew ray through a system using direction cosines (k,l,m) relative to x, y, and z respectively, in the right-handed coordinate system, as shown in Figure 12.3.

The products of the index of refraction and the direction cosines are called "optical direction cosines" (K_{-1}, L_{-1}, M_{-1}), where they are defined in the space between j_{-1} and j surfaces.

12.2.1 Transfer equations between spherical surfaces, j_{-1} to j

Figure 12.4 shows a ray transferring between two surfaces, from surface j_{-1} to surface j. This is a ray with optical direction cosines (K_{-1}, L_{-1}, and M_{-1}) starting at the x_{-1}, y_{-1}, z_{-1} position on surface (j_{-1}). An object will typically be flat, lying in the x–y plane only, not on a curved surface as shown in Figure 12.4.

Snell's law is applied at the location at which the skew ray intersects the spherical surface (j) to produce the new optical direction cosines. The tangent

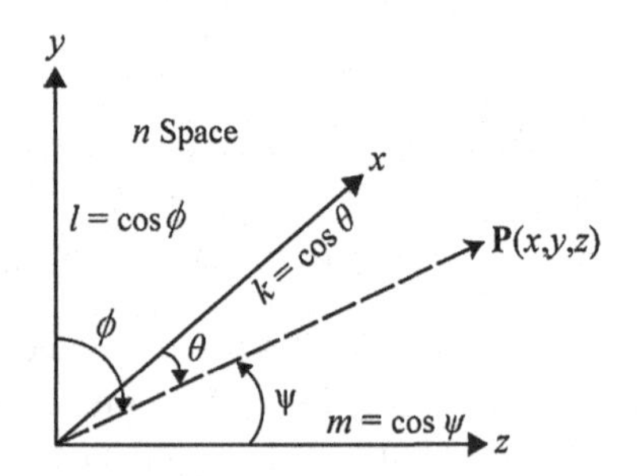

Figure 12.3 Vector **P** has direction cosines (k, l, m) for a skew ray.

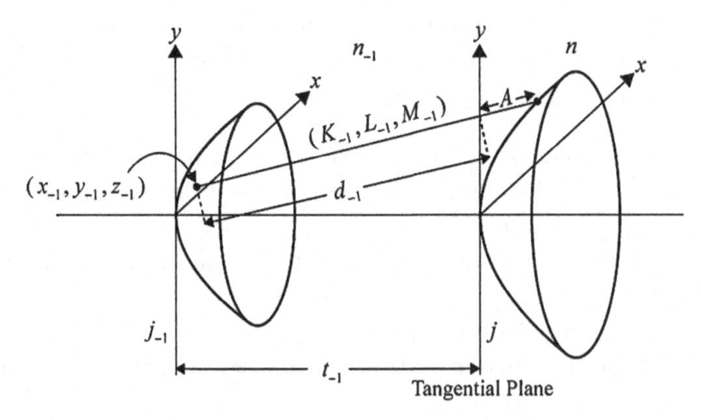

Figure 12.4 Skew ray trace from surface j_{-1} to the jth surface.

plane is the first location at which this computation is made, and the process continues at all successive surfaces until the image point is reached.

The tangent plane coordinates are (x_T, y_T, 0) for all points on the plane. The new value of x_T is the value of x_{-1} plus the change introduced in the transfer (Δx). This Δx is the projection in the x axis over the length, d_{-1}, onto the x axis:

$$x_T = x_{-1} + \Delta x = x_{-1} + d_{-1}\frac{K_{-1}}{n_{-1}}, \tag{12.1}$$

$$y_T = y_{-1} + \Delta y = y_{-1} + d_{-1}\frac{L_{-1}}{n_{-1}}, \tag{12.2}$$

where K_{-1}/n_{-1} and L_{-1}/n_{-1} are the direction cosines with respect to x and y.

The separation between the surface j_{-1} and the tangent plane at the jth surface is not known, but must be calculated from the initial bounding conditions. The change in the z coordinate from z_{-1} to z_T is taken from Figure 12.4:

$$\Delta z = t_{-1} - z_{-1}. \tag{12.3}$$

The change in the z position (Δz) equals the projection of the ray length, d_{-1}, along the z axis:

$$\Delta z = d_{-1}\frac{M_{-1}}{n_{-1}}. \tag{12.4}$$

The skew ray intersection with the tangent plane of the next surface (j), is calculated using

$$x_T = x_{-1} + d_{-1}\frac{K_{-1}}{n_{-1}}, \tag{12.5}$$

$$y_T = y_{-1} + d_{-1}\frac{L_{-1}}{n_{-1}}, \tag{12.6}$$

$$z_T = 0, \tag{12.7}$$

where $d_{-1}/n_{-1} = (t_{-1} - z_{-1})(1/M_{-1})$ from Equations (12.3) and (12.4).

Thus far, we have taken a skew ray on surface j_{-1} with coordinates of (x_{-1}, y_{-1}, z_{-1}) along with direction cosines to locate the intersection of the skew ray with a plane tangent to the vertex of the next optical surface. Thus, Equations (12.5), (12.6), and (12.7) give the x_T, y_T, z_T coordinates of the skew ray. We need to calculate these quantities for the actual physical surface. Since the tangent plane is non-refracting, the direction cosines do not change for the last segment, A, of the ray to the spherical surface, as shown in Figure 12.4. That surface's coordinates, x, y, z, are calculated as

$$x = x_T + \frac{A}{n_{-1}} K_{-1}\,, \tag{12.8}$$

$$y = y_T + \frac{A}{n_{-1}} L_{-1}\,, \tag{12.9}$$

$$z = \frac{A}{n_{-1}} M_{-1}\,. \tag{12.10}$$

The critical value in the above equations is A, which depends on the direction cosines, the coordinates in the tangent plane (x_T, y_T, z_T), and the curvature of the spherical surface.

From the equation of a sphere with the coordinate origin at the vertex,

$$x^2 + y^2 + (z - R)^2 = R^2\,, \tag{12.11}$$

$$x^2 + y^2 + z^2 - 2RZ = 0\,. \tag{12.12}$$

Recalling that curvature $C = 1/R$,

$$C^2(x^2 + y^2 + z^2) - 2CZ = 0\,. \tag{12.13}$$

Substituting Equations (12.8)–(12.10) into (12.13) and collecting terms,

$$C^2\left[\left(x_T + \frac{A}{n_{-1}} K_{-1}\right)^2 + \left(y_T + \frac{A}{n_{-1}} L_{-1}\right)^2 + \left(\frac{A}{n_{-1}} M_{-1}\right)^2\right] - 2C\frac{A}{n_{-1}} M_{-1} = 0$$

$$C^2\left[x_T^2 + 2\frac{A}{n_{-1}} K_{-1} x_T + \left(\frac{A}{n_{-1}}\right)^2 K_{-1}^2 + y_T^2 + 2\frac{A}{n_{-1}} L_{-1} y_T + \left(\frac{A}{n_{-1}}\right)^2 L_{-1}^2 + \left(\frac{A}{n_{-1}}\right)^2 M_{-1}^2\right]$$
$$-2C\frac{A}{n_{-1}} M_{-1} = 0$$

$$C^2(x_T^2 + y_T^2) + 2\frac{A}{n_{-1}} C^2\left[K_{-1} x_T + L_{-1} y_T - \frac{M_{-1}}{C}\right]$$
$$+ \left(\frac{A}{n_{-1}}\right)^2 C^2\left[K_{-1}^2 + L_{-1}^2 + M_{-1}^2\right]$$
$$= 0\,.$$

Rearranging terms,

$$\left(\frac{A}{n_{-1}}\right)^2 C[K_{-1}^2 + L_{-1}^2 + M_{-1}^2] - 2\left(\frac{A}{n_{-1}}\right)[M_{-1} - C(K_{-1}x_T + L_{-1}y_T)] + C(x_T^2 + y_T^2) = 0 . \tag{12.14}$$

The sum of the direction cosines squared is unity. Optical direction cosines simplify to the index of refraction squared. In addition, new terms may be defined to simplify the quadratic equation:

$$Cn_{-1}^2\left(\frac{A}{n_{-1}}\right)^2 - 2B\frac{A}{n_{-1}} + H = 0 , \tag{12.15}$$

where

$$H = C(x_T^2 + y_T^2) ,$$
$$B = [M_{-1} - C(K_{-1}x_T + L_{-1}y_T)] .$$

Solving Equation (12.15) for A/n_{-1},

$$\frac{A}{n_{-1}} = \frac{2B \pm \sqrt{4B^2 - 4Cn_{-1}^2 H}}{2Cn_{-1}^2} , \tag{12.16}$$

where obviously for a plane surface ($A = 0$ or $C = 0$), the tangent plane coordinates would prevail and $(A/n_{-1}) = 0$. In order for the mathematics to represent the physical situation, the sign in front of the radical in Equation (12.16) has to be negative. Rewriting,

$$\frac{A}{n_{-1}} = \frac{B - n_{-1}\sqrt{(B/n_{-1})^2 - CH}}{Cn_{-1}^2} . \tag{12.17}$$

The extension, A, of the ray from the tangent plane to the physical surface is shown in the plane of incidence in Figure 12.5.

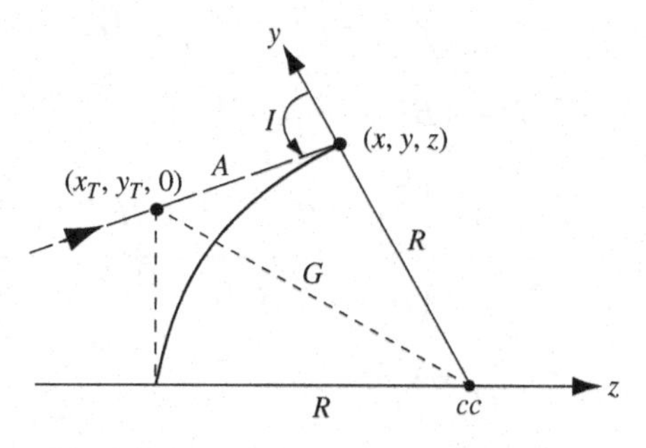

Figure 12.5 Plane of incidence of skew ray at a spherical surface.

From the law of cosines:

$$G^2 = A^2 + R^2 + 2AR\cos I. \tag{12.18}$$

Also, applying the Pythagorean Theorem to the line segments in Figure 12.5:

$$G^2 = R^2 + \left(x_T^2 + y_T^2\right) = A^2 + R^2 + 2AR\cos I. \tag{12.19}$$

Solving for cos I gives

$$\cos I = \frac{(H/C) - A^2}{2AR},$$

where H/C has been substituted for $\left(x_T^2 + y_T^2\right)$. Rearranging terms,

$$n_{-1}\cos I = \frac{H - CA^2}{2(A/n_{-1})}. \tag{12.20}$$

Substituting Equation (12.17) for A/n_{-1} into Equation (12.20),

$$n_{-1}\cos I = \frac{H - Cn_{-1}^2\left[\dfrac{B - n_{-1}\sqrt{(B/n)^2 - CH}}{Cn_{-1}^2}\right]^2}{2\left[\dfrac{B - n_{-1}\sqrt{(B/n)^2 - CH}}{Cn_{-1}^2}\right]}$$

$$= \frac{Cn_{-1}^2 H - \left[B - \left(B^2 - n_{-1}^2 CH\right)^{1/2}\right]^2}{2\left[B - \left(B^2 - n_{-1}^2 CH\right)^{1/2}\right]}$$

$$= \frac{Cn_{-1}^2 H - \left[B^2 + \left(B^2 - n_{-1}^2 CH\right) - 2B\left(B^2 - n_{-1}^2 CH\right)^{1/2}\right]}{2\left[B - \left(B^2 - n_{-1}^2 CH\right)^{1/2}\right]}$$

$$= \frac{-2\left(B^2 - n_{-1}^2 CH\right) + 2B\left(B^2 - n_{-1}^2 CH\right)^{1/2}}{2\left[B - \left(B^2 - n_{-1}^2 CH\right)^{1/2}\right]}$$

$$= \frac{2\left(B^2 - n_{-1}^2 CH\right)^{1/2}\left[-\left(B^2 - n_{-1}^2 CH\right)^{1/2} + B\right]}{2\left[B - \left(B^2 - n_{-1}^2 CH\right)^{1/2}\right]}$$

$$= \left(B^2 - n_{-1} CH\right)^{1/2}.$$

So

$$n_{-1} \cos I = n_{-1} \left[(B/n_{-1})^2 - CH \right]^{1/2}. \quad (12.21)$$

Using Equation (12.21) in Equation (12.17),

$$\frac{A}{n_{-1}} = \frac{B - n_{-1} \cos I}{Cn_{-1}^2}. \quad (12.22)$$

Squaring Equation (12.21) and rewriting,

$$\cos^2 I = \frac{B^2}{n_{-1}^2} - CH$$

$$n_1^2 \cos^2 I = B^2 - n_{-1}^2 CH$$

$$Cn_{-1}^2 = \frac{B^2 - n_{-1}^2 \cos^2 I}{H}$$

$$= \frac{(B + n_{-1} \cos I)(B - n_{-1} \cos I)}{H}.$$

Substituting this into Equation (12.22) to get the length of ray segment A:

$$\frac{A}{n_{-1}} = \frac{H}{B + n_{-1} \cos I}, \quad (12.23)$$

where

$$H = C(x_T^2 + y_T^2), \quad (12.24)$$

$$B = M_{-1} - C(y_T L_{-1} + x_T K_{-1}), \quad (12.25)$$

$$n_{-1} \cos I = n_{-1} \left[\left(\frac{B}{n_{-1}} \right)^2 - CH \right]^{\frac{1}{2}}. \quad (12.26)$$

Now we can calculate the values for the coordinates x, y, z, on the spherical surface using Equations (12.8), (12.9), and (12.10).

12.3 Refraction at the spherical surface

The position of the intersection of the ray on the jth spherical surface with optical direction cosines (K_{-1}, L_{-1}, M_{-1}) has been determined. We now need to apply Snell's law in vector form to find the direction cosines of the refracted ray into the next optical space. If vectors are drawn from the x, y, z point on the

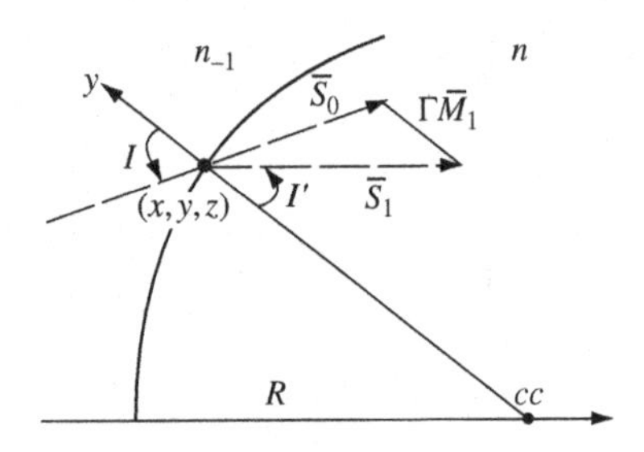

Figure 12.6 Refraction for a spherical surface, using Snell's law.

spherical surface, then the incident ray, $\bar{S}_0$, and the refracted ray, $\bar{S}_1$ will have amplitudes of n_{-1} and n for the jth surface. The layout is shown in Figure 12.6.

The resulting vector $\Gamma\bar{M}_1$ is parallel to the normal to the surface.

$$\bar{S}_1 - \bar{S}_0 = \Gamma\bar{M}_1 . \tag{12.27}$$

The directions of $\bar{S}_0$ and $\bar{S}_1$ are the directions of the incident and refracted rays, respectively. The length Γ is just the difference in the vector components of $\bar{S}_1$ and $\bar{S}_0$ projected onto the radius of curvature:

$$\Gamma = n \cos I' - n_{-1} \cos I , \tag{12.28}$$

$$\Gamma = \left[(n_{-1} \cos I)^2 - n_{-1}^2 + n^2 \right]^{\frac{1}{2}} - n \cos I , \tag{12.29}$$

which is parallel to the normal of the surface.

Figure 12.6 shows the plane of incidence with the radius of curvature in it. The unit vector $\bar{M}_1$ is the vector parallel to the normal times the curvature:

$$\bar{M}_1 = C\left[-x\bar{i} - y\bar{j} + (R - z)\bar{k} \right] , \tag{12.30}$$

where $\bar{i}$, $\bar{j}$, $\bar{k}$ are unit vectors along the coordinate axes.

Using Equation (12.27):

$$\bar{S}_1 - \bar{S}_0 = -Cx\Gamma\bar{i} - C\Gamma y\bar{j} + C\Gamma(R - z)\bar{k} . \tag{12.31}$$

Also,

$$\bar{S}_0 = K_{-1}\bar{i} + L_{-1}\bar{j} + M_{-1}\bar{k} ,$$
$$\bar{S}_1 = K\bar{i} + L\bar{j} + M\bar{k} ,$$

so

$$\bar{S}_1 - \bar{S}_0 = (K - K_{-1})\bar{i} + (L - L_{-1})\bar{j} + (M - M_{-1})\bar{k} . \tag{12.32}$$

Rewriting the equation needed to find the optical direction cosines from the initial data:

$$n \cos I' = n\left[\left(\frac{n_{-1}}{n}\cos I\right)^2 - \left(\frac{n_{-1}}{n}\right)^2 + 1\right]^{\frac{1}{2}}, \tag{12.33}$$

$$\Gamma = n \cos I' - n_{-1} \cos I, \tag{12.34}$$

$$K = K_{-1} - xC\Gamma, \tag{12.35}$$

$$L = L_{-1} - yC\Gamma, \tag{12.36}$$

$$M = M_{-1} - (zC - 1)\Gamma. \tag{12.37}$$

Summarizing thus far, the above equations develop the skew ray trace for a ray from spherical surface j_{-1} to spherical surface j in the space of refractive index n, and using the initial data (x_{-1}, y_{-1}, z_{-1}) and the axial distance, t_{-1}, as shown in Figure 12.7. These equations are similar to, although more complex than, the transfer and refraction equations for paraxial optics. Finding x, y, z and the new optical direction cosines (K, L, M) in the space with refractive index n is the same as using the refraction equations.

Table 12.1 shows the sequence of equations required to trace a real skew ray.

The calculated values for x, y, z and K, L, M are now used as the initial starting point to repeat the steps of (a)–(o) in Table 12.1. The process is performed iteratively until the ray is traced through the system.

The real ray trace procedure is shown step by step in worksheet format in Figure 12.8. The concept is similar to that developed in Chapter 10 for paraxial ray tracing. The given quantities are shown above the bold line. The given and

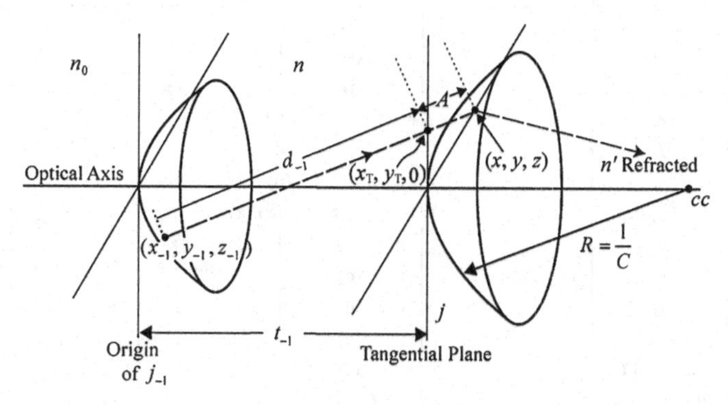

Figure 12.7 Skew ray trace between two spherical surfaces.

Table 12.1. *Equations used to trace a real skew ray*

$\dfrac{d_{-1}}{n_{-1}} = (t_{-1} - z_{-1})\dfrac{1}{M_{-1}}$	(a)
$x_T = x_{-1} + \dfrac{d_{-1}}{n_{-1}} K_{-1}$	(b)
$y_T = y_{-1} + \dfrac{d_{-1}}{n_{-1}} L_{-1}$	(c)
$H = C(x_T^2 + y_T^2)$	(d)
$B = M_{-1} - C(y_T L_{-1} + x_T K_{-1})$	(e)
$n_{-1} \cos I = n_{-1} \left[\left(\dfrac{B}{n_{-1}} \right)^2 - CH \right]^{\frac{1}{2}}$	(f)
$\dfrac{A}{n_{-1}} = \dfrac{H}{B + n_{-1} \cos I}$	(g)
$x = x_T + \dfrac{A}{n_{-1}} K_{-1}$	(h)
$y = y_T + \dfrac{A}{n_{-1}} L_{-1}$	(i)
$z = \dfrac{A}{n_{-1}} M_{-1}$	(j)
$n \cos I' = n \left[\left(\dfrac{n_{-1}}{n} \cos I \right)^2 - \left(\dfrac{n_{-1}}{n} \right)^2 + 1 \right]^{\frac{1}{2}}$	(k)
$\Gamma = \left[(n_{-1} \cos I)^2 - n_{-1}^2 + n^2 \right]^{\frac{1}{2}} - n_{-1} \cos I$	(l)
$K = K_{-1} - xC\Gamma$	(m)
$L = L_{-1} - yC\Gamma$	(n)
$M = M_{-1} - (zC - 1)\Gamma$	(o)

Surface	O (object)		1		2		3
C	C_0		C_1		C_2		
t		t_0		t_1		t_2	
n		n_0		n_1		n_2	
x	X_0		(h)				
y	Y_0		(i)				
z	Z_0		(j)				
K		K_0		(m)			
L		L_0		(n)			
M		M_0		(o)			
d_{-1}/n_{-1}		(a)		Next Step (a)			
x_T			(b)				
y_T			(c)				
H			(d)				
B			(e)				
$n_{-1} \cos I$			(f)				
$B + n_{-1} \cos I$			(e+f)				
A/n_{-1}			(g)				
$n \cos I'$			(k)				
Γ			(l)				

Figure 12.8 Real ray trace worksheet.

calculated values follow the format of staggered columns as shown, where the object surface is labeled as the zero (0) surface.

An advantage of this approach is its ability to be adapted into a spreadsheet computer program.

Example 12.1

Trace a real ray starting at the object (0.1, 0.2, 0) with direction cosines (0.1, 0.05, 0.993) through a lens located 15 cm away with radii 5 cm and –8 cm, respectively, index 1.5 and axial thickness 2 cm. The paraxial image plane is 10.18 cm from the last surface.

Real ray trace (calculations)

Parameter	Object	Object space	Surface 1	Lens space	Surface 2	Image space	Image
C	0		0.2		– 0.125		0
t		15		2		10.18	
n		1		1.5		1	1
x	0.1		1.647 792		1.572 996		– 0.393 986
y	0.2		0.973 896		0.921 122		– 0.332 26
z	0		0.380 876		– 0.210 442		0
K		0.1		– 0.079 477		– 0.184 71	
L		0.05		– 0.056 076		– 0.1177	
M		0.99373		1.496 843		0.975 72	
d–1/n–1		15.09464		1.081693		10.649	
xt			1.609 464		1.561 822		– 0.393 986
yt			0.954 732		0.913 239		– 0.332 26
H			0.700 377		– 0.409 162		0
B			0.951 994		1.474 926		0.975 72
n–1 cos I			0.875 338		1.435 385		0.975 72
B + n–1 cos i			1.827 332		2.91 031		1.951 439
A/n–1			0.383 279		– 0.140 59		0
n cos i′			1.419 935		0.900 183		0.975 72
RR			0.544 597		– 0.535 202		0
CRR			0.108 919		0.066 9		0
xnew			1.647 792		1.572 996		–0.393986
ynew			1.973 896		1.921 122		–0.33226
znew			0.380 876		0.210 442		0
			1.419 935				

12.4 Meridional real ray trace

For a real ray in the meridional plane (y–z plane) there are two ways to accomplish a real ray trace: (1) use the equation developed for the skew ray but force the x component of the ray to zero; or (2) the Q–U approach (O'Shea, 1985).

The meridional plane ray trace utilizes a simplified version of the skew ray equations. To ray trace a real ray in the meridional plane, the corresponding equations given in Table 12.1, which were developed for the skew ray, are simplified by letting $x = 0$ and $K = 0$:

$$\frac{d_{-1}}{n_{-1}} = (t_{-1} - z_{-1})\frac{1}{M_{-1}}, \tag{12.38}$$

$$y_T = y_{-1} + \frac{d_{-1}}{n_{-1}}L_{-1}, \tag{12.39}$$

$$H = Cy_{\mathrm{T}}^2, \tag{12.40}$$

$$B = M_{-1} - Cy_{\mathrm{T}}L_{-1}, \tag{12.41}$$

$$n_{-1}\cos I = n_{-1}\left[\left(\frac{B}{n_{-1}}\right)^2 - CH\right]^{\frac{1}{2}}, \tag{12.42}$$

$$\frac{A}{n_{-1}} = \frac{H}{B + n_{-1}\cos I}, \tag{12.43}$$

$$y = y_{\mathrm{T}} + \frac{A}{n_{-1}}L_{-1}, \tag{12.44}$$

$$z = \frac{A}{n_{-1}}M_{-1}, \tag{12.45}$$

$$n\cos I' = n\left[\left(\frac{n_{-1}}{n}\cos I\right)^2 - \left(\frac{n_{-1}}{n}\right)^2 + 1\right]^{\frac{1}{2}}, \tag{12.46}$$

$$\Gamma = n\cos I' - n_{-1}\cos I, \tag{12.47}$$

$$L = L_{-1} - yC\Gamma, \tag{12.48}$$

$$M = M_{-1} - \Gamma(1 - C^2y^2)^{\frac{1}{2}}. \tag{12.49}$$

12.5 Q–U method of real ray trace

The Q–U method of ray tracing specifies angles relative to the optical axis and perpendicular distances from the spherical surface vertex to the ray. The initial ray must have two values given: its angle in relation to the optical axis (U) and its ray–vertex perpendicular distance (Q). Figure 12.9 depicts the values of U' and Q with the corresponding refracted ray values of U' and Q'. From the construction lines shown in Figure 12.9:

$$Q = R \sin I = +R \sin(-U) \; . \tag{12.50}$$

Therefore from Snell's law we can solve for sin I':

$$\sin I' = \frac{n}{n'} \sin I \; . \tag{12.51}$$

So the values of I, I', and U are known. We can find the characteristics of the refracted ray from the geometry shown in Figure 12.10.

From Figure 12.10, for the refracted ray after the spherical surface, U' and I' are the refracted ray slope angle and the angle of refraction after the surface, respectively, so

$$U' = U - (I - I') \; . \tag{12.52}$$

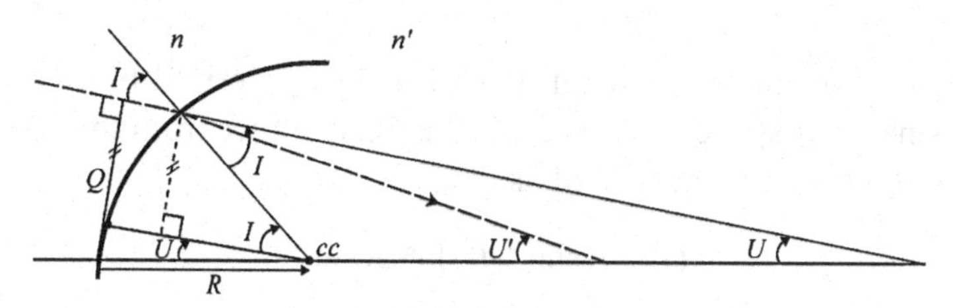

Figure 12.9 Q–U ray trace for a real ray.

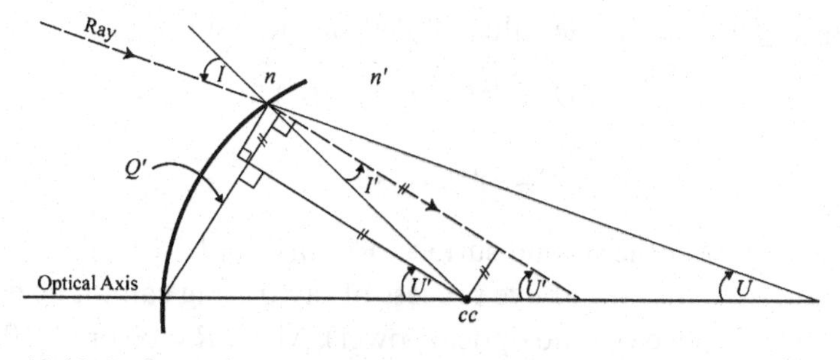

Figure 12.10 Refracted ray at a spherical surface.

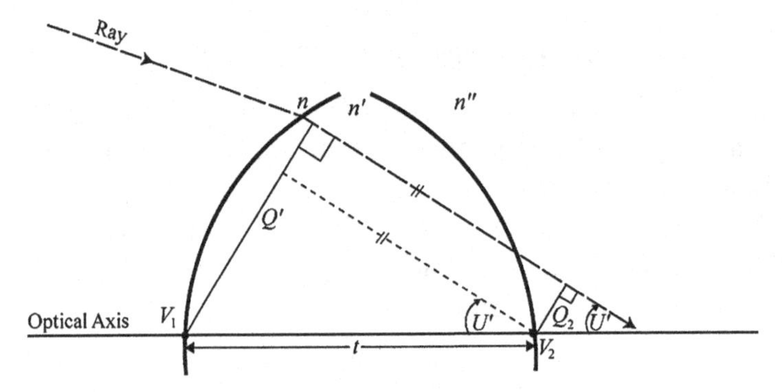

Figure 12.11 Transfer of the exact ray to the next surface.

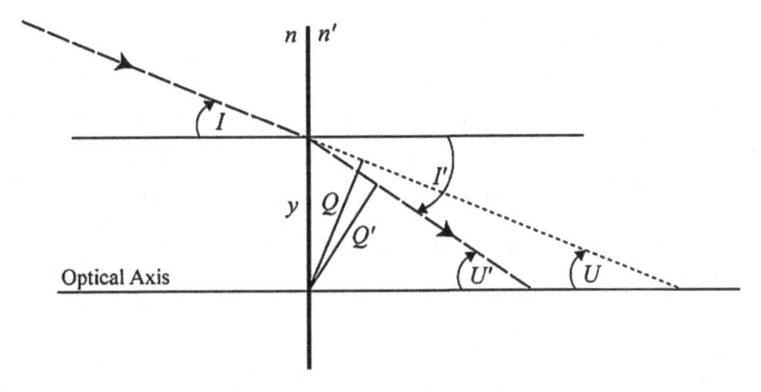

Figure 12.12 Transfer across a plane surface.

Therefore, we now have a new refracted ray, characterized by U' values just inside the surface in space n'. The perpendicular distance from the vertex to the refracted ray, Q', can be expressed as

$$\begin{aligned} Q' &= R\sin(-U') + R\sin I' \\ &= R(\sin I' - \sin U') \ . \end{aligned} \tag{12.53}$$

The axial transfer of an exact ray to the next surface for an axial thickness (t) is shown in Figure 12.11. The value of Q at the next surface is

$$Q_2 = Q' - t\sin(-U') \tag{12.54}$$

$$= Q' + t\sin U' \ . \tag{12.55}$$

Now we have Q–U at the second surface. Recall $U = U'$.

One should note that the above process of ray tracing cannot be done for a plane surface (a surface with no optical power), where $R = \infty$ or $C = 0$. A plane surface refraction of the ray is shown in Figure 12.12:

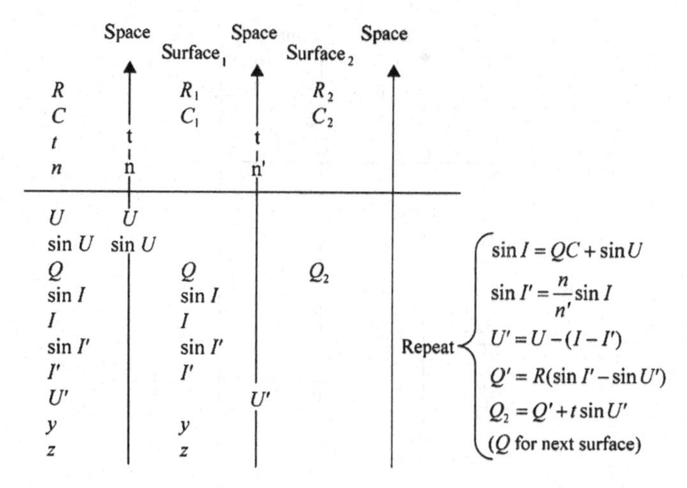

Figure 12.13 *Q–U* trace worksheet.

$$y = \frac{Q}{\cos U} = \frac{Q'}{\cos U'} , \tag{12.56}$$

$$Q' = \frac{\cos U'}{\cos U} Q . \tag{12.57}$$

The five-equation sequence listed below must be applied repetitively to transfer a ray from surface to surface throughout the optical system. This approach presumes that the initial values of *Q and U* are given.

(1) $\sin I = Q/R + \sin U = QC + \sin U$.
(2) $\sin I' = (n/n')\sin I$.
(3) $U' = U - (I - I')$.
(4) $Q' = R(\sin I' - \sin U')$.
(5) $Q_2 = Q' + t\sin U'$.

Figure 12.13 illustrates the sequence in worksheet format.

Example 12.2

Ray trace an exact ray from object to image for the thick lens shown.

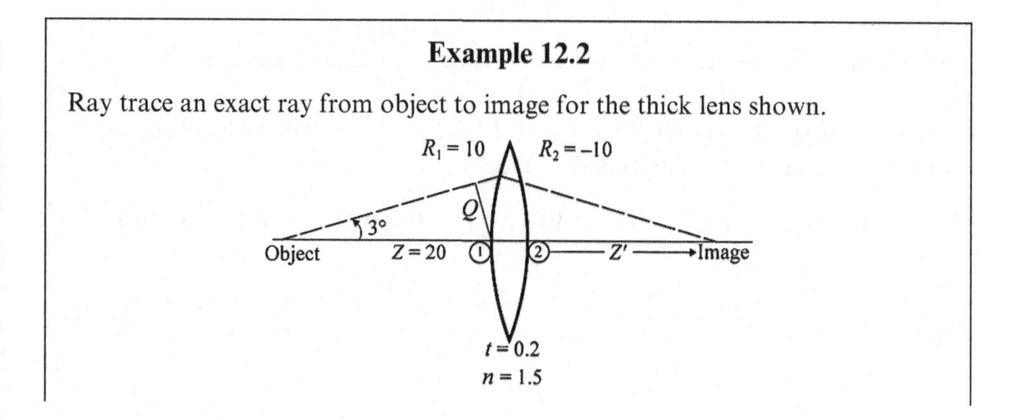

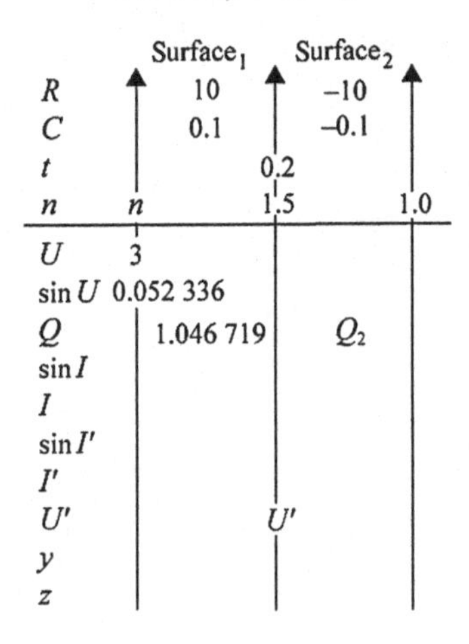

Open/initial knowledge

$$Q = t \sin U = +20 \sin 3$$
$$Q = 1.046\,719$$

Exact ray trace (calculations)

Parameter	Object space	Surface 1	Lens space	Surface 2	Image space
R		10		− 10	
C		0.1		− 0.1	
t	20		0.2		BFD = ??
n	1		1.5		1
U	3°				
sin U	0.052 336				
Q		1.046 719			
sin I	0.157 008				
I	9.033 258°				
sin I'			0.104 672		
I'			6.008 26		
U'			− 0.025 001		
sin U'			− 0.000 436		
Q'			1.05108		
Q_2				1.050 995	

From the table, $Q_2 = 1.050\,995$, $C_2 = -0.1$, and sin $U' = -0.000\,436$. Using these values, we can solve for the second surface:

$$\sin I_2 = Q_2 C_2 + \sin U' = (1.059\,95)(-0.1) - 0.000\,436 = -0.106\,431 \rightarrow I_2 = -6.058\,020°$$

(note $U' = U_2$).

$$\sin I_2' = \frac{n}{n'}\sin I_2 = \frac{1.5}{1}(-0.105\,536) = -0.158\,303 \to I_2' = -9.108\,425°\ ,$$

$$U_2' = U_2 - I_2 + I_2' = -0.025\,001 + 6.058\,020 - 9.108\,425 = -3.075\,406°\ ,$$

$$\sin U_2' = -0.053\ 650,$$

$$Q_2' = R_2(\sin I_2' - \sin U_2') = -10(-0.158\ 303 + 0.053\ 650) = 1.046\ 531,$$

$$Q_3 = Q_2' + t\sin U_2' = 0 = 1.046\ 531 + BFD(-0.053\ 650).$$

$$\to BFD = 19.50\ 655.$$

As a check, you can use the Gaussian equation and/or perform a paraxial ray trace as shown below. Remember, however, that paraxial approximations are used and the answers will not match perfectly.

Gaussian method

$$\phi_1 = \phi_2 = \frac{n' - n}{R} = 0.05.$$

$$\phi_T = \phi_1 + \phi_2 - \phi_1\phi_2\frac{t}{n} = 2(0.05) - (0.05^2)\frac{0.2}{1.5} = 0.099\ 667.$$

$$\frac{1}{z'} - \frac{1}{z} = \phi \to \frac{1}{z'} - \frac{1}{-20} = 0.099\ 667 \to z' = 20.1341\ .$$

Paraxial method

$$\phi_1 = \phi_2 = 0.05,\ u = 3° = 0.052\ 36\ \text{rad},\ z = -20\,.$$

Approximate

$$u = \frac{y_1}{-z} \to y_1 = 1.0472\ ,$$

$$n'u' = nu - y\phi_1 = (1)(0.05236) - (1.0472)(0.05) = 0,$$

$$y_2 = y_1 + \frac{t'}{n'}(n'u') = 1.0472 + 0 = 1.0472,$$

$$n''u'' = n'u' - y_2\phi_2 = 0 - 1.0472(0.05) = -0.05236,$$

$$y_3 = y_2 + \frac{t''}{n''}(n''u'') = 0 = 1.0472 + BFD(-0.052\ 36) \to BFD = 20\ .$$

One problem with these ray traces is the computational inaccuracy caused by the differences of small and large numbers. Hence, a very large number of significant figures such as six or eight places must be carried.

Problems

12.1 Do a real ray trace for the lens below for a ray at $y = 10$ from an object at infinity. What is the *BFD*?

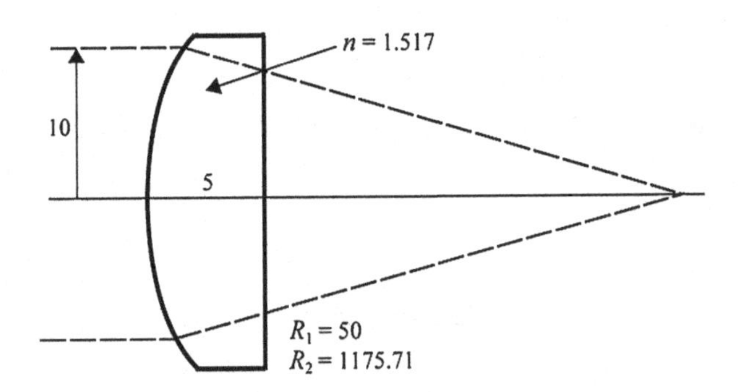

12.2 Derive Equation (12.52).

12.3 What is the difference between direction cosines and optical direction cosines?

12.4 Exact ray trace the doublet shown below for an object at infinity.

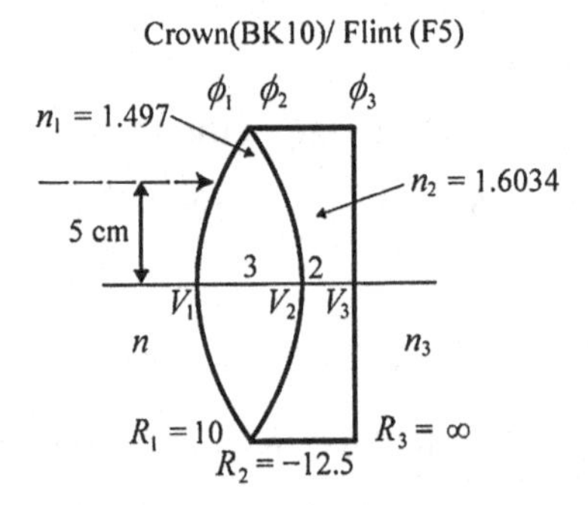

Bibliography

Hopkins, R., Hanau, R., Osenberg, H., *et al.* (1962). *Military Standardized Handbook 141 (MIL HDBK-141)*. US Government Printing Office.

O'Shea, D. C. (1985). *Elements of Modern Optical Design*. New York: Wiley.

Appendix A
Linear prism dispersion design

This appendix describes a first-order technique for designing a linearized-dispersion prism. Such a prism is characterized by linear dispersion, similar to the dispersion associated with a grating.

A.1 Wedge approximation

We begin by approximating a prism as a thin wedge. In that case, the deviation angle, δ, is related to the apex angle,α, and the index of refraction, n, by

$$\delta = \alpha(n-1). \tag{A.1}$$

The angular separation between the shortest and the longest wavelength, i.e. the angular width of the spectrum, is given by (see Figure A.1)

$$\Delta = \delta_F - \delta_C = \alpha[(n_F - 1) - (n_C - 1)] = \frac{\delta_d}{V\#}, \tag{A.2}$$

where $V^{\#}$ is the Abbe number associated with the prism material.[1] Note that although the use of F, d, C, and the Abbe number implies the visible part of the electromagnetic spectrum, analogous quantities can be determined for any part of the spectrum including the short-wavelength infrared and medium-wavelength infrared.

A.2 Two-material prism

When two prisms are placed in series and are separated by an air gap, as shown in Figure A.2, equations similar to Equations (A.1) and (A.2), must be linearly combined or added. The total deviation angle is:

$$\bar{\delta}_i = \delta_{1,i} + \delta_{2,i}, \tag{A.3}$$

where the subscript, i, can denote the short, the center, or the long wavelength in the instrument bandwidth, or

[1] $$V\# = \frac{n_d - 1}{n_F - n_C}.$$

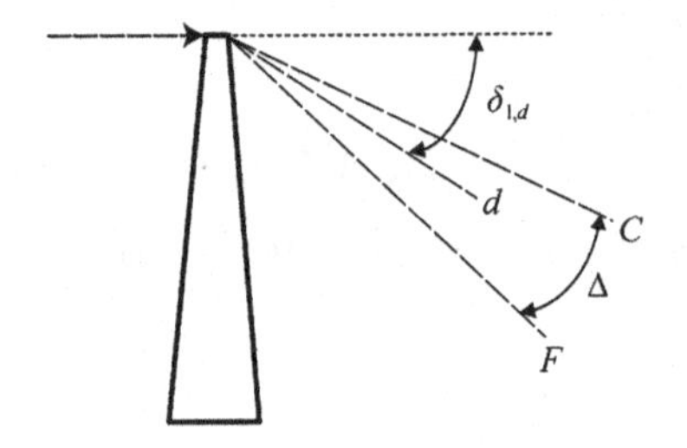

Figure A.1 Single-prism nomenclature.

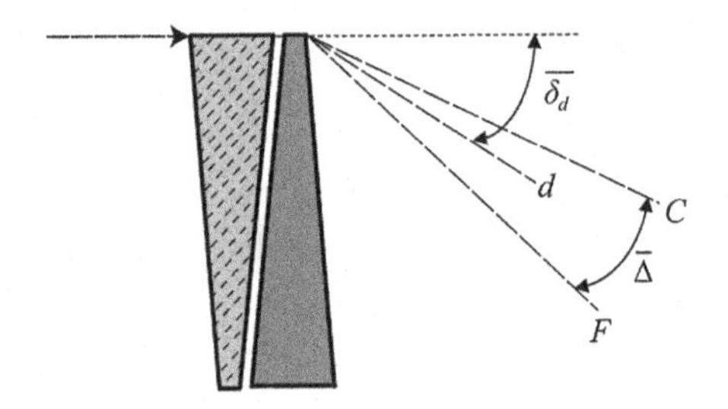

Figure A.2 Two-material prism nomenclature.

$$\overline{\delta_F} = \delta_{1,F} + \delta_{2,F}, \qquad \overline{\delta_d} = \delta_{1,d} + \delta_{2,d}, \qquad \overline{\delta_C} = \delta_{1,C} + \delta_{2,C}.$$

The angular width of the spectrum produced by the prism pair is:

$$\begin{aligned}\overline{\Delta} &= \overline{\delta_F} - \overline{\delta_C} = \delta_{1,F} + \delta_{2,F} - \left(\delta_{1,C} + \delta_{2,C}\right)\\ &= \left(\delta_{1,F} - \delta_{1,C}\right) + \left(\delta_{2,F} - \delta_{2,C}\right),\\ &= \frac{\delta_{1,d}}{V_1} + \frac{\delta_{2,d}}{V_2}.\end{aligned} \tag{A.4}$$

The third equation constrains the angular separation of the center wavelength and the long wavelength to equal one-half of the total angular width of the spectrum. Specifically,

$$\begin{aligned}\overline{\delta_d} - \overline{\delta_C} &= \delta_{1,d} + \delta_{2,d} - \delta_{1,C} - \delta_{2,C}\\ &= \delta_{1,d} - \delta_{1,C} + \delta_{2,d} - \delta_{2,C}\\ &= \alpha_1(n_{d1} - 1) - \alpha_1(n_{C1} - 1) + \alpha_2(n_{d2} - 1) - \alpha_2(n_{C2} - 1)\\ &= \alpha_1(n_{d1} - n_{C1}) + \alpha_2(n_{d2} - n_{C2})\\ &= \alpha_1(n_{d1} - n_{C1})\left(\frac{n_{d1} - 1}{n_{d1} - 1}\right) + \alpha_2(n_{d2} - n_{C2})\left(\frac{n_{d2} - 1}{n_{d2} - 1}\right)\\ &= \delta_{1,d} \times \left(\frac{n_{d1} - n_{C1}}{n_{d1} - 1}\right) \times \left(\frac{n_{F1} - n_{C1}}{n_{F1} - n_{C1}}\right) + \delta_{2,d} \times \left(\frac{n_{d2} - n_{C2}}{n_{d2} - 1}\right) \times \left(\frac{n_{F2} - n_{C2}}{n_{F2} - n_{C2}}\right)\\ &= \delta_{1,d} \times \left(\frac{P_1}{V_1}\right) + \delta_{2,d}\left(\frac{P_2}{V_2}\right)\\ &= \frac{\overline{\Delta}}{2}.\end{aligned}$$

$$\frac{\overline{\Delta}}{2} = \delta_{1,d}\frac{P_1}{V_1} + \delta_{2,d}\frac{P_2}{V_2}, \tag{A.5}$$

where P is the relative partial dispersion associated with each prism material.[2] At this point, we have three equations and two unknowns, namely the center-wavelength (d) deviations of each prism.

A.3 Least-squares solution

The three equations (A.3), (A.4), and (A.5) and two unknown deviations ($\delta_{1,d}$ and $\delta_{2,d}$) can be compactly expressed as a matrix–vector product:

$$\begin{bmatrix} \overline{\delta_d} \\ \Delta \\ \frac{\Delta}{2} \end{bmatrix} = \begin{bmatrix} 1 & 1 \\ \frac{1}{V_1} & \frac{1}{V_2} \\ \frac{P_1}{V_1} & \frac{P_1}{V_2} \end{bmatrix} \begin{bmatrix} \delta_{1,d} \\ \delta_{2,d} \end{bmatrix} = \mathbf{H} \begin{bmatrix} \delta_{1,d} \\ \delta_{2,d} \end{bmatrix}. \tag{A.6}$$

Once estimates for the two center-wavelength deviations are calculated, the deviations can be related to prism apex angles via Equation (A.1), thus completing the first-order design of a linear prism. The center-wavelength deviations are found using a pseudo inverse of the matrix **H**:

$$\begin{bmatrix} \delta_{1,d} \\ \delta_{2,d} \end{bmatrix} = \left(\mathbf{H}^{\mathrm{T}}\mathbf{H}\right)^{-1}\mathbf{H}^{\mathrm{T}} \begin{bmatrix} \overline{\delta_d} \\ \Delta \\ \frac{\Delta}{2} \end{bmatrix}. \tag{A.7}$$

A.4 Three-material prism

The total deviation, angular spectrum width, and linear-dispersion specifications can be satisfied exactly if three materials can be employed in the construction of the prism. In that case, by extension of the equations for the two-material prism, we have the matrix–vector equation:

$$\begin{bmatrix} \overline{\delta_d} \\ \Delta \\ \frac{\Delta}{2} \end{bmatrix} = \begin{bmatrix} 1 & 1 & 1 \\ \frac{1}{V_1} & \frac{1}{V_2} & \frac{1}{V_3} \\ \frac{P_1}{V_1} & \frac{P_2}{V_2} & \frac{P_3}{V_3} \end{bmatrix} \begin{bmatrix} \delta_{1,d} \\ \delta_{2,d} \\ \delta_{3,d} \end{bmatrix} = \mathbf{H}_3 \begin{bmatrix} \delta_{1,d} \\ \delta_{2,d} \\ \delta_{3,d} \end{bmatrix}. \tag{A.8}$$

The center-wavelength deviations and therefore the apex angles of the three prisms can be calculated upon inverting the matrix $\mathbf{H}_3$.

A.5 Concluding remarks

The Matlab program below shows examples of two-material prism design. This first-order design technique should be considered as a starting point in selecting likely material pairs (or triplets) for further optimization in a ray-tracing lens design

[2] $$P = \frac{n_d - n_C}{n_F - n_C}.$$

program such as Zemax. In that sense, the linear prism design technique presented here is analogous to selecting glasses for a two-glass achromat or a three-glass apochromat lens.

```
MATLAB Command Window
To get started, select "MATLAB Help" from the Help menu.

>> p1=(1.5168-1.51432)/.008054
p2=(1.61564-1.61165)/.017064
nd1=1.5168
v1=64.17
nd2=1.61659
v2=36.63
H=[1 1 ;1/v1 1/v2 ;p1/v1 p2/v2]
dispersion=[10*pi/180; 4*pi/180; 2*pi/180]
Hpi=inv(H'*H)*H'
dev=Hpi*dispersion
% for prism 1
dev(1)
alpha1=dev(1)/(n1-1)
alpha1=dev(1)/(nd1-1)
dispersion=[0; 4*pi/180; 2*pi/180]
dev=Hpi*dispersion
dispersion=[10; 4; 2]
dev=Hpi*dispersion
dispersion=[0; 1*pi/180; .5*pi/180]
dev=Hpi*dispersion
p1
dispersion=[0; .001; .0005]
dev=Hpi*dispersion
alpha=dev[1]/(nd1-1)
alpha=dev(1)/(nd1-1)
alpha1=dev(1)/(nd1-1)
alpha2=dev(2)/(nd2-1)
```

Appendix B
Linear mixing model

An imaging system will necessarily have limited field of view and spatial resolution. This limitation is imposed by such factors as pixel size, detector-array format, the number of data collected, etc. The corresponding instantaneous field of view (IFOV) is therefore likely to encompass several "patches" of materials that possess different reflectance and/or emissivity properties. If we are lucky, the combined signal from each IFOV is a linear mixture of weighted radiances from each "pure" material within the IFOV.

Given sufficient signal to noise ratio (SNR), sub-pixel traces of a particular material may be detected based on the presence of distinctive spectral features in the combined signature. A simpler technique relies on a single spectral channel within which the target and the background exhibit different radiance. For example, the 3–5 μm window that may be used to detect smoldering fires in a natural background and the detection of narrow roads of concrete surrounded by vegetation is accomplished in the 0.6–0.7 μm band. Compare the spectra of healthy vegetation and soil to concrete or asphalt in this spectral region to see why.

Take a look at Figures B.1(a) and B.1(b) for some examples of a geometric interpretation of linearly mixed pixels.[1]

A data cube from a space-borne spectrometer provides spectral data on spatial locations of interest, as shown in Figure B.2. The signal measured in each IFOV is radiometrically made up of the constituents, i.e. $[A(x)+B(x)+C(x)+D(x)]$, along with their fractional area. The signal consists of the following from each constituent:

(1) reflectance (ρ) as a function of wavelength;
(2) fractional area.

In order to solve for these constituents it is assumed that only positive materials are present (the materials have to be present or exist, so no negative values are possible), and that topography variation, i.e. shadowing, is ignored.

To represent what each resolution element (resel) of data consists of, use matrix algebra to get the measurements. Consider a series of measurements at different wavelength intervals over a spectrum of m intervals for a given constituent:

[1] J. Boardman (1993), Automating spectral unmixing of AVIRIS data using convex geometry concepts, *Summaries of the Fourth Annual JPL Airborne Geoscience Workshop.*

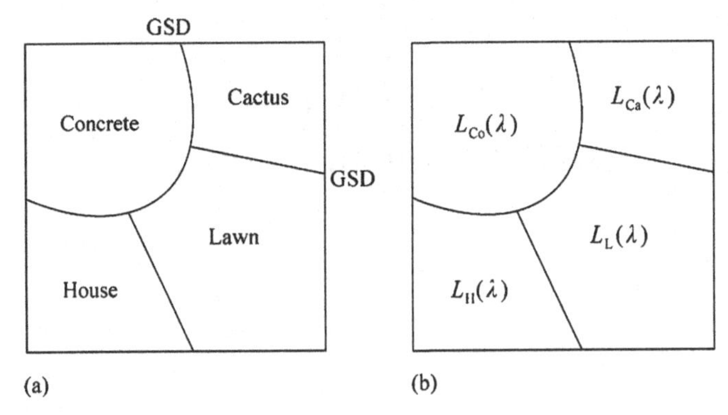

Figure B.1 A resolution region of an object that contains several materials within an instantaneous field of view (IFOV): (a) ground sampling distance with various constituents, with IFOV; (b) radiance from various constituents in IFOV.

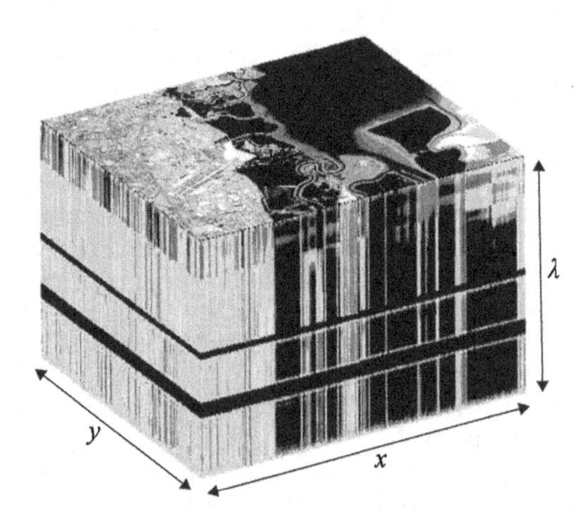

Figure B.2 Data cube of aviris data (JPL).

$$\begin{bmatrix} \text{value} \\ \text{measured} \\ \text{for a given} \\ \text{wavelength}(m) \end{bmatrix} = \begin{bmatrix} h_1^1 & h_1^2 & h_1^3 & \cdots & h_n^n \\ h_2^1 & h_2^2 & h_2^3 & \cdots & h_n^n \\ \vdots & \vdots & \vdots & & \\ h_m^1 & h_m^2 & h_m^3 & \cdots & h_m^n \end{bmatrix} \begin{bmatrix} \text{abundance} \\ \text{of each} \\ \text{contstituent} \\ (n) \end{bmatrix}$$

As shown in the mathematical representation, each of the m measurements represents the sum of the contributions from each constituent for a given wavelength of the spectrum. There are m wavelength band measurements for the unknown constituents that are n in number, rewritten in equation form:

$$\bar{g}_m = \mathbf{H}\bar{f}_n,$$

where $\mathbf{H}$ is an m by n matrix representing the possible constituents of the given measurement, h_m^n.

The object of interest for this example is a concrete pad within the resolution element, but it is not resolved. For each of the five spatial resels shown in

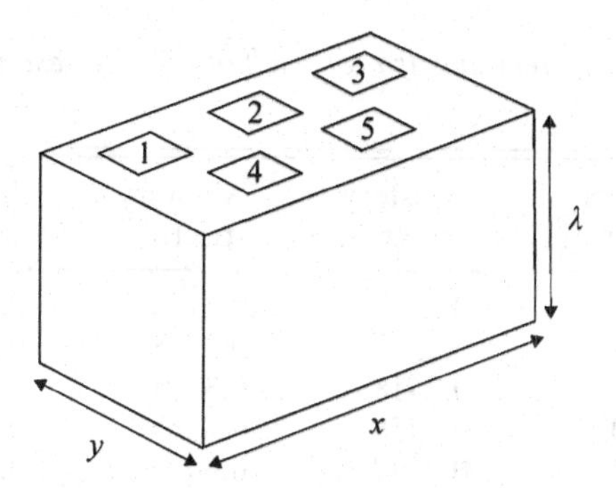

Figure B.3 Data cube with the five resels of interest.

Figure B.3, determine the abundances of the various constituents (i.e. conifer, gypsum, dry grass, concrete, maple tree, blackbush, calcite, sagebrush, fir tree, $n=9$). Solve for the abundances in the five resels with the given spectra in the mystery spectra.xls files provided in Table B.1. Which resel has concrete in it?

Approach to solution

Think of the spectrum at a resel, I, as a vector, $\bar{g}_m$, which is made up of the abundances of each of the various constituents which have a known value at the same spectral region.

Mathematically,

$$\bar{g}_m = \mathbf{H}\bar{f}_n, \tag{B.1}$$

where $\overline{f_n}$ is the partial abundance of constituent f_n and $\sum_{n=1}^{a} f_n \equiv 1$.

$$\bar{f}_n = \begin{bmatrix} f_1 \\ f_2 \\ f_3 \\ f_4 \\ f_5 \\ f_6 \\ f_7 \\ f_8 \\ f_9 \end{bmatrix}.$$

$$\mathbf{H} = \begin{bmatrix} h_1^1 & h_1^2 & h_1^3 & h_1^4 & h_1^5 & h_1^6 & h_1^7 & h_1^8 & h_1^9 \\ h_2^1 & h_2^2 & h_2^3 & h_2^4 & h_2^5 & h_2^6 & h_2^7 & h_2^8 & h_2^9 \\ h_3^1 & h_3^2 & h_3^3 & h_3^4 & h_3^5 & h_3^6 & h_3^7 & h_3^8 & h_3^9 \\ h_4^1 & h_4^2 & h_4^3 & h_4^4 & h_4^5 & h_4^6 & h_4^7 & h_4^8 & h_4^9 \\ h_5^1 & h_5^2 & h_5^3 & h_5^4 & h_5^5 & h_5^6 & h_5^7 & h_5^8 & h_5^9 \\ h_6^1 & h_6^2 & h_6^3 & h_6^4 & h_6^5 & h_6^6 & h_6^7 & h_6^8 & h_6^9 \\ h_7^1 & h_7^2 & h_7^3 & h_7^4 & h_7^5 & h_7^6 & h_7^7 & h_7^8 & h_7^9 \\ h_8^1 & h_8^2 & h_8^3 & h_8^4 & h_8^5 & h_8^6 & h_8^7 & h_8^8 & h_8^9 \\ h_9^1 & h_9^2 & h_9^3 & h_9^4 & h_9^5 & h_9^6 & h_9^7 & h_9^8 & h_9^9 \end{bmatrix}.$$

Table B.1. *Reflectance mystery spectra for each of the five resolution points in Figure B.3*

Wavelength (microns)	Mystery spectra 1	Mystery spectra 2	Mystery spectra 3	Mystery spectra 4	Mystery spectra 5
0.5	0.3181	0.2339	0.3199	0.3597	0.9371
0.505	0.322	0.2378	0.3248	0.3636	0.9391
0.51	0.3261	0.2418	0.3301	0.3677	0.9419
0.515	0.3303	0.2458	0.3355	0.3721	0.9446
0.52	0.3342	0.2497	0.3405	0.3759	0.9435
0.525	0.3398	0.2548	0.3442	0.3795	0.9422
0.53	0.3455	0.2599	0.348	0.3832	0.9409
0.535	0.351	0.2649	0.3511	0.3863	0.9402
0.54	0.3567	0.27	0.355	0.3901	0.9395
0.545	0.3613	0.2746	0.356	0.3926	0.9396
0.55	0.3655	0.2791	0.3562	0.3943	0.9415
0.555	0.3699	0.2837	0.3569	0.3965	0.9433
0.56	0.3744	0.2882	0.3577	0.3986	0.9435
0.565	0.3794	0.2927	0.3568	0.4003	0.9423
0.57	0.3837	0.2972	0.3542	0.4001	0.9411
0.575	0.3883	0.3017	0.3525	0.4009	0.9415
0.58	0.393	0.3062	0.3508	0.4016	0.9424
0.585	0.3968	0.3096	0.3506	0.4024	0.9434
0.59	0.4011	0.3131	0.3516	0.4045	0.9438
0.595	0.405	0.3165	0.3517	0.4056	0.9442
0.6	0.409	0.32	0.352	0.407	0.9444
0.605	0.4134	0.3231	0.3521	0.4081	0.9434
0.61	0.4176	0.3262	0.3517	0.4087	0.9424
0.615	0.4217	0.3293	0.351	0.4091	0.9438
0.62	0.426	0.3324	0.3508	0.4098	0.9456
0.625	0.431	0.3356	0.3513	0.411	0.9473
0.63	0.436	0.3388	0.3518	0.4121	0.9488
0.635	0.441	0.342	0.3524	0.4135	0.9494
0.64	0.4459	0.3453	0.3528	0.4145	0.9476
0.645	0.4502	0.3481	0.3527	0.4152	0.9457
0.65	0.4546	0.3509	0.353	0.4162	0.9451
0.655	0.4588	0.3537	0.353	0.417	0.9461
0.66	0.4631	0.3565	0.3531	0.4178	0.947
0.665	0.4685	0.3599	0.3549	0.4197	0.9484
0.67	0.4739	0.3632	0.357	0.4219	0.95
0.675	0.4791	0.3666	0.3583	0.4232	0.9516
0.68	0.4844	0.37	0.36	0.4251	0.9527
0.685	0.49	0.3735	0.3699	0.4308	0.951
0.69	0.4956	0.3769	0.3797	0.4367	0.9494
0.695	0.5013	0.3804	0.39	0.4429	0.9477
0.7	0.507	0.3839	0.4001	0.4489	0.946
0.705	0.5132	0.3877	0.4231	0.4614	0.947
0.71	0.5195	0.3915	0.4463	0.474	0.948
0.715	0.5259	0.3953	0.4696	0.4867	0.9491

Table B.1. (*cont.*)

Wavelength (microns)	Mystery spectra 1	Mystery spectra 2	Mystery spectra 3	Mystery spectra 4	Mystery spectra 5
0.72	0.5321	0.3991	0.4927	0.4992	0.9501
0.725	0.5352	0.401	0.5133	0.5102	0.9512
0.73	0.5381	0.4029	0.5337	0.5208	0.9516
0.735	0.5411	0.4049	0.5542	0.5316	0.9513
0.74	0.5439	0.4068	0.5741	0.5418	0.9509
0.745	0.5465	0.4084	0.5813	0.5459	0.951
0.75	0.549	0.4101	0.588	0.5495	0.9513
0.755	0.5517	0.4118	0.5951	0.5535	0.9516
0.76	0.5543	0.4134	0.602	0.5573	0.952
0.765	0.5567	0.4149	0.6043	0.5589	0.9523
0.77	0.5589	0.4163	0.6064	0.5603	0.9525
0.775	0.5613	0.4178	0.6085	0.5617	0.9528
0.78	0.5637	0.4193	0.6108	0.5634	0.9531
0.785	0.5665	0.4208	0.6124	0.5647	0.9543
0.79	0.5691	0.4224	0.6135	0.5656	0.9558
0.795	0.5718	0.424	0.6149	0.5668	0.9572
0.8	0.5738	0.4255	0.6145	0.5662	0.9586
0.805	0.5762	0.4269	0.6156	0.5671	0.96
0.81	0.5785	0.4283	0.6166	0.568	0.9615
0.815	0.5809	0.4296	0.6177	0.569	0.9618
0.82	0.5831	0.431	0.6185	0.5696	0.9616
0.825	0.5862	0.4327	0.62	0.5709	0.9615
0.83	0.5892	0.4344	0.6212	0.5719	0.9613
0.835	0.5922	0.4362	0.6223	0.5729	0.9612
0.84	0.5951	0.4379	0.6234	0.5738	0.961
0.845	0.5974	0.4393	0.6242	0.5745	0.9608
0.85	0.5998	0.4407	0.6255	0.5756	0.9607
0.855	0.6021	0.4421	0.6264	0.5763	0.9605
0.86	0.6044	0.4435	0.6274	0.5772	0.9603
0.865	0.6067	0.4451	0.6279	0.5779	0.96
0.87	0.609	0.4467	0.6286	0.5787	0.9597
0.875	0.6113	0.4483	0.6292	0.5795	0.9594
0.88	0.6136	0.4499	0.6297	0.5802	0.9591
0.885	0.6166	0.4519	0.6304	0.5811	0.9588
0.89	0.6196	0.4539	0.6311	0.582	0.9585
0.895	0.6225	0.4558	0.6315	0.5826	0.9582
0.9	0.6254	0.4578	0.6318	0.5832	0.9579
0.905	0.628	0.4594	0.6323	0.5838	0.9576
0.91	0.6306	0.461	0.6327	0.5844	0.9573
0.915	0.6333	0.4627	0.6332	0.5851	0.957
0.92	0.6358	0.4643	0.6334	0.5854	0.9567
0.925	0.6385	0.466	0.6329	0.5857	0.9564
0.93	0.6412	0.4678	0.6325	0.5861	0.9561
0.935	0.6438	0.4696	0.6318	0.5862	0.9563
0.94	0.6463	0.4714	0.6308	0.5859	0.957

Table B.1. (*cont.*)

Wavelength (microns)	Mystery spectra 1	Mystery spectra 2	Mystery spectra 3	Mystery spectra 4	Mystery spectra 5
0.945	0.6472	0.472	0.6279	0.5845	0.9577
0.95	0.648	0.4726	0.625	0.583	0.9584
0.955	0.6488	0.4732	0.6221	0.5814	0.9591
0.96	0.6495	0.4739	0.619	0.5797	0.9598
0.965	0.6502	0.4746	0.6181	0.579	0.9605
0.97	0.6508	0.4753	0.6171	0.5783	0.9613
0.975	0.6514	0.4761	0.6161	0.5775	0.962
0.98	0.652	0.4768	0.6152	0.5767	0.9627
0.985	0.653	0.4778	0.6156	0.5768	0.963
0.99	0.6542	0.4788	0.6165	0.5774	0.9629
0.995	0.6555	0.4797	0.6177	0.5782	0.9629
1	0.657	0.4807	0.6193	0.5795	0.9629
1.005	0.6586	0.4815	0.6226	0.5821	0.9629
1.01	0.66	0.4823	0.6253	0.5842	0.9629
1.015	0.6614	0.4832	0.628	0.5863	0.9629
1.02	0.6628	0.484	0.6308	0.5884	0.9628
1.025	0.664	0.4848	0.6332	0.5901	0.963
1.03	0.6653	0.4855	0.6356	0.5919	0.9635
1.035	0.6666	0.4863	0.6382	0.5938	0.9641
1.04	0.6678	0.4871	0.6407	0.5956	0.9646
1.045	0.6693	0.4881	0.6428	0.5971	0.9651
1.05	0.6706	0.4892	0.6445	0.5982	0.9657
1.055	0.6719	0.4902	0.6463	0.5995	0.9662
1.06	0.6733	0.4912	0.6484	0.601	0.9665
1.065	0.6741	0.4919	0.6488	0.6014	0.9661
1.07	0.6747	0.4925	0.6491	0.6018	0.9657
1.075	0.6754	0.4931	0.6494	0.6022	0.9653
1.08	0.676	0.4938	0.6494	0.6022	0.9649
1.085	0.6767	0.4945	0.649	0.6022	0.9645
1.09	0.6773	0.4951	0.6483	0.602	0.9641
1.095	0.6778	0.4958	0.6474	0.6017	0.964
1.1	0.6783	0.4965	0.6465	0.6012	0.9644
1.105	0.6786	0.497	0.6449	0.6006	0.9648
1.11	0.6788	0.4975	0.6428	0.5995	0.9653
1.115	0.679	0.498	0.6409	0.5986	0.9657
1.12	0.6793	0.4985	0.6391	0.5978	0.9661
1.125	0.6792	0.4987	0.6328	0.5945	0.9666
1.13	0.679	0.499	0.6263	0.5912	0.9668
1.135	0.6787	0.4992	0.6194	0.5874	0.9666
1.14	0.6782	0.4995	0.6123	0.5834	0.9665
1.145	0.6776	0.4997	0.6075	0.5803	0.9663
1.15	0.6768	0.5	0.6021	0.5767	0.9661
1.155	0.6756	0.5002	0.5958	0.5721	0.966
1.16	0.6742	0.5005	0.5891	0.5671	0.9658
1.165	0.672	0.5002	0.5842	0.5631	0.9655

Table B.1. (*cont.*)

Wavelength (microns)	Mystery spectra 1	Mystery spectra 2	Mystery spectra 3	Mystery spectra 4	Mystery spectra 5
1.17	0.67	0.5	0.5801	0.5599	0.965
1.175	0.6684	0.4997	0.5767	0.5573	0.9646
1.18	0.667	0.4995	0.5742	0.5556	0.9641
1.185	0.6681	0.5002	0.5749	0.5562	0.9636
1.19	0.669	0.501	0.5753	0.5565	0.9631
1.195	0.6698	0.5018	0.5757	0.5568	0.9626
1.2	0.6707	0.5026	0.576	0.557	0.9622
1.205	0.6735	0.5043	0.578	0.5586	0.9621
1.21	0.6767	0.506	0.5813	0.5615	0.962
1.215	0.68	0.5077	0.5846	0.5643	0.9619
1.22	0.6828	0.5095	0.5868	0.5662	0.9618
1.225	0.683	0.5097	0.5883	0.5671	0.9617
1.23	0.6831	0.51	0.5898	0.5682	0.9616
1.235	0.6836	0.5103	0.5921	0.5701	0.9615
1.24	0.6843	0.5106	0.595	0.5725	0.9617
1.245	0.6848	0.5108	0.5968	0.5741	0.9618
1.25	0.6851	0.511	0.5982	0.5753	0.9619
1.255	0.6854	0.5112	0.5997	0.5765	0.962
1.26	0.6855	0.5115	0.6003	0.577	0.9621
1.265	0.687	0.5124	0.6003	0.5774	0.9623
1.27	0.6884	0.5134	0.6	0.5777	0.9624
1.275	0.6898	0.5143	0.5998	0.578	0.9625
1.28	0.6912	0.5153	0.5995	0.5782	0.9627
1.285	0.691	0.5153	0.5971	0.5771	0.9628
1.29	0.6906	0.5153	0.5942	0.5756	0.9629
1.295	0.6901	0.5154	0.5911	0.5738	0.9631
1.3	0.6896	0.5154	0.5882	0.5722	0.9632
1.305	0.6883	0.5151	0.5824	0.5689	0.9632
1.31	0.6868	0.5148	0.5761	0.5651	0.963
1.315	0.6853	0.5144	0.5697	0.5611	0.9627
1.32	0.6837	0.5141	0.563	0.5569	0.9624
1.325	0.6812	0.5132	0.556	0.5526	0.9622
1.33	0.6785	0.5124	0.5488	0.548	0.9619
1.335	0.6756	0.5115	0.5409	0.5426	0.9616
1.34	0.6725	0.5106	0.5325	0.5368	0.9615
1.345	0.6683	0.509	0.5245	0.5313	0.9614
1.35	0.664	0.5074	0.5164	0.5257	0.9614
1.355	0.66	0.5058	0.5088	0.5206	0.9614
1.36	0.6559	0.5041	0.501	0.5153	0.9614
1.365	0.6512	0.5019	0.487	0.5068	0.9614
1.37	0.6466	0.4996	0.4732	0.4984	0.9613
1.375	0.6421	0.4973	0.4597	0.4904	0.9613
1.38	0.6377	0.4951	0.4465	0.4827	0.9614
1.385	0.6313	0.4913	0.4332	0.4744	0.9615
1.39	0.6247	0.4876	0.4194	0.4656	0.9616

Table B.1. (*cont.*)

Wavelength (microns)	Mystery spectra 1	Mystery spectra 2	Mystery spectra 3	Mystery spectra 4	Mystery spectra 5
1.395	0.6179	0.4838	0.4049	0.4561	0.9616
1.4	0.6103	0.48	0.3886	0.4449	0.9617
1.405	0.602	0.4762	0.3786	0.4363	0.9618
1.41	0.593	0.4725	0.367	0.4261	0.9617
1.415	0.5839	0.4687	0.3549	0.4154	0.9611
1.42	0.5749	0.4649	0.3433	0.4051	0.9605
1.425	0.5686	0.4632	0.3354	0.3979	0.9599
1.43	0.5617	0.4615	0.326	0.3892	0.9592
1.435	0.5538	0.4598	0.3142	0.3781	0.9586
1.44	0.5464	0.4581	0.3037	0.3682	0.958
1.445	0.545	0.4578	0.3034	0.368	0.9578
1.45	0.5462	0.4576	0.3094	0.3741	0.9585
1.455	0.5472	0.4574	0.3153	0.38	0.9592
1.46	0.5474	0.4571	0.319	0.3838	0.9599
1.465	0.5491	0.4582	0.3229	0.3871	0.9606
1.47	0.5505	0.4592	0.3262	0.3896	0.9613
1.475	0.5517	0.4602	0.3287	0.3913	0.962
1.48	0.5521	0.4612	0.3296	0.3915	0.9623
1.485	0.5543	0.4634	0.3308	0.3917	0.9618
1.49	0.5578	0.4656	0.335	0.395	0.9613
1.495	0.5625	0.4677	0.3425	0.4016	0.9608
1.5	0.5674	0.4699	0.3504	0.4086	0.9603
1.505	0.5692	0.4706	0.3565	0.4137	0.9598
1.51	0.5701	0.4713	0.3605	0.4168	0.9593
1.515	0.5706	0.4719	0.3633	0.4186	0.9592
1.52	0.5708	0.4726	0.3653	0.4198	0.9598
1.525	0.5696	0.4727	0.3664	0.4196	0.9605
1.53	0.5685	0.4727	0.3676	0.4198	0.9612
1.535	0.5679	0.4728	0.3703	0.4213	0.9618
1.54	0.5681	0.4729	0.3749	0.4248	0.9625
1.545	0.5704	0.4737	0.3815	0.4305	0.9632
1.55	0.5729	0.4745	0.3882	0.4363	0.9638
1.555	0.5752	0.4753	0.3948	0.442	0.9639
1.56	0.5774	0.4761	0.4009	0.4471	0.9636
1.565	0.5822	0.4785	0.4065	0.4521	0.9633
1.57	0.5868	0.4809	0.4119	0.4567	0.963
1.575	0.5914	0.4832	0.4169	0.461	0.9627
1.58	0.5959	0.4856	0.422	0.4654	0.9624
1.585	0.5991	0.4873	0.4265	0.4692	0.9621
1.59	0.6022	0.489	0.431	0.4729	0.9618
1.595	0.6055	0.4907	0.4356	0.4769	0.9615
1.6	0.6088	0.4924	0.4405	0.481	0.9612
1.605	0.6103	0.4932	0.4446	0.4844	0.9607

Table B.1. (*cont.*)

Wavelength (microns)	Mystery spectra 1	Mystery spectra 2	Mystery spectra 3	Mystery spectra 4	Mystery spectra 5
1.61	0.6117	0.4939	0.4483	0.4874	0.9603
1.615	0.6131	0.4947	0.452	0.4905	0.9598
1.62	0.6145	0.4955	0.4558	0.4936	0.9594
1.625	0.6151	0.4959	0.4585	0.4959	0.959
1.63	0.6157	0.4962	0.461	0.498	0.9585
1.635	0.6164	0.4966	0.464	0.5006	0.9581
1.64	0.617	0.497	0.4667	0.5028	0.9576
1.645	0.6169	0.497	0.4684	0.5044	0.9572
1.65	0.6166	0.497	0.4699	0.5057	0.9566
1.655	0.6163	0.497	0.4714	0.5069	0.956
1.66	0.6159	0.4971	0.4726	0.508	0.9553
1.665	0.6134	0.4958	0.4727	0.5079	0.9547
1.67	0.6107	0.4946	0.4723	0.5073	0.9541
1.675	0.6078	0.4934	0.4714	0.5062	0.9534
1.68	0.6049	0.4922	0.4706	0.5053	0.9528
1.685	0.603	0.4917	0.4685	0.5034	0.9521
1.69	0.6008	0.4912	0.4659	0.5009	0.9515
1.695	0.5983	0.4907	0.4625	0.4976	0.9509
1.7	0.5954	0.4902	0.4582	0.4935	0.9502
1.705	0.593	0.49	0.4524	0.4883	0.9492
1.71	0.5902	0.4898	0.4457	0.4822	0.9481
1.715	0.5872	0.4897	0.4385	0.4757	0.9471
1.72	0.5842	0.4895	0.4314	0.4692	0.946
1.725	0.5819	0.4897	0.4256	0.4635	0.945
1.73	0.5797	0.4898	0.4201	0.4581	0.944
1.735	0.5779	0.49	0.4155	0.4536	0.9429
1.74	0.5766	0.4901	0.4119	0.4501	0.9419
1.745	0.577	0.4906	0.4097	0.4488	0.9408
1.75	0.578	0.4911	0.4088	0.4489	0.9398
1.755	0.5793	0.4916	0.409	0.45	0.9387
1.76	0.5807	0.4921	0.4091	0.451	0.9388
1.765	0.5819	0.4927	0.4109	0.4528	0.939
1.77	0.5831	0.4933	0.4127	0.4545	0.9393
1.775	0.5844	0.4939	0.4147	0.4564	0.9396
1.78	0.586	0.4945	0.4176	0.4594	0.9399
1.785	0.5883	0.4953	0.4193	0.462	0.9402
1.79	0.5906	0.4961	0.4211	0.4647	0.9405
1.795	0.5926	0.4969	0.4223	0.4669	0.9408
1.8	0.5945	0.4977	0.4231	0.4687	0.9411
1.805	0.5959	0.4981	0.4212	0.4687	0.9414
1.81	0.5972	0.4986	0.4191	0.4685	0.9417
1.815	0.5982	0.4991	0.4165	0.4679	0.942
1.82	0.5991	0.4996	0.4134	0.4668	0.9385

Table B.1. (*cont.*)

Wavelength (microns)	Mystery spectra 1	Mystery spectra 2	Mystery spectra 3	Mystery spectra 4	Mystery spectra 5
1.825	0.5989	0.4996	0.4049	0.4626	0.9351
1.83	0.5985	0.4996	0.3962	0.4582	0.9317
1.835	0.598	0.4995	0.387	0.4534	0.9283
1.84	0.5972	0.4995	0.3769	0.4477	0.9248
1.845	0.593	0.4974	0.3658	0.4404	0.9214
1.85	0.5884	0.4954	0.3537	0.4322	0.918
1.855	0.5834	0.4934	0.3406	0.423	0.9146
1.86	0.5781	0.4913	0.3265	0.4127	0.9111
1.865	0.5628	0.4828	0.3135	0.4003	0.9077
1.87	0.547	0.4743	0.2994	0.3866	0.9043
1.875	0.5311	0.4658	0.2851	0.3728	0.9009
1.88	0.5153	0.4574	0.2708	0.359	0.8974
1.885	0.5009	0.4471	0.2594	0.3458	0.894
1.89	0.4864	0.4368	0.2476	0.3322	0.8955
1.895	0.4719	0.4266	0.2356	0.3184	0.8969
1.9	0.4573	0.4163	0.2236	0.3047	0.8984
1.905	0.4488	0.4108	0.2142	0.2939	0.8998
1.91	0.4404	0.4053	0.205	0.2834	0.9013
1.915	0.432	0.3998	0.196	0.273	0.9027
1.92	0.4238	0.3944	0.1873	0.2629	0.9041
1.925	0.4236	0.3962	0.1824	0.2579	0.9056
1.93	0.424	0.3981	0.1786	0.2542	0.907
1.935	0.4249	0.3999	0.1765	0.252	0.9085
1.94	0.4267	0.4018	0.1765	0.2519	0.9099
1.945	0.4325	0.4058	0.1799	0.2554	0.9114
1.95	0.4389	0.4098	0.1849	0.2606	0.9128
1.955	0.4454	0.4139	0.1904	0.2661	0.9143
1.96	0.452	0.4179	0.1959	0.2718	0.9154
1.965	0.4564	0.4209	0.2007	0.2765	0.9165
1.97	0.4606	0.4239	0.2051	0.2809	0.9176
1.975	0.465	0.4268	0.2098	0.2855	0.9188
1.98	0.4697	0.4298	0.2156	0.2912	0.9199
1.985	0.4732	0.4316	0.223	0.2987	0.921
1.99	0.4772	0.4334	0.2319	0.3077	0.9222
1.995	0.4818	0.4352	0.2422	0.318	0.9233
2	0.4863	0.437	0.2526	0.3284	0.9244

$\bar{f}_n$ are the abundances of each known constituent library, and h_m^n are the spectral library values of the known materials at resel I forming a column of the $\mathbf{H}$ matrix.

For this problem, I goes from 1 to 5, representing the five measured mystery spectra. They contain a linear mixing of the nine spectral signatures of various possible constituents as shown in the data library in Table B.2, with graphs of their reflectance following the table.

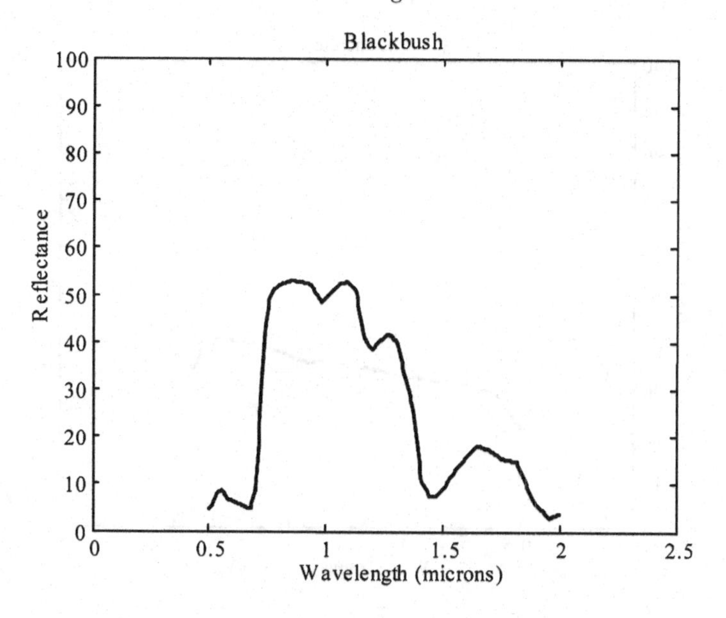
Blackbush
Reflectance
Wavelength (microns)
0
10
20
30
40
50
60
70
80
90
100
0
0.5
1
1.5
2
2.5

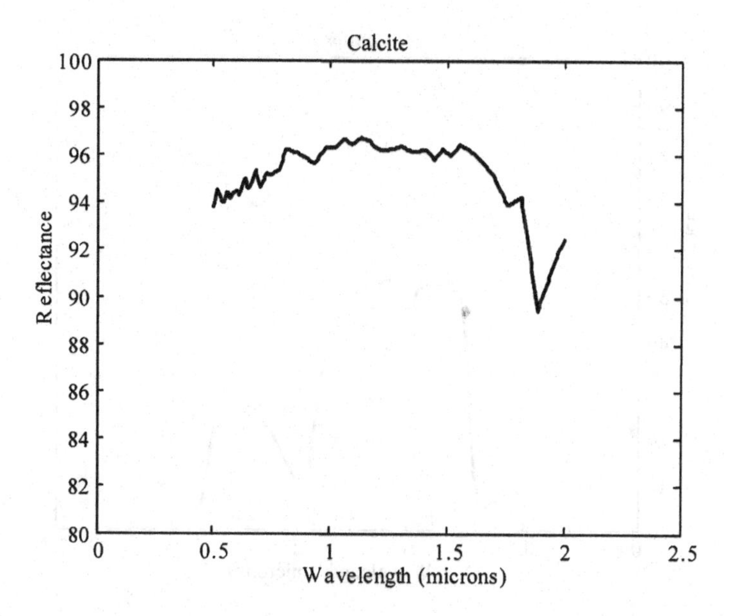
Calcite
Reflectance
Wavelength (microns)
80
82
84
86
88
90
92
94
96
98
100
0
0.5
1
1.5
2
2.5

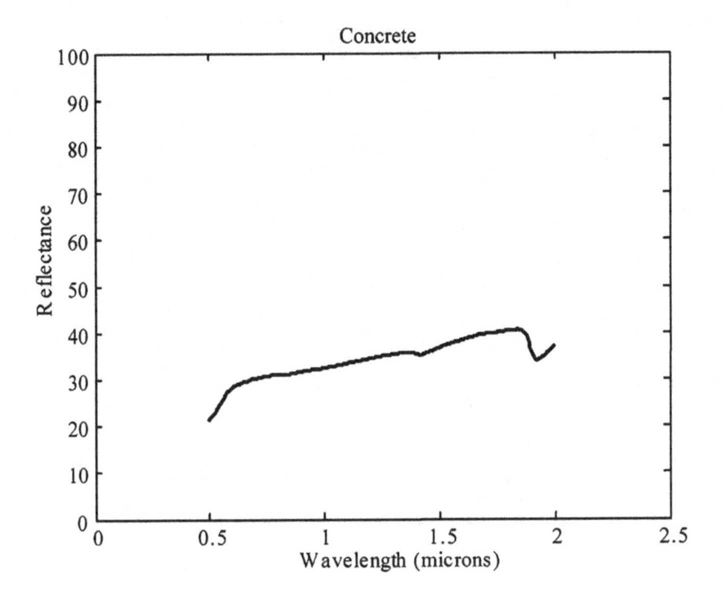
Concrete
100
90
80
70
60
50
40
30
20
10
0
Reflectance
0
0.5
1
1.5
2
2.5
Wavelength (microns)

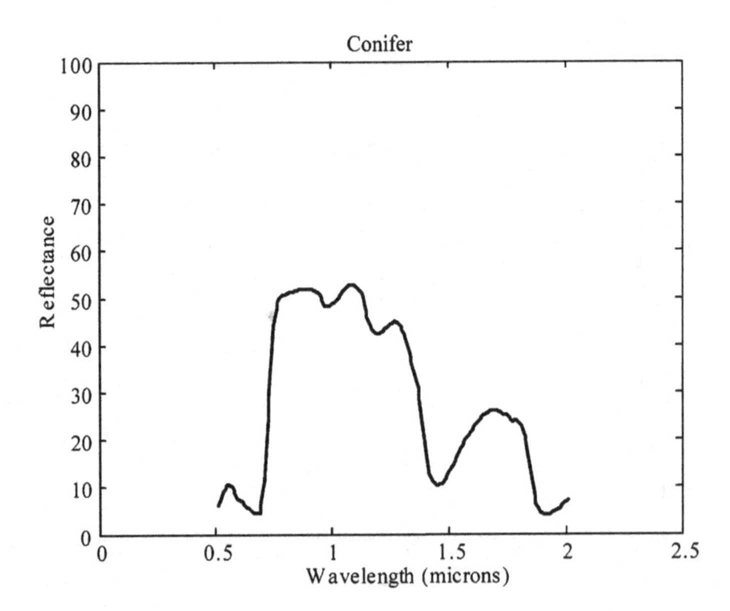
Conifer
100
90
80
70
60
50
40
30
20
10
0
Reflectance
0
0.5
1
1.5
2
2.5
Wavelength (microns)

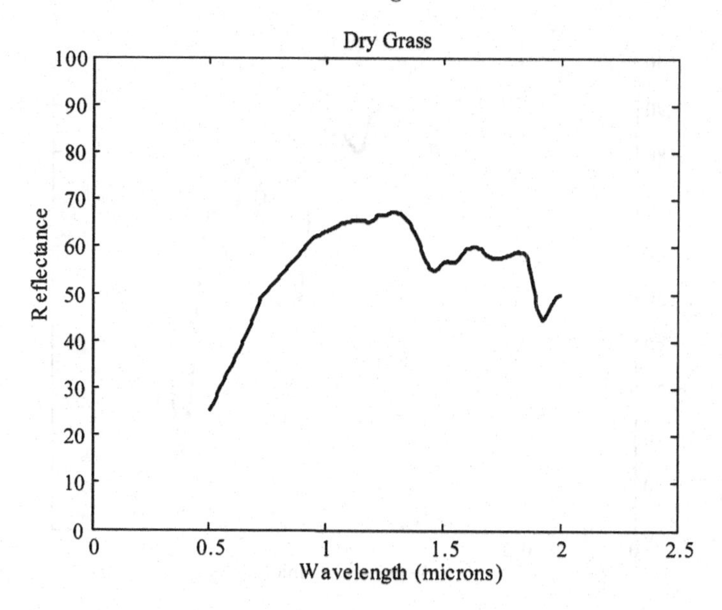
Dry Grass
100
90
80
70
60
50
40
30
20
10
0
Reflectance
0
0.5
1
1.5
2
2.5
Wavelength (microns)

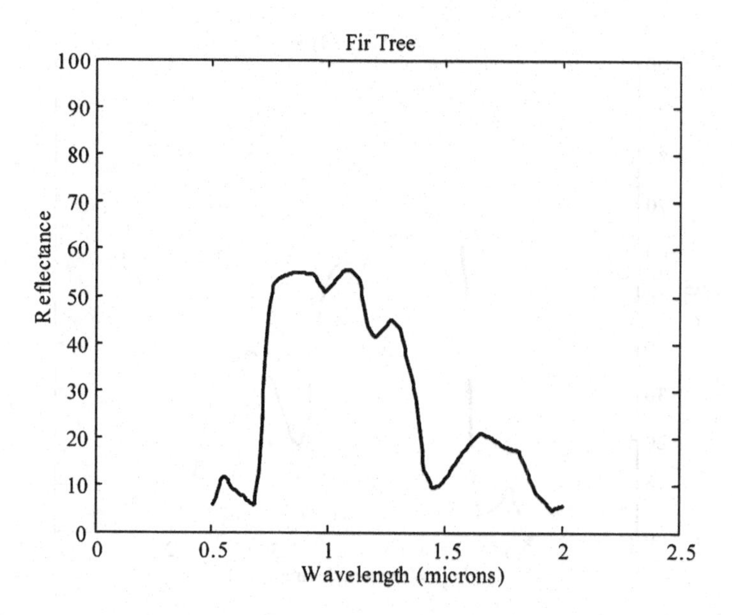
Fir Tree
100
90
80
70
60
50
40
30
20
10
0
Reflectance
0
0.5
1
1.5
2
2.5
Wavelength (microns)

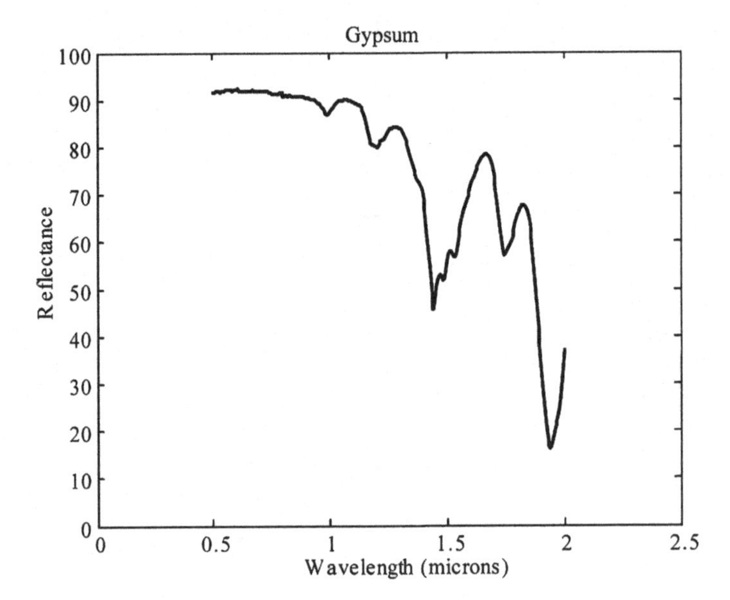
Gypsum
Reflectance
Wavelength (microns)
100
90
80
70
60
50
40
30
20
10
0
0
0.5
1
1.5
2
2.5

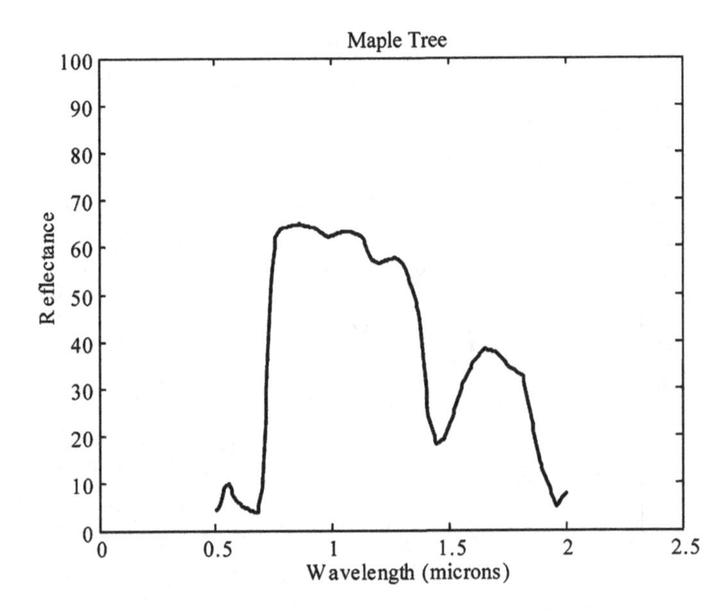
Maple Tree
Reflectance
Wavelength (microns)
100
90
80
70
60
50
40
30
20
10
0
0
0.5
1
1.5
2
2.5

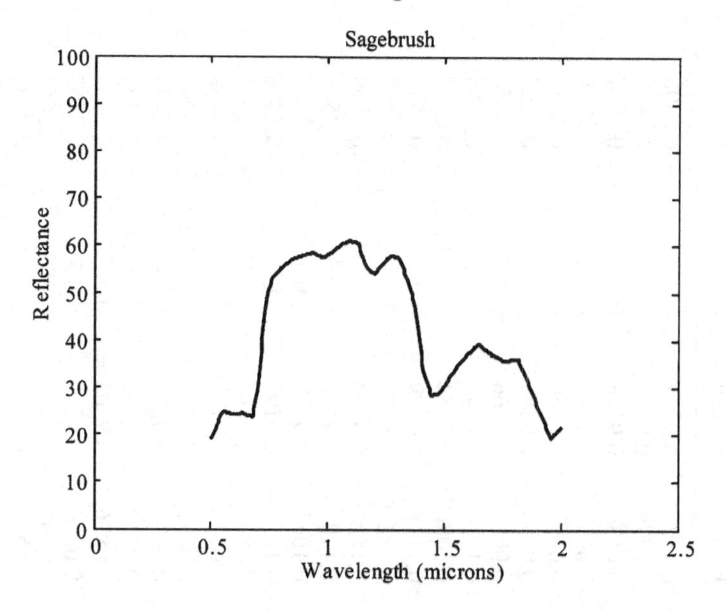

We can arrange the nine library spectral signatures in alphabetical order, as given in Table B.2:

(1) blackbush,
(2) calcite,
(3) concrete,
(4) conifer,
(5) dry grass,
(6) fir tree,
(7) gypsum,
(8) maple tree,
(9) sagebrush.

Each column of the **H** is a library spectra signature of one of these constituents. **H** has 301 rows and 9 columns, so each $\bar{g}_m$ value is a vector that has a length of 301 rows by one column.

To solve for $\bar{f}_n$, we must invert Equation (B.1) somehow, but since **H** is not square, we must use the least squares procedure (or singular value decomposition, SVD) to find a pseudo inverse $\mathbf{H}^+$ of **H**, such that:

$$\hat{\bar{f}}_n = \mathbf{H}^+ \bar{g}_m,$$

where $\hat{\bar{f}}_n$ is an estimated value of the abundance vector and has a length of 9 (for the nine possible constituents).

Table B.2. *Reflectance spectral signature library*

Wavelength (micrometers)	Blackbush	Calcite	Concrete	Conifer	Dry grass	Fir tree	Gypsum	Maple tree	Sagebrush
0.5	0.0459	0.9371	0.2159	0.0565	0.2518	0.0537	0.9147	0.0391	0.1908
0.505	0.0484	0.9391	0.2194	0.0644	0.2563	0.0579	0.9142	0.0421	0.1946
0.51	0.0517	0.9419	0.2229	0.0724	0.2607	0.0636	0.9148	0.0462	0.1997
0.515	0.0551	0.9446	0.2264	0.0803	0.2651	0.0693	0.9165	0.0503	0.2048
0.52	0.0616	0.9435	0.2299	0.0882	0.2696	0.0793	0.9159	0.061	0.2125
0.525	0.0683	0.9422	0.2338	0.0924	0.2758	0.0896	0.916	0.0722	0.2203
0.53	0.0745	0.9409	0.2377	0.0966	0.2821	0.0991	0.9166	0.0825	0.2278
0.535	0.0779	0.9402	0.2415	0.1008	0.2884	0.1038	0.9145	0.0873	0.2335
0.54	0.0813	0.9395	0.2454	0.105	0.2946	0.1086	0.9154	0.0921	0.2392
0.545	0.0839	0.9396	0.2497	0.1034	0.2994	0.1118	0.9179	0.0955	0.244
0.55	0.0847	0.9415	0.254	0.1018	0.3042	0.112	0.917	0.0962	0.2469
0.555	0.0855	0.9433	0.2583	0.1002	0.309	0.1123	0.9183	0.0969	0.2499
0.56	0.0839	0.9435	0.2626	0.0987	0.3138	0.1102	0.9195	0.0933	0.2506
0.565	0.0803	0.9423	0.2664	0.0926	0.319	0.1062	0.923	0.086	0.2492
0.57	0.0766	0.9411	0.2703	0.0866	0.3242	0.1022	0.9194	0.0788	0.2478
0.575	0.0732	0.9415	0.2741	0.0806	0.3293	0.0979	0.9195	0.0728	0.2467
0.58	0.0699	0.9424	0.2779	0.0746	0.3345	0.0934	0.9194	0.0673	0.2458
0.585	0.0667	0.9434	0.2803	0.0729	0.3389	0.0889	0.9176	0.0618	0.2448
0.59	0.0658	0.9438	0.2828	0.0712	0.3434	0.087	0.9205	0.0599	0.2449
0.595	0.0651	0.9442	0.2853	0.0695	0.3478	0.0853	0.9199	0.0582	0.2452
0.6	0.0642	0.9444	0.2877	0.0678	0.3522	0.0836	0.9203	0.0565	0.2453
0.605	0.0629	0.9434	0.2891	0.0651	0.357	0.0814	0.9212	0.0542	0.2452
0.61	0.0617	0.9424	0.2905	0.0625	0.3618	0.0793	0.92	0.052	0.2451
0.615	0.0601	0.9438	0.2919	0.0598	0.3666	0.077	0.9179	0.0497	0.2449
0.62	0.0585	0.9456	0.2933	0.0572	0.3714	0.0747	0.9173	0.0474	0.2447
0.625	0.0576	0.9473	0.2942	0.0554	0.3769	0.0736	0.9173	0.0463	0.2453
0.63	0.0571	0.9488	0.2952	0.0536	0.3825	0.0733	0.9173	0.0458	0.2464

0.635	0.0564	0.9494	0.2961	0.0518	0.388	0.0723	0.9179	0.045	0.2469
0.64	0.0553	0.9476	0.297	0.0501	0.3936	0.0696	0.9173	0.0433	0.2459
0.645	0.0541	0.9457	0.2978	0.0479	0.3984	0.0668	0.9167	0.0416	0.2448
0.65	0.0529	0.9451	0.2987	0.0458	0.4032	0.0643	0.9173	0.0402	0.2438
0.655	0.0517	0.9461	0.2995	0.0437	0.408	0.0621	0.9167	0.0392	0.2426
0.66	0.0504	0.947	0.3003	0.0416	0.4128	0.0598	0.9167	0.0382	0.2415
0.665	0.0493	0.9484	0.3011	0.042	0.4186	0.0581	0.9168	0.0374	0.2408
0.67	0.0482	0.95	0.302	0.0425	0.4245	0.0566	0.9184	0.0368	0.2402
0.675	0.0471	0.9516	0.3028	0.043	0.4305	0.0551	0.9167	0.0361	0.2397
0.68	0.0476	0.9527	0.3036	0.0435	0.4363	0.0565	0.9168	0.0376	0.2414
0.685	0.0567	0.951	0.3043	0.0604	0.4426	0.073	0.9161	0.0501	0.2547
0.69	0.0659	0.9494	0.305	0.0773	0.4489	0.0894	0.9156	0.0626	0.268
0.695	0.075	0.9477	0.3057	0.0942	0.4552	0.1059	0.9167	0.075	0.2813
0.7	0.0841	0.946	0.3064	0.1111	0.4614	0.1223	0.9169	0.0875	0.2946
0.705	0.1152	0.947	0.307	0.1539	0.4684	0.1646	0.9162	0.1357	0.3202
0.71	0.1472	0.948	0.3076	0.1968	0.4754	0.2079	0.9162	0.1853	0.3463
0.715	0.1825	0.9491	0.3081	0.2397	0.4825	0.2521	0.9167	0.2367	0.3729
0.72	0.2308	0.9501	0.3087	0.2826	0.4895	0.2996	0.9162	0.2954	0.4012
0.725	0.2791	0.9512	0.3092	0.3219	0.4928	0.3472	0.9167	0.354	0.4294
0.73	0.3252	0.9516	0.3098	0.3612	0.4961	0.3893	0.9162	0.4104	0.4543
0.735	0.3681	0.9513	0.3103	0.4005	0.4994	0.4231	0.9161	0.4632	0.474
0.74	0.411	0.9509	0.3108	0.4399	0.5028	0.4569	0.9139	0.516	0.4937
0.745	0.4388	0.951	0.3111	0.4527	0.5057	0.4785	0.9139	0.5507	0.5062
0.75	0.4566	0.9513	0.3115	0.4655	0.5087	0.4918	0.9122	0.5734	0.514
0.755	0.4745	0.9516	0.3119	0.4784	0.5117	0.5052	0.9121	0.5961	0.5217
0.76	0.491	0.952	0.3122	0.4912	0.5146	0.5175	0.911	0.617	0.529
0.765	0.4962	0.9523	0.3125	0.4942	0.5172	0.5207	0.9116	0.6217	0.5324
0.77	0.5014	0.9525	0.3129	0.4972	0.5198	0.524	0.9111	0.6264	0.5357
0.775	0.5066	0.9528	0.3132	0.5003	0.5224	0.5272	0.9111	0.6311	0.5391
0.78	0.5118	0.9531	0.3135	0.5033	0.525	0.5305	0.9116	0.6358	0.5424
0.785	0.5142	0.9543	0.3137	0.5043	0.528	0.5322	0.9128	0.6372	0.5451
0.79	0.5159	0.9557	0.3139	0.5053	0.5310	0.5336	0.9106	0.6378	0.5477

Table B.2. (*cont.*)

Wavelength (micrometers)	Blackbush	Calcite	Concrete	Conifer	Dry grass	Fir tree	Gypsum	Maple tree	Sagebrush
0.795	0.5176	0.9572	0.3140	0.5064	0.5339	0.5351	0.9084	0.6383	0.5502
0.8	0.5193	0.9586	0.3142	0.5074	0.5369	0.5365	0.9062	0.6389	0.5528
0.805	0.521	0.96	0.3143	0.5082	0.5395	0.5379	0.9064	0.6395	0.5554
0.81	0.5227	0.9615	0.3145	0.5089	0.5421	0.5393	0.9066	0.64	0.558
0.815	0.5238	0.9618	0.3146	0.5096	0.5447	0.5402	0.9069	0.6407	0.56
0.82	0.5246	0.9616	0.3147	0.5104	0.5473	0.5409	0.9059	0.6414	0.5617
0.825	0.5254	0.9615	0.3148	0.5112	0.5506	0.5417	0.9071	0.642	0.5634
0.83	0.5262	0.9613	0.315	0.512	0.5539	0.5424	0.9068	0.6427	0.5651
0.835	0.527	0.9612	0.3151	0.5127	0.5573	0.5431	0.9065	0.6434	0.5668
0.84	0.5278	0.961	0.3152	0.5135	0.5606	0.5438	0.9058	0.6441	0.5685
0.845	0.5286	0.9608	0.3154	0.5143	0.5632	0.5445	0.905	0.6448	0.5702
0.85	0.5294	0.9607	0.3156	0.5151	0.5658	0.5452	0.9058	0.6454	0.5719
0.855	0.5302	0.9605	0.3158	0.5159	0.5684	0.5459	0.9053	0.6461	0.5736
0.86	0.5308	0.9603	0.316	0.5167	0.571	0.5464	0.9053	0.6465	0.5751
0.865	0.5302	0.96	0.3166	0.5168	0.5736	0.5461	0.9045	0.6459	0.5758
0.87	0.5297	0.9597	0.3173	0.5168	0.5762	0.5458	0.9045	0.6452	0.5765
0.875	0.5291	0.9594	0.3179	0.5168	0.5788	0.5454	0.9045	0.6445	0.5772
0.88	0.5286	0.9591	0.3185	0.5169	0.5814	0.5451	0.9039	0.6439	0.5779
0.885	0.528	0.9588	0.3191	0.5165	0.5847	0.5448	0.904	0.6432	0.5786
0.89	0.5275	0.9585	0.3197	0.5162	0.588	0.5444	0.9041	0.6426	0.5793
0.895	0.527	0.9582	0.3203	0.5158	0.5914	0.5441	0.9031	0.6419	0.58
0.9	0.5264	0.9579	0.3209	0.5154	0.5947	0.5438	0.9016	0.6412	0.5807
0.905	0.5259	0.9576	0.3212	0.5151	0.5977	0.5434	0.9014	0.6406	0.5814
0.91	0.5253	0.9573	0.3215	0.5147	0.6006	0.5431	0.9007	0.6399	0.5821
0.915	0.5248	0.957	0.3217	0.5143	0.6036	0.5428	0.9008	0.6393	0.5828
0.92	0.5242	0.9567	0.322	0.5139	0.6065	0.5425	0.8992	0.6386	0.5835
0.925	0.5237	0.9564	0.3226	0.5113	0.6095	0.5421	0.8994	0.6379	0.5842

0.93	0.5231	0.9561	0.3231	0.5088	0.6125	0.5418	0.8999	0.6373	0.5849
0.935	0.5212	0.9563	0.3237	0.5062	0.6154	0.54	0.8993	0.6361	0.5849
0.94	0.5172	0.957	0.3243	0.5036	0.6184	0.536	0.8974	0.6341	0.584
0.945	0.5132	0.9577	0.3244	0.4982	0.6195	0.532	0.8957	0.6321	0.5831
0.95	0.5091	0.9584	0.3246	0.4928	0.6207	0.528	0.8938	0.6301	0.5822
0.955	0.5051	0.9591	0.3247	0.4874	0.6218	0.5239	0.8918	0.6282	0.5813
0.96	0.5011	0.9598	0.3249	0.48445	0.6229	0.5199	0.88865	0.6262	0.58035
0.965	0.4971	0.9605	0.3251	0.4815	0.624	0.5159	0.8855	0.6242	0.5794
0.97	0.493	0.9613	0.3255	0.4809	0.6252	0.5119	0.8816	0.6222	0.5785
0.975	0.489	0.962	0.3259	0.4804	0.6263	0.5079	0.8773	0.6203	0.5776
0.98	0.485	0.9627	0.3262	0.4799	0.6274	0.5039	0.8735	0.6183	0.5767
0.985	0.4849	0.963	0.3267	0.4818	0.6289	0.504	0.87	0.6182	0.5773
0.99	0.4874	0.9629	0.3271	0.4837	0.6304	0.5069	0.8682	0.6193	0.5789
0.995	0.49	0.9629	0.3275	0.4856	0.6319	0.5099	0.8678	0.6204	0.5805
1	0.4925	0.9629	0.328	0.4875	0.6334	0.5128	0.869	0.6215	0.5821
1.005	0.495	0.9629	0.3285	0.4905	0.6345	0.5157	0.8747	0.6227	0.5836
1.01	0.4975	0.9629	0.329	0.4935	0.6357	0.5186	0.8786	0.6238	0.5852
1.015	0.5001	0.9629	0.3295	0.4964	0.6368	0.5215	0.8824	0.6249	0.5868
1.02	0.5026	0.9628	0.33	0.4994	0.6379	0.5244	0.8862	0.626	0.5884
1.025	0.5052	0.963	0.3305	0.5026	0.6391	0.5274	0.8883	0.627	0.5902
1.03	0.508	0.9635	0.3309	0.5058	0.6402	0.5307	0.8908	0.6276	0.5923
1.035	0.5107	0.9641	0.3313	0.5089	0.6413	0.5339	0.8937	0.6283	0.5945
1.04	0.5135	0.9646	0.3318	0.5121	0.6425	0.5371	0.8962	0.6289	0.5966
1.045	0.5162	0.9651	0.3323	0.515	0.6439	0.5403	0.8972	0.6295	0.5987
1.05	0.519	0.9657	0.3329	0.5179	0.6454	0.5436	0.8966	0.6302	0.6009
1.055	0.5218	0.9662	0.3334	0.5209	0.6469	0.5468	0.8967	0.6308	0.603
1.06	0.5239	0.9665	0.334	0.5238	0.6484	0.5492	0.8976	0.6311	0.6048
1.065	0.5244	0.9661	0.3345	0.524	0.6492	0.5496	0.8979	0.6305	0.6059
1.07	0.5249	0.9657	0.335	0.5243	0.65	0.5501	0.8977	0.6299	0.607
1.075	0.5254	0.9653	0.3356	0.5246	0.6507	0.5505	0.8979	0.6294	0.6081
1.08	0.5259	0.9649	0.3361	0.5249	0.6515	0.5509	0.8964	0.6288	0.6092
1.085	0.5264	0.9645	0.3367	0.5236	0.6522	0.5514	0.8965	0.6282	0.6102

Table B.2. (*cont.*)

Wavelength (micrometers)	Blackbush	Calcite	Concrete	Conifer	Dry grass	Fir tree	Gypsum	Maple tree	Sagebrush
1.09	0.5269	0.9641	0.3373	0.5223	0.653	0.5518	0.8956	0.6276	0.6113
1.095	0.5264	0.964	0.3379	0.521	0.6538	0.5513	0.8941	0.6269	0.6119
1.1	0.5236	0.9644	0.3385	0.5197	0.6545	0.5486	0.8921	0.6256	0.6112
1.105	0.5208	0.9648	0.3391	0.5163	0.6549	0.5458	0.8921	0.6243	0.6106
1.11	0.5181	0.9653	0.3396	0.513	0.6553	0.5431	0.89	0.623	0.61
1.115	0.5153	0.9657	0.3402	0.5096	0.6557	0.5403	0.8889	0.6218	0.6094
1.12	0.5125	0.9661	0.3408	0.5062	0.6561	0.5376	0.8879	0.6205	0.6087
1.125	0.5097	0.9666	0.3413	0.4942	0.6561	0.5349	0.8864	0.6192	0.6081
1.13	0.5036	0.9668	0.3418	0.4822	0.6562	0.5286	0.8845	0.6164	0.6055
1.135	0.4895	0.9666	0.3423	0.4702	0.6562	0.5143	0.881	0.6099	0.5984
1.14	0.4755	0.9665	0.3428	0.4582	0.6562	0.5	0.8764	0.6034	0.5913
1.145	0.4615	0.9663	0.3433	0.4518	0.6562	0.4857	0.87	0.5968	0.5841
1.15	0.4475	0.9661	0.3437	0.4453	0.6563	0.4713	0.8615	0.5903	0.577
1.155	0.4335	0.966	0.3442	0.4388	0.6563	0.457	0.8493	0.5838	0.5699
1.16	0.4194	0.9658	0.3446	0.4323	0.6563	0.4427	0.8353	0.5773	0.5628
1.165	0.4086	0.9655	0.3452	0.4296	0.6552	0.4316	0.8224	0.5723	0.5571
1.17	0.405	0.965	0.3458	0.4268	0.6542	0.4282	0.8128	0.5707	0.5549
1.175	0.4015	0.9646	0.3463	0.4241	0.6531	0.4247	0.8056	0.5692	0.5528
1.18	0.3979	0.9641	0.3469	0.4213	0.652	0.4213	0.8022	0.5676	0.5506
1.185	0.3944	0.9636	0.3473	0.4221	0.6532	0.4178	0.8021	0.5661	0.5484
1.19	0.3909	0.9631	0.3478	0.4229	0.6543	0.4144	0.801	0.5645	0.5463
1.195	0.3873	0.9626	0.3482	0.4237	0.6554	0.4109	0.7997	0.5629	0.5441
1.2	0.3857	0.9622	0.3486	0.4245	0.6565	0.4094	0.7982	0.5622	0.5435
1.205	0.3884	0.9621	0.3491	0.4267	0.6595	0.4123	0.7992	0.5633	0.5465
1.21	0.3911	0.962	0.3495	0.4288	0.6625	0.4152	0.8052	0.5645	0.5494
1.215	0.3938	0.9619	0.35	0.431	0.6654	0.4181	0.8109	0.5656	0.5524
1.22	0.3966	0.9618	0.3505	0.4331	0.6684	0.421	0.8126	0.5667	0.5554

1.225	0.3993	0.9617	0.351	0.4355	0.6684	0.4239	0.8136	0.5679	0.5584
1.23	0.402	0.9616	0.3516	0.4379	0.6685	0.4268	0.815	0.569	0.5613
1.235	0.4045	0.9615	0.3521	0.4402	0.6685	0.4296	0.8195	0.57	0.5641
1.24	0.4064	0.9617	0.3526	0.4426	0.6685	0.4321	0.8262	0.5705	0.5665
1.245	0.4083	0.9618	0.353	0.4439	0.6685	0.4347	0.831	0.5711	0.5689
1.25	0.4102	0.9619	0.3534	0.4452	0.6686	0.4372	0.8339	0.5716	0.5712
1.255	0.4122	0.962	0.3539	0.4465	0.6686	0.4397	0.837	0.5722	0.5736
1.26	0.4141	0.9621	0.3543	0.4478	0.6686	0.4422	0.8371	0.5728	0.576
1.265	0.416	0.9623	0.3547	0.4462	0.6701	0.4447	0.8386	0.5733	0.5784
1.27	0.4167	0.9624	0.3552	0.4446	0.6716	0.4458	0.8393	0.5733	0.5799
1.275	0.4146	0.9625	0.3556	0.443	0.6731	0.4435	0.8402	0.5718	0.5795
1.28	0.4125	0.9627	0.356	0.4414	0.6746	0.4413	0.8407	0.5703	0.5791
1.285	0.4104	0.9628	0.3564	0.4364	0.6743	0.439	0.8413	0.5688	0.5787
1.29	0.4083	0.9629	0.3567	0.4314	0.6739	0.4367	0.8401	0.5673	0.5783
1.295	0.4062	0.9631	0.3571	0.4264	0.6736	0.4345	0.8381	0.5658	0.578
1.3	0.4041	0.9632	0.3575	0.4215	0.6733	0.4322	0.8366	0.5643	0.5776
1.305	0.3994	0.9632	0.358	0.4119	0.6722	0.4274	0.8336	0.5611	0.5754
1.31	0.3889	0.963	0.3584	0.4024	0.6711	0.4167	0.8284	0.5542	0.569
1.315	0.3783	0.9627	0.3589	0.3929	0.67	0.4061	0.8228	0.5473	0.5626
1.32	0.3678	0.9624	0.3593	0.3834	0.669	0.3954	0.8161	0.5403	0.5562
1.325	0.3573	0.9622	0.3597	0.3735	0.6668	0.3847	0.8105	0.5334	0.5498
1.33	0.3467	0.9619	0.3601	0.3635	0.6646	0.3741	0.8037	0.5264	0.5434
1.335	0.3362	0.9616	0.3605	0.3536	0.6624	0.3634	0.794	0.5195	0.537
1.34	0.3259	0.9615	0.3609	0.3436	0.6603	0.3529	0.7824	0.512	0.5301
1.345	0.316	0.9614	0.3613	0.3341	0.6566	0.3427	0.7732	0.5032	0.5219
1.35	0.3061	0.9614	0.3617	0.3245	0.653	0.3325	0.7636	0.4945	0.5138
1.355	0.2962	0.9614	0.3622	0.315	0.6493	0.3224	0.7558	0.4857	0.5056
1.36	0.2863	0.9614	0.3626	0.3054	0.6457	0.3122	0.7476	0.4769	0.4974
1.365	0.2765	0.9614	0.3624	0.2834	0.6413	0.302	0.7401	0.4682	0.4892
1.37	0.2666	0.9613	0.3623	0.2613	0.6369	0.2919	0.7332	0.4594	0.481
1.375	0.2532	0.9613	0.3622	0.2392	0.6325	0.2781	0.7277	0.4442	0.4695
1.38	0.2315	0.9614	0.362	0.2172	0.6282	0.256	0.7234	0.4138	0.45

Table B.2. (*cont.*)

Wavelength (micrometers)	Blackbush	Calcite	Concrete	Conifer	Dry grass	Fir tree	Gypsum	Maple tree	Sagebrush
1.385	0.2098	0.9615	0.361	0.1961	0.6216	0.2339	0.7189	0.3835	0.4305
1.39	0.1881	0.9616	0.3601	0.1751	0.615	0.2118	0.7123	0.3531	0.4111
1.395	0.1665	0.9616	0.3592	0.1541	0.6084	0.1897	0.7028	0.3228	0.3916
1.4	0.1448	0.9617	0.3582	0.1331	0.6019	0.1676	0.6865	0.2924	0.3721
1.405	0.1231	0.9618	0.3576	0.1269	0.5949	0.1455	0.6656	0.262	0.3526
1.41	0.1066	0.9617	0.3569	0.1207	0.588	0.1284	0.6387	0.2382	0.3367
1.415	0.102	0.9611	0.3563	0.1146	0.581	0.1229	0.6096	0.2294	0.3291
1.42	0.0974	0.9605	0.3557	0.1084	0.5741	0.1174	0.5824	0.2207	0.3215
1.425	0.0929	0.9599	0.3567	0.1067	0.5697	0.1119	0.5584	0.212	0.3139
1.43	0.0883	0.9592	0.3577	0.1049	0.5653	0.1064	0.5289	0.2032	0.3063
1.435	0.0837	0.9586	0.3586	0.1032	0.5609	0.1009	0.4895	0.1945	0.2987
1.44	0.0792	0.958	0.3596	0.1015	0.5565	0.0954	0.4552	0.1857	0.2911
1.445	0.076	0.9578	0.3606	0.1021	0.5551	0.0918	0.4541	0.1802	0.2859
1.45	0.0759	0.9585	0.3616	0.1027	0.5537	0.0924	0.4785	0.182	0.2863
1.455	0.0759	0.9592	0.3625	0.1034	0.5522	0.0931	0.5021	0.1838	0.2867
1.46	0.0759	0.9599	0.3635	0.104	0.5508	0.0937	0.517	0.1856	0.2871
1.465	0.0759	0.9606	0.3644	0.1079	0.5519	0.0944	0.524	0.1874	0.2875
1.47	0.0759	0.9613	0.3653	0.1119	0.553	0.095	0.5281	0.1892	0.2879
1.475	0.0758	0.962	0.3662	0.1158	0.5542	0.0957	0.5292	0.191	0.2883
1.48	0.0768	0.9623	0.3671	0.1197	0.5553	0.0972	0.5237	0.1944	0.2898
1.485	0.08	0.9618	0.3681	0.1244	0.5586	0.1009	0.5158	0.2015	0.2938
1.49	0.0832	0.9613	0.3691	0.1291	0.562	0.1045	0.5199	0.2085	0.2977
1.495	0.0864	0.9608	0.3702	0.1338	0.5653	0.1082	0.5369	0.2155	0.3017
1.5	0.0897	0.9603	0.3712	0.1385	0.5686	0.1118	0.556	0.2226	0.3057
1.505	0.0929	0.9598	0.3721	0.1432	0.569	0.1155	0.5704	0.2296	0.3097
1.51	0.0961	0.9593	0.3731	0.1479	0.5694	0.1191	0.5767	0.2367	0.3137
1.515	0.0995	0.9592	0.374	0.1526	0.5698	0.1229	0.578	0.2439	0.3176

1.52	0.1034	0.9598	0.375	0.1573	0.5702	0.1268	0.5766	0.2517	0.3213
1.525	0.1073	0.9605	0.3759	0.1627	0.5695	0.1307	0.5706	0.2595	0.325
1.53	0.1112	0.9612	0.3767	0.1681	0.5688	0.1347	0.5655	0.2673	0.3288
1.535	0.115	0.9618	0.3776	0.1735	0.5681	0.1386	0.566	0.2751	0.3325
1.54	0.1189	0.9625	0.3784	0.1789	0.5674	0.1425	0.5745	0.2828	0.3362
1.545	0.1228	0.9632	0.3792	0.1833	0.5681	0.1464	0.5911	0.2906	0.34
1.55	0.1267	0.9638	0.3801	0.1878	0.5689	0.1504	0.6085	0.2984	0.3437
1.555	0.1301	0.9639	0.3809	0.1922	0.5697	0.154	0.6253	0.3047	0.3469
1.56	0.1331	0.9636	0.3817	0.1966	0.5704	0.1573	0.6398	0.3101	0.3498
1.565	0.1362	0.9633	0.3824	0.2002	0.5745	0.1607	0.6512	0.3154	0.3526
1.57	0.1392	0.963	0.3831	0.2038	0.5786	0.1641	0.6613	0.3207	0.3555
1.575	0.1422	0.9627	0.3839	0.2074	0.5826	0.1675	0.6701	0.326	0.3584
1.58	0.1453	0.9624	0.3846	0.211	0.5867	0.1709	0.6791	0.3314	0.3612
1.585	0.1483	0.9621	0.3854	0.2146	0.5893	0.1742	0.6875	0.3367	0.3641
1.59	0.1513	0.9618	0.3862	0.2183	0.5919	0.1776	0.6954	0.342	0.367
1.595	0.1544	0.9615	0.3869	0.2219	0.5945	0.181	0.7042	0.3474	0.3698
1.6	0.1572	0.9612	0.3877	0.2255	0.5971	0.184	0.7139	0.352	0.3725
1.605	0.1597	0.9607	0.3885	0.229	0.5979	0.1864	0.7224	0.3551	0.3748
1.61	0.1621	0.9603	0.3892	0.2325	0.5986	0.1888	0.7293	0.3581	0.377
1.615	0.1645	0.9598	0.39	0.2361	0.5994	0.1911	0.7367	0.3612	0.3793
1.62	0.167	0.9594	0.3908	0.2396	0.6001	0.1935	0.7439	0.3642	0.3815
1.625	0.1694	0.959	0.3916	0.2419	0.6002	0.1958	0.7498	0.3673	0.3838
1.63	0.1718	0.9585	0.3923	0.2443	0.6002	0.1982	0.7553	0.3704	0.386
1.635	0.1743	0.9581	0.3931	0.2467	0.6002	0.2005	0.7623	0.3734	0.3883
1.64	0.1767	0.9576	0.3938	0.249	0.6002	0.2029	0.7683	0.3765	0.3905
1.645	0.1791	0.9572	0.3945	0.2506	0.5995	0.2052	0.7729	0.3796	0.3928
1.65	0.1799	0.9566	0.3953	0.2522	0.5988	0.2059	0.7766	0.3808	0.3929
1.655	0.1792	0.956	0.396	0.2537	0.5981	0.2049	0.7799	0.3803	0.3909
1.66	0.1784	0.9553	0.3967	0.2553	0.5974	0.2039	0.7826	0.3798	0.3888
1.665	0.1776	0.9547	0.3972	0.2564	0.5945	0.2029	0.7834	0.3792	0.3868
1.67	0.1768	0.9541	0.3977	0.2575	0.5916	0.2019	0.7825	0.3787	0.3847
1.675	0.176	0.9534	0.3981	0.2587	0.5887	0.201	0.7795	0.3782	0.3827

Table B.2. (*cont.*)

Wavelength (micrometers)	Blackbush	Calcite	Concrete	Conifer	Dry grass	Fir tree	Gypsum	Maple tree	Sagebrush
1.68	0.1752	0.9528	0.3986	0.2598	0.5858	0.2	0.777	0.3777	0.3807
1.685	0.1744	0.9521	0.3991	0.2597	0.5843	0.199	0.7705	0.3771	0.3786
1.69	0.1736	0.9515	0.3995	0.2595	0.5829	0.198	0.7617	0.3766	0.3766
1.695	0.1729	0.9509	0.3999	0.2594	0.5815	0.197	0.7498	0.3761	0.3746
1.7	0.1721	0.9502	0.4004	0.2593	0.58	0.196	0.7341	0.3755	0.3725
1.705	0.1703	0.9492	0.4008	0.2571	0.5793	0.1943	0.716	0.3727	0.3712
1.71	0.1686	0.9481	0.4011	0.2549	0.5786	0.1926	0.6943	0.3698	0.3699
1.715	0.1668	0.9471	0.4014	0.2527	0.5779	0.1908	0.6709	0.3669	0.3686
1.72	0.1651	0.946	0.4018	0.2505	0.5772	0.1891	0.6474	0.364	0.3673
1.725	0.1633	0.945	0.4021	0.2504	0.5772	0.1874	0.6243	0.3611	0.3661
1.73	0.1616	0.944	0.4024	0.2503	0.5772	0.1857	0.6024	0.3582	0.3648
1.735	0.1598	0.9429	0.4027	0.2502	0.5773	0.1839	0.5841	0.3553	0.3635
1.74	0.158	0.9419	0.403	0.2502	0.5773	0.1822	0.57	0.3524	0.3622
1.745	0.1563	0.9408	0.4032	0.2466	0.5781	0.1805	0.5673	0.3495	0.3609
1.75	0.1545	0.9398	0.4034	0.2431	0.5788	0.1788	0.5703	0.3466	0.3596
1.755	0.1528	0.9387	0.4036	0.2396	0.5796	0.177	0.5772	0.3437	0.3583
1.76	0.1521	0.9388	0.4038	0.236	0.5803	0.1762	0.5839	0.3418	0.3583
1.765	0.1516	0.939	0.4043	0.2367	0.5811	0.1755	0.5891	0.3401	0.3586
1.77	0.1512	0.9393	0.4048	0.2374	0.5819	0.1748	0.5941	0.3385	0.359
1.775	0.1507	0.9396	0.4052	0.238	0.5826	0.1742	0.5999	0.3368	0.3593
1.78	0.1503	0.9399	0.4057	0.2387	0.5834	0.1735	0.6098	0.3351	0.3596
1.785	0.1498	0.9402	0.4061	0.2353	0.5845	0.1728	0.6221	0.3334	0.3599
1.79	0.1493	0.9405	0.4066	0.232	0.5856	0.1722	0.6347	0.3318	0.3603
1.795	0.1489	0.9408	0.407	0.2286	0.5868	0.1715	0.6453	0.3301	0.3606
1.8	0.1484	0.9411	0.4074	0.2252	0.5879	0.1708	0.6542	0.3284	0.3609
1.805	0.148	0.9414	0.4076	0.2177	0.5887	0.1702	0.6608	0.3267	0.3612

1.81	0.1475	0.9417	0.4079	0.2101	0.5894	0.1695	0.6667	0.3251	0.3616
1.815	0.1471	0.942	0.4081	0.2025	0.5902	0.1689	0.6707	0.3234	0.3619
1.82	0.1412	0.9385	0.4083	0.195	0.5909	0.1626	0.6728	0.3105	0.3559
1.825	0.1353	0.9351	0.4086	0.1778	0.5906	0.1564	0.6734	0.2975	0.3499
1.83	0.1294	0.9317	0.4088	0.1607	0.5903	0.1501	0.6729	0.2846	0.3439
1.835	0.1236	0.9283	0.4091	0.1436	0.5899	0.1439	0.6708	0.2716	0.3379
1.84	0.1177	0.9248	0.4094	0.1264	0.5896	0.1376	0.6653	0.2587	0.3319
1.845	0.1118	0.9214	0.4089	0.1105	0.5859	0.1314	0.6563	0.2458	0.3259
1.85	0.1059	0.918	0.4085	0.0945	0.5823	0.1251	0.6434	0.2328	0.3199
1.855	0.1	0.9146	0.4081	0.0786	0.5787	0.1189	0.6266	0.2199	0.3139
1.86	0.0942	0.9111	0.4076	0.0627	0.575	0.1127	0.6056	0.207	0.3079
1.865	0.0883	0.9077	0.4045	0.0577	0.5611	0.1064	0.5778	0.194	0.302
1.87	0.0824	0.9043	0.4015	0.0527	0.5472	0.1002	0.545	0.1811	0.296
1.875	0.0765	0.9009	0.3984	0.0477	0.5332	0.0939	0.5117	0.1682	0.29
1.88	0.0706	0.8974	0.3954	0.0427	0.5193	0.0877	0.4786	0.1552	0.284
1.885	0.0648	0.894	0.3873	0.0417	0.5069	0.0814	0.4473	0.1423	0.278
1.89	0.0622	0.8955	0.3793	0.0407	0.4944	0.0788	0.4144	0.1356	0.2719
1.895	0.0596	0.8969	0.3712	0.0397	0.482	0.0761	0.3809	0.1288	0.2659
1.9	0.057	0.8984	0.3631	0.0388	0.4695	0.0734	0.3474	0.1221	0.2599
1.905	0.0544	0.8998	0.358	0.0391	0.4637	0.0707	0.3149	0.1154	0.2539
1.91	0.0518	0.9013	0.3529	0.0395	0.4578	0.068	0.2833	0.1086	0.2478
1.915	0.0493	0.9027	0.3477	0.0398	0.452	0.0653	0.2524	0.1019	0.2418
1.92	0.0467	0.9041	0.3426	0.0401	0.4461	0.0626	0.2228	0.0951	0.2358
1.925	0.0441	0.9056	0.3437	0.0414	0.4487	0.0599	0.1979	0.0884	0.2298
1.93	0.0415	0.907	0.3448	0.0427	0.4513	0.0573	0.1778	0.0817	0.2237
1.935	0.0389	0.9085	0.3459	0.044	0.4539	0.0546	0.1642	0.0749	0.2177
1.94	0.0363	0.9099	0.347	0.0453	0.4565	0.0519	0.1588	0.0682	0.2117
1.945	0.0338	0.9114	0.3492	0.047	0.4624	0.0492	0.163	0.0615	0.2057
1.95	0.0312	0.9128	0.3514	0.0488	0.4683	0.0465	0.1738	0.0547	0.1996
1.955	0.0286	0.9143	0.3535	0.0505	0.4742	0.0438	0.1862	0.048	0.1936
1.96	0.0296	0.9154	0.3557	0.0523	0.4801	0.0449	0.1989	0.0513	0.1963

Table B.2. (*cont.*)

Wavelength (micrometers)	Blackbush	Calcite	Concrete	Conifer	Dry grass	Fir tree	Gypsum	Maple tree	Sagebrush
1.965	0.0306	0.9165	0.358	0.0547	0.4838	0.046	0.2093	0.0547	0.1991
1.97	0.0315	0.9176	0.3602	0.0572	0.4875	0.047	0.2186	0.058	0.2018
1.975	0.0325	0.9188	0.3625	0.0596	0.4912	0.0481	0.2288	0.0613	0.2045
1.98	0.0335	0.9199	0.3647	0.0621	0.4949	0.0492	0.2431	0.0647	0.2073
1.985	0.0345	0.921	0.3672	0.0644	0.496	0.0502	0.2671	0.068	0.21
1.99	0.0355	0.9222	0.3697	0.0666	0.4972	0.0513	0.2973	0.0713	0.2128
1.995	0.0365	0.9233	0.3721	0.0689	0.4983	0.0524	0.3328	0.0747	0.2155
2	0.0375	0.9244	0.3746	0.0712	0.4994	0.0534	0.3685	0.078	0.2182

Using least squares,

$$\mathbf{H}^{+} = \left(\mathbf{H}^{\mathrm{T}}\mathbf{H}\right)^{-1}\mathbf{H}^{\mathrm{T}},$$

where $\mathbf{H}^{\mathrm{T}}$ is the transpose of $\mathbf{H}$:

$$\hat{\bar{f}}_n = \left(\mathbf{H}^{\mathrm{T}}\mathbf{H}\right)^{-1}\mathbf{H}^{\mathrm{T}}\bar{g}_m.$$

The least-squares fit to the spectrum $\bar{g}_m$ for a given pixel is given by[2]

$$\hat{\bar{f}}_n = \left(\mathbf{H}^{\mathrm{T}}\mathbf{H}\right)^{-1}\mathbf{H}^{\mathrm{T}}\bar{g}_m,$$

where "^" above the vector indicates *estimate*. However, a potential problem is that the weights (abundances) may become *negative*. This is physically impossible, so that constituent must be thrown out.

Putting these matrices into a Matlab program for this evaluation,

Matlab program – unmixing

```
function y = unmixing
% Appendix B
% Solution to the spectral unmixing problem
clear all
clc
% read in the libraries
spectrums = xlsread ('06spectralLibrary.xls');
%This makes a matrix of the various materials that make up the library
   of elements
mystery = xlsread('06mysterySpectra.xls');
%Makes a matrix of the mystery spectrum values
%Get rid of the wavelength column
[rows cols] = size(spectrums);
S = spectrums(:,2:cols);
[rows cols] = size(mystery);
M = mystery(:,2:cols);
%Abundances are yi
yi = inv(S'*S)*S'*M;
% The abundances are from the equation (Ht*H)o -1*Ht*xi = yi in notes
% Where H is the given library spectrums, xi is the mystery spectra and
% yi are the relative amounts of materials in the mystery spectrum.
% The various abundances of blackbush, calcite, concrete, conifer, dry grass,
% Fir tree, gypsum, Maple tree, and sage brush are listed for each mystery
   material yi(:,1),yi(:,2),yi(:,3),yi(:,4),yi(:,5)
```

These are the abundances for each vector:

[2] H. H. Barrett and K. J. Myers (2004). *Foundations of Image Science*. New York: John Wiley & Sons, Inc., p. 56.

Constituent

	$\bar{f}_1$	$\bar{f}_2$	$\bar{f}_3$	$\bar{f}_4$	$\bar{f}_5$
Blackbush	0.00	0.00	0.00	0.00	0.00
Calcite	0.00	0.00	0.00	0.00	1.00
Concrete	0.00	*0.50*	0.00	*0.25*	0.00
Conifer	0.00	0.00	0.50	0.25	0.00
Dry grass	0.90	0.50	0.25	0.25	0.00
Fir tree	0.00	0.00	0.00	0.00	0.00
Gypsum	0.10	0.00	0.25	0.25	0.00
Maple tree	0.00	0.00	0.00	0.00	0.00
Sagebrush	0.00	0.00	0.00	0.00	0.00

The concrete target is in pixels 2 and 4.

Appendix C
Nature's optical phenomena

Many optical phenomena that appear in the sky may be explained using the geometrical ray tracing described in this book. A few that will be discussed here are rainbows, halos, sun dogs, and the optical illusion of mirages in the desert.

C.1 Rainbows

A rainbow is an optical phenomenon that causes a nearly continuous spectrum of light to appear in the sky. It is only present when the rain location is opposite the sun's direction from the observer – in fact, when they are 180° apart. As white light from the sun shines onto drops of water, a polychromatic bow is formed, with red in the outermost arc and blue/violet on the inside of the bow. The infrared color regions beyond the red are ignored; however, they do exist.[1] This is the most traditional rainbow, in which sunlight is spread out into its spectrum of colors and diverted to the eye of the observer by spherical water drops. The "bow" part of the word comes from the fact that rainbows contain a group of nearly circular arcs of color (blue to red), all having a common center of curvature. A second rainbow sometimes appears as a fainter arc, with colors in the opposite order, i.e., with violet on the outside, and red on the inside.

The rainbow's appearance is caused by the dispersive properties of water as sunlight is refracted by the raindrops. The light is refracted twice: once as it enters the surface of the raindrop, and again as it leaves the drop. A rainbow does not actually have a pot of gold at the end, but is an optical phenomenon whose apparent position and location depends on the observer's location. All water drops reflect sunlight, but only the light from certain raindrops are seen and perceived to form a rainbow by the observer.

The location of the observer in relation to the sun is critical. A rainbow is never observed at noon. The geometry of the sun and the observer's eye forms an optical axis which is the center of the bow of arc as shown in Figure C.1.

The effect of the rays is shown in Figure C.2. The rays are reflected back until the critical angle (see Section 2.4.2) is reached and causes a boundary, which is the spread of the rainbow. At greater angles, the dispersion is simply overshadowed by the

[1] Greenler, R. (1980). *Rainbows, Halos, and Glories*. Cambridge: Cambridge University Press.

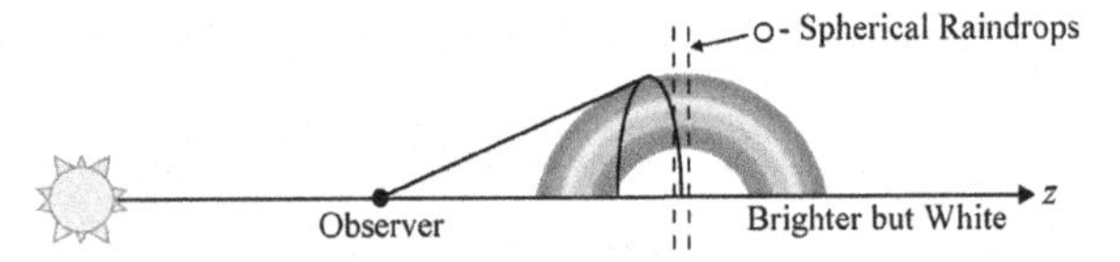

Figure C.1 Sun–observer orientation forming an optical axis.

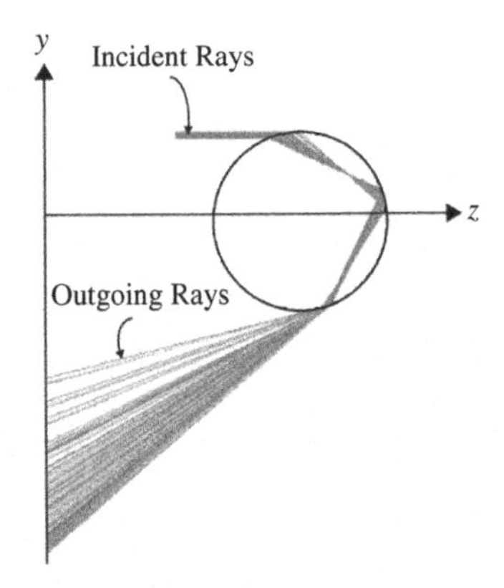

Figure C.2 Chromatic spread of the sun's rays from a raindrop.

superposition of all the other colors, but at the critical angle, the transmission stops and the spectrum appears as shown in Figure C.2.

This critical angle is dependent on the index of refraction of water, which varies with wavelength. The spectrum from the water drop thus causes maximum intensity. Blue light is refracted at a higher angle than red, but because the eye is the aperture stop of the system, to the observer, blue light comes in at a shallower angle, and thus appears at the bottom of the bow.

The rainbow appears at an angle of approximately 41° to the optical axis, the line from the observer's eye to the sun. Thus, if the sun is higher than 42° above the horizon, the rainbow is below the horizon and cannot be seen. However, if the observer is within a sprinkler system, a rainbow may be seen while looking at the ground. One has the opportunity to see the whole circle of the rainbow from an airplane or a high rise in Hawaii,[2] if the geometry is correct. A photograph showing a bull's-eye rainbow is very difficult to capture, and would require a wide angle lens of at least 84°, such as an inverse telephoto lens. A wide angle lens is a lens whose focal length is much shorter than the diagonal of the recording medium. Normally, the focal length of a lens is about equal to the diagonal of the detector array or film. For a 1 cm CCD/CMOS array to record a rainbow, it would require a focal length of 6 mm!

The angle of the rainbow from the observer's line of sight can be explained by modeling the raindrop as a sphere, as shown in Figure C.3. The parallel rays from the sun are impinging on the sphere from all positions, as shown in Figure C.2; however,

[2] Capps, R. (1981) private communication.

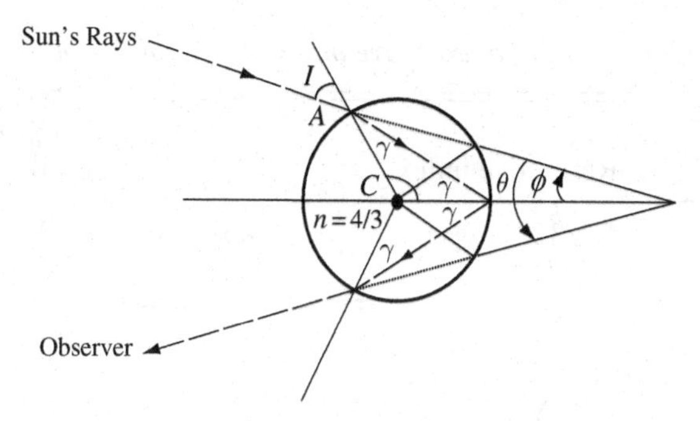

Figure C.3 Raindrop effect on the sun's rays.

the refracted rays are limited by the critical angle that can allow energy through the sphere. To find the maximum deviation of the ray, one must look at the geometry of the ray propagation through the drop in a symmetrical manner.

Using Snell's law for the angle of incidence, I, in air:

$$\sin I = n \sin \gamma, \tag{C.1}$$

consider the triangle AOC:

$$\phi + I + (180 - 2\gamma) = 180,$$
$$\phi = 2\gamma - I,$$
$$\theta = 4\gamma - 2I,$$

where θ is the total angle of deviation of the sun's ray, as shown in Figure C.3. For the primary rainbow dispersed rays, a single reflection internal to the drop occurs at the back surface.

To find the maximum angle θ, set its derivative with respect to incidence angle, I, equal to zero:

$$\frac{\mathrm{d}\theta}{\mathrm{d}I} = 4\frac{\mathrm{d}\gamma}{\mathrm{d}I} - 2 = 0. \tag{C.2}$$

Taking the full differential of Equation (C.1):

$$\cos I \, \mathrm{d}I = n \cos \gamma \, \mathrm{d}\gamma,$$

so,

$$\frac{\mathrm{d}\gamma}{\mathrm{d}I} = \frac{\cos I}{n \cos \gamma}. \tag{C.3}$$

Substituting Equation (C.3) into Equation (C.2),

$$\frac{4 \cos I}{n \cos \gamma} - 2 = 0, \tag{C.4}$$

so,

Table C.1. *Approximate refractive index values for various colors*

Color	Refractive index	I	Angle (θ) of total deviation
Red	1.3312	59.5°	42.3°
Blue	1.3371	59.2°	41.5°

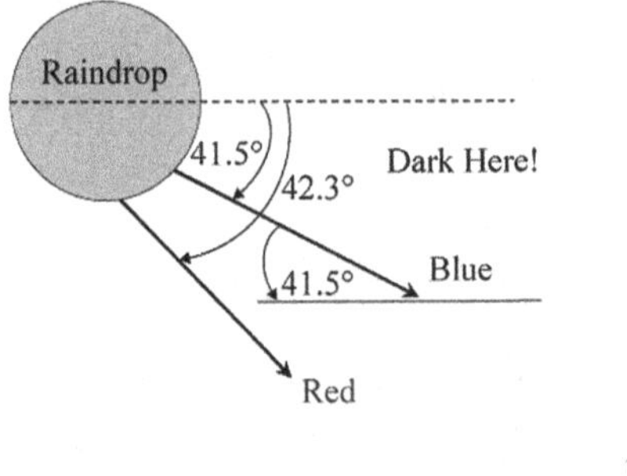

Figure C.4 Spread of colors from a raindrop.

$$\begin{aligned} 2\cos I - n\cos\gamma &= 0 \\ 2\cos I - n\left(1 - \sin^2\gamma\right)^{1/2} &= 0 \\ 2\cos I - \left(n^2 - n^2\sin^2\gamma\right)^{1/2} &= 0 \\ 4\cos^2 I = n^2 - \sin^2 I &= n^2 - \left(1 - \cos^2 I\right) \\ \cos I &= \sqrt{\frac{n^2 - 1}{3}}. \end{aligned} \tag{C.5}$$

So, the maximum angle of incidence to allow energy through the water drop is a function of the refractive index, which depends on wavelength. For water, the variation in refractive index is slight. Table C.1 lists approximate refractive index values for various colors.

As shown in Figure C.4, the red rays are deviated at the largest angle. However, from the observer's point of view, with the eye being the stop of the optical system, the angles of the rays entering the entrance pupil give the illusion that the red light is above the blue light. From the line of sight formed by the sun–observer, shown in Figure C.1, the rainbow appears at about 42° in a radial pattern, as shown in Figure C.5.

C.2 Secondary rainbows

When a ray is reflected twice inside the water sphere, a secondary rainbow is formed beyond the 42° angle at about 52°, as shown in Figure C.6. As a result of the double reflections, the colors of the secondary rainbow are inverted, with blue on the outside and red on the inside of the bow. The region of the sky between the two

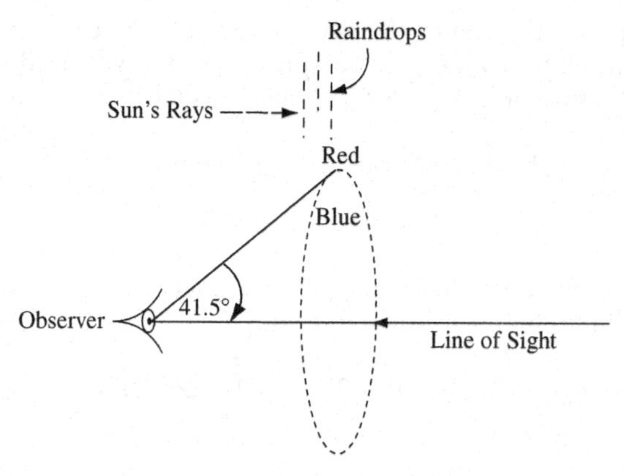

Figure C.5 Observation of a rainbow with the Sun's rays parallel to line of sight.

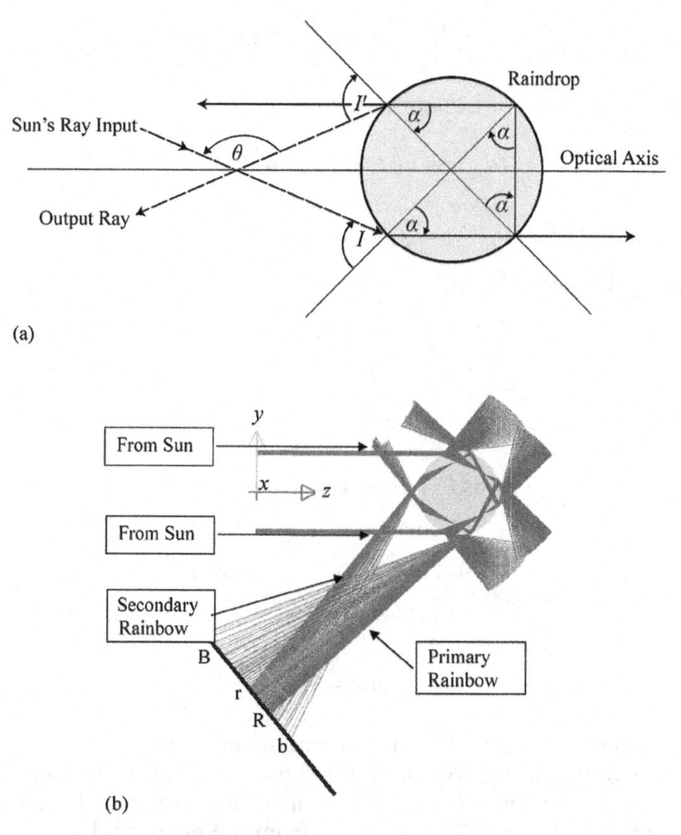

Figure C.6 (a) The creation of a secondary rainbow due to two reflections inside the raindrops. (b) The integrated effect to produce both primary and secondary rainbows with inverse color (b to r).

rainbows is darker than the rest of the sky background. The geometry that explains the secondary rainbow follows the same approach that was used in the case of the primary rainbow, as shown in Figure C.6, where, as before, the incident ray is at an angle of I.

From Snell's law, for the incident ray:

$$\sin I = n \sin \alpha. \tag{C.6}$$

From Figure C.6(a), for a symmetrical optical axis, the total deviation, θ, from the sun's input rays (all parallel), as shown in Figure C.6(b) is:

$$\begin{aligned} \theta &= I - \alpha + 180 - 2\alpha + 180 - 2\alpha + I' - \alpha \\ &= I + I' + 360 - 6\alpha. \end{aligned}$$

From symmetry, $I = I'$:

$$\theta = 2I - 6\alpha + 360. \tag{C.7}$$

To find the maximum deviation, take the derivative of θ with respect to I:

$$\frac{\mathrm{d}\theta}{\mathrm{d}I} = 2 - 6\frac{\mathrm{d}\alpha}{\mathrm{d}I} = 0. \tag{C.8}$$

From the total differential of Equation (C.6)

$$\begin{aligned} \cos I \, \mathrm{d}I &= n \cos \alpha \, \mathrm{d}\alpha \\ \frac{\mathrm{d}\alpha}{\mathrm{d}I} &= \frac{\cos I}{n \cos \alpha}. \end{aligned}$$

Substituting into Equation (C.8),

$$2 - 6\frac{\cos I}{n \cos \alpha} = 0.$$

Rearranging,

$$\begin{aligned} 3 \cos I = n \cos \alpha &= \sqrt{n^2 - \sin^2 I} \\ 9 \cos^2 I &= n^2 - \sin^2 I \\ 9\left(1 - \sin^2 I\right) &= n^2 - \sin^2 I \\ 8 \sin^2 I &= 9 - n^2 \\ \sin I &= \sqrt{\frac{9 - n^2}{8}}. \end{aligned}$$

This is the incident angle that gives the maximum deviation.

The geometry of the secondary rainbow is shown in Figure C.7. Now blue is on top of or above red from the observer's eye position and at a larger angle than the primary rainbow, as shown in Figure C.8. The angle from the horizontal is 50.2° for the red light and 52.2° for the blue light. Detailed calculations for red and blue colors are given in Table C.2.

Table C.2. *Deviation angles for various colors for the secondary rainbow*

Color	Refractive index	I	α	Total deviation angle θ	Angle with line of sight ϕ
Red	1.3312	71.9°	45.6°	230.4°	50.4°
Blue	1.3371	71.7°	45.2°	232.0°	52.0°

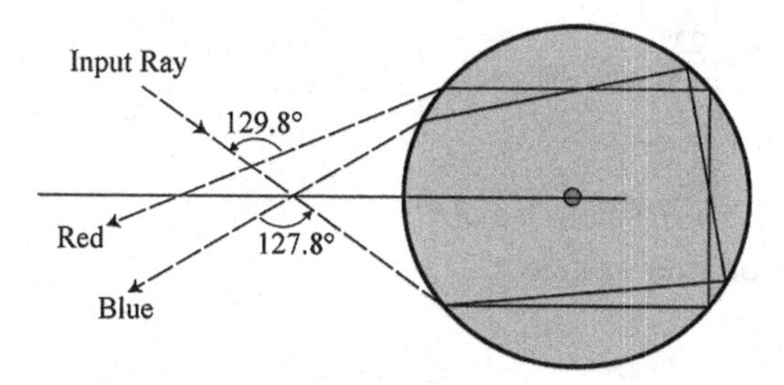

Figure C.7 Secondary rainbow geometry.

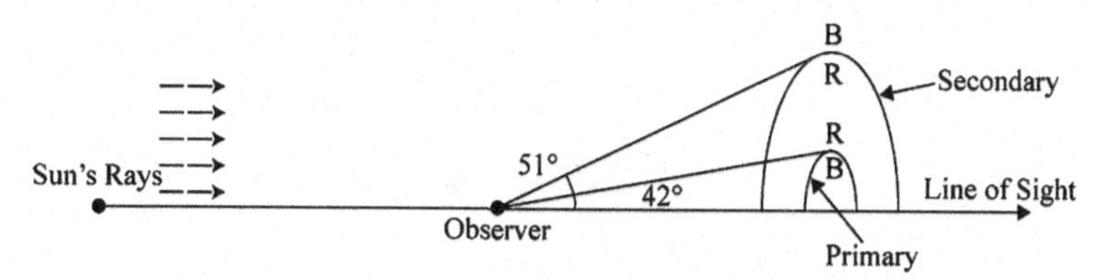

Figure C.8. Geometry of the observer and the rainbows.

The geometry of the observer and the rainbows is shown in Figure C.8 as two arcs at approximately 42° and 51° from the line of sight defined by the sun to eye.

C.3 Halos

During the winter months, halos are often observed around the moon, appearing as a ring at approximately 24° from the moon. This is due to ice crystals in the atmosphere. These ice crystals grow in a hexagonal form, as shown in Figure C.9.

If two sides of the hexagon are extended, as shown in Figure C.9, the extension resembles a prism with an apex angle of 60°, which can deviate light, as was discussed in Section 4.3.3. The minimum deviation angle and the index of refraction were given by Equation (4.26) as:

$$\delta_{\text{min}} = A - 2\sin^{-1}\left(n \sin\frac{A}{2}\right),$$

$$A = \text{apex angle} = 60^\circ,$$

$$n = \text{index of refraction of ice.}$$

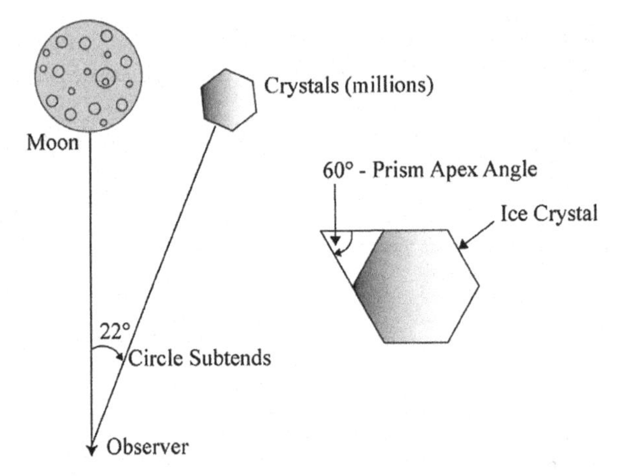

Figure C.9 Moon ring or halo.

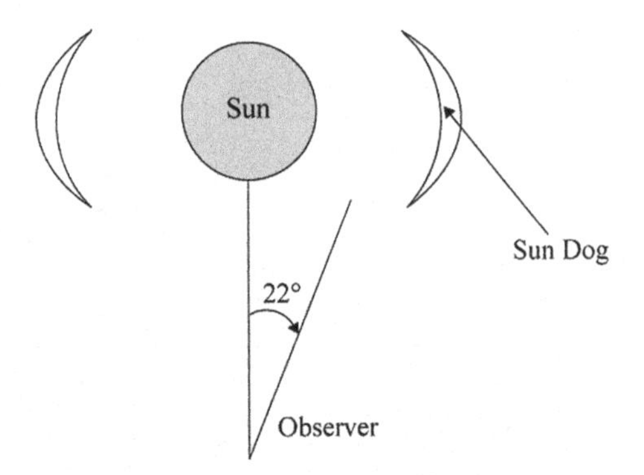

Figure C.10 Sun dogs at 22°.

Substituting in the refractive index of 4/3 for ice, one finds that the deviation angle is about 24°, with the outer edge having a slight blue color since the refractive index there is slightly higher.

A similar effect with the same angle (about 22°) is seen with sun dogs. However, sun dogs are only observed when the sun is near the horizon, and are horizontally located on either side of the sun – not in a circle. This is because the ice crystals that are between the observer and the sun are floating down like a leaf falling from a tree, and therefore are oriented with the apex angle of the ice prisms pointing to the sun. This unique orientation of the crystals produces only two bright regions, one on each side of the sun; these are called sun dogs, and are shown in Figure C.10.

C.4 Mirages

Mirages are often seen in the desert. White Sands, New Mexico, is one of the best places to observe them. Mirages occur when the heat from the earth has caused the air's index of refraction to vary with height above ground. This variation in index of refraction causes the light to bend in a curve such that the light is reflected back from

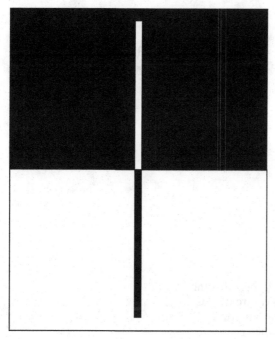

Figure C.11 Edge response to observe rainbow.

the Earth forming a mirage that looks like a lake. The variation in the refractive index of air due to temperature also causes the twinkling of star light.

In order to make your own rainbow of colors using a prism, Figure C.11 is provided to give you both the ordinary rainbow of colors and complementary colors. This requires you to have a prism.

Appendix D
Nomenclature for equations

Upper case:

BFD	Back focal distance
C	Curvature of lens
CA	Clear aperture
D	Diameter
D_{ent}	Diameter of entrance pupil
D_{ex}	Diameter of exit pupil
E	Electric field
E_{i}	Incident radiant power
E_{r}	Reflected radiant power
E_{t}	Transmitted radiant power
F	Front focal point
F^*	Back focal point
FFD	Front focal distance
FOV	Field of view
$(F/\#)_{\infty}$	F-number infinity
$(F/\#)_{w}$	F-number working
I, I_{i}	Angle of incidence
I'	Angle of refraction
I_{r}	Angle of reflection
I_{c}	Critical angle
$IFOV$	Instantaneous Field of View
K_{e}	Extinction coefficient
K_{o}	Dielectric constant
KK	Conic constant
L	Distance
L'	Reduced distance
M_{t}	Transverse magnification
M_{z}	Axial/longitundinal magnification
M_{α}	Angular magnification
N	Front nodal point
N^*	Rear nodal point
NA	Numerical aperture

OPD	Optical path distance
OPL	Optical path length
P	Front principal point/plane
P^*	Rear principal point/plane
R	Radius of curvature
U	Wavefront
$V^{\#}$	Abbe number
W	Wavefront error (subscript indicates type)

Lower case:

c	Speed of light
cc	Center of curvature
efl	Effective focal length
f	Front focal length
f^*	Back focal length
f_C^*, f_d^*, f_F^*	Focal length for C, d, and F-light
h	Planck's constant, object height
h'	Image height
k	Wave number
n	Refractive index
n_C, n_d, n_F	Refractive index of C, d, and F light
n_g	Refractive index of glass
p	Partial dispersion
$\bar{u}$	Chief ray angle
u	Marginal ray angle
v	Velocity
x	Object to front focal point distance (Newtonian form)
x'	Image to back focal point distance (Newtonian form)
$\bar{y}$	Chief ray height
y	Marginal ray height
z	Object distance (measured from P)
z'	Image distance (measured from P^*)
z_H	Hyperfocal distance
z_s	Sag of a spherical surface
Δz	Depth of field

Greek letters

α_a	Aperture scaling factor
α_f	Field angle scaling factor
α_r	Rayleigh criterion
α_w	Whole system scaling factor
δ	Distance from vertex to front principal plane
$\delta_C, \delta_d, \delta_F$	Deviation of C, d, and F light
δ_p	Prism deviation
δ^*	Distance from back vertex to back principal plane
ϵ	Eccentricity
ϵ_m	Permittivity of a medium
ϵ_r	Relative permittivity

η	Surface normal
λ	Wavelength
$\lambda_C, \lambda_d, \lambda_F$	Wavelength of *C*, *d*, and *F* light
μ_{m}	Permeability of a medium
μ_{r}	Relative permeability
υ	Frequency (Hz)
ρ	Fresnel reflectance
τ	Fresnel transmittance
ϕ	Optical power (numerical subscripts indicate surfaces)
ϕ_C, ϕ_d, ϕ_F	Power of *C*, *d* and *F* light
θ	Angle (typically in the pupil plane)
ψ	Angle
ω	Frequency (radians)
Ω	Steradians (solid angle)

Other characters:

$\overline{PF}$	Directed distance between front principal plane and front focal point
$\overline{P^*F^*}$	Directed distance between back principal plane and back focal point
$\overline{PP^*}$	Directed distance between principal planes
$\overline{PN}$	Directed distance between front principal plane and front nodal point
$\overline{P^*N^*}$	Directed distance between back principal plane and back nodal point
₱	Position factor
$	Shape factor
ж	Lagrange invariant
$\overline{V_1F}$	Directed distance between front vertex and front focal point
$\overline{V_2F^*}$	Directed distance between back vertex and back focal point
$\overline{V_1P}$	Directed distance between front vertex and front principal plane
$\overline{V_2P^*}$	Directed distance between back vertex and back principal plane

Appendix E
Fundamental physical constants and trigonometric identities

Constants (10^6 accuracy)

Quantity	Symbol	Value	Units
Speed of light in a vacuum	c	299 792 458	$m\ s^{-1}$
Permeability of vacuum	μ_c	$4\pi\ (10^{-7}) = 12.566\ 370\ 614\ldots$	$10^{-7}\ m^{-3}\ kg\ s^{-2}\ C^{-2}$
Permittivity of vacuum	ϵ_0	8.854 187 817	$10^{-12}\ m^{-5}\ kg^{-1}\ s^4\ C$
Elementary charge	e	1.602 176 462	$10^{-19}\ C$
Planck's constant	h	6.626 068 76	$10^{-34}\ J\ s$
		4.135 667 43	$10^{-15}\ eV\ s$
Dirac's constant	$\hbar$	$h/2\pi = 1.054\ 571\ 596$	$10^{-34}\ J\ s$
		$= 6.582\ 119\ 15$	$10^{-16}\ eV\ s$
Boltzmann constant	k	1.380 650 3	$10^{-23}\ J\ K^{-1}$
Wien displacement constant	b	2.897 768 5	$10^{-3}\ m\ K$
Stefan Boltzmann constant	σ	5.670 400	$10^{-8}\ W\ m^{-2}\ K^{-4}$
Electron mass	m_e	9.109 381 88	$10^{-31}\ kg$
Proton mass	m_p	1.672 621 58	$10^{-27}\ kg$
Neutron mass	m_n	1.674 927 16	$10^{-27}\ kg$
Gravitational constant	G	6.674 200	$10^{-11}\ m^3\ kg^{-1}\ s^{-2}$
F light wavelength	λ_F	486.10	$10^{-9}\ m$
C light wavelength	λ_C	656.30	$10^{-9}\ m$
d light wavelength	λ_d	587.60	$10^{-9}\ m$
pi	π	3.141 592 653 5	
Lumens per watt conversion		683	$lm\ W^{-1}$
Electron volt	eV	$1.602\ 177\ (10^{-19})$	J

Common trigonometric identities

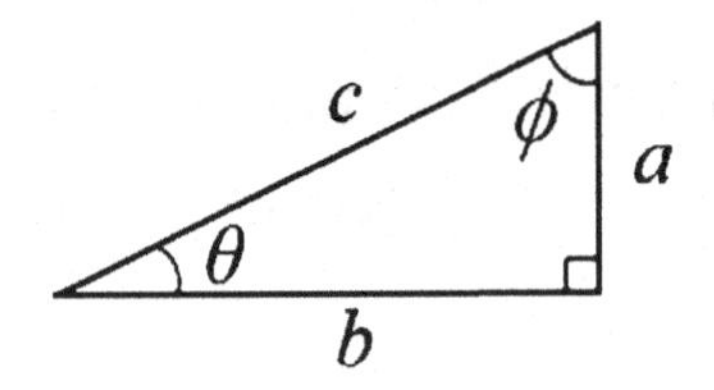

$$\sin\theta = a/c,$$

$$\cos\theta = b/c,$$

$$\tan\theta = \frac{\sin\theta}{\cos\theta} = \frac{a}{b}.$$

Reciprocal and quotient identities

$$\cos u = \sin(u + \pi/2), \qquad \sec u = \frac{1}{\cos u},$$

$$\tan u = \frac{\sin u}{\cos u}, \qquad \text{cosec } u = \frac{1}{\sin u},$$

$$\cot u = \frac{\cos u}{\sin u} = \frac{1}{\tan u}.$$

Pythagorean identities

$$\sin^2 u + \cos^2 u = 1,$$

$$\tan^2 u + 1 = \sec^2 u,$$

$$1 + \cot^2 u = \text{cosec}^2 u.$$

Cofunction identities

$$\sin(\pi/2 - u) = \cos u, \qquad \sec(\pi/2 - u) = \text{cosec } u,$$

$$\cos(\pi/2 - u) = \sin u, \qquad \operatorname{cosec}(\pi/2 - u) = \sec u,$$

$$\tan(\pi/2 - u) = \cot u, \qquad \cot(\pi/2 - u) = \tan u.$$

Even–odd identities

$$\sin(-u) = -\sin u, \qquad \operatorname{cosec}(-u) = -\operatorname{cosec} u,$$

$$\cos(-u) = \cos u, \qquad \sec(-u) = \sec u,$$

$$\tan(-u) = -\tan u, \qquad \cot(-u) = -\cot u.$$

Sum–difference formulas

$$\sin(u \pm v) = \sin u \cos v \pm \cos u \sin v,$$

$$\cos(u \pm v) = \cos u \cos v \mp \sin u \sin v,$$

$$\tan(u \pm v) = \frac{\tan u \pm \tan v}{1 \mp \tan u \tan v}.$$

Double-angle formulas

$$\sin(2u) = 2 \sin u \cos u,$$

$$\cos(2u) = \cos^2 u - \sin^2 u = 2\cos^2 u - 1 = 1 - 2\sin^2 u,$$

$$\tan(2u) = \frac{2 \tan u}{1 - \tan^2 u}.$$

Power-reducing/half angle formulas

$$\sin^2 u = \frac{1 - \cos(2u)}{2},$$

$$\cos^2 u = \frac{1 + \cos(2u)}{2},$$

$$\tan^2 u = \frac{1 - \cos(2u)}{1 + \cos(2u)}.$$

Sum-to-product formulas

$$\sin u + \sin v = 2\sin\left(\frac{u+v}{2}\right)\cos\left(\frac{u-v}{2}\right),$$

$$\sin u - \sin v = 2\cos\left(\frac{u+v}{2}\right)\sin\left(\frac{u-v}{2}\right),$$

$$\cos u + \cos v = 2\cos\left(\frac{u+v}{2}\right)\cos\left(\frac{u-v}{2}\right),$$

$$\cos u - \cos v = -2\sin\left(\frac{u+v}{2}\right)\sin\left(\frac{u-v}{2}\right).$$

Product-to-sum formulas

$$\sin u \sin v = \frac{1}{2}[\cos(u-v) - \cos(u+v)],$$

$$\cos u \cos v = \frac{1}{2}[\cos(u-v) + \cos(u+v)],$$

$$\sin u \cos v = \frac{1}{2}[\sin(u+v) + \sin(u+v)],$$

$$\cos u \sin v = \frac{1}{2}[\sin(u+v) - \sin(u-v)].$$

Law of sines

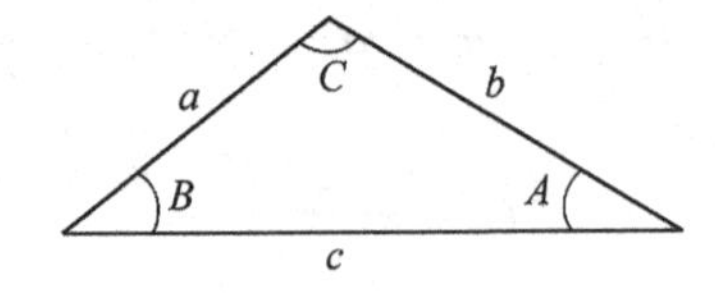

$$\frac{a}{\sin A} = \frac{b}{\sin B} = \frac{c}{\sin C}.$$

Law of cosines

$$c^2 = a^2 + b^2 - 2ab \cos C,$$

$$b^2 = a^2 + c^2 - 2ac \cos B,$$

$$a^2 = b^2 + c^2 - 2bc \cos A.$$

Law of tangents

$$\frac{a-b}{a+b} = \frac{\tan\left(\frac{1}{2}(A-B)\right)}{\tan\left(\frac{1}{2}(A+B)\right)}.$$

Euler's relations

$$\mathrm{e}^{\mathrm{i}x} = \cos x + \mathrm{i} \sin x,$$

$$\mathrm{e}^{-\mathrm{i}x} = \cos x - \mathrm{i} \sin x,$$

$$\cos x = \frac{\mathrm{e}^{\mathrm{i}x} + \mathrm{e}^{-\mathrm{i}x}}{2},$$

$$\sin x = \frac{\mathrm{e}^{\mathrm{i}x} - \mathrm{e}^{-\mathrm{i}x}}{2\mathrm{i}},$$

where $\mathrm{i}^2 = -1$.

Glossary

Abbe number (glass factor, $V^{\#}$) A quantitative measure of the average slope of the dispersion curve (refractive index versus wavelength curve).

Afocal system An optical system which changes angular magnification but has a plane wavefront input and a plane wavefront output.

Angle of incidence (I_i) The angle at which light first interacts with a surface, measured at the point of incidence relative to the normal of the surface; forms a plane of action.

Angle of reflection (I_r) The angle at which light is reflected after interacting with a surface, equal in magnitude to but opposite in sign from the angle of incidence, measured at the point of incidence relative to the normal of the surface; lies in plane of action.

Angle of refraction (I') Angle that light bends after interacting with a surface, measured at the point of interaction to the normal of the surface; lies in plane of action.

Angular magnification (M_α) The ratio of the apparent angular size of the object observed through the optical system to that of the object viewed by the unaided eye.

Aperture stop The limiting diameter of a lens, mirror, or baffle within a system that determines the amount of light that enters by excluding rays farther from the optical axis than the aperture stop diameter.

Axial magnification Also known as longitudinal magnification; occurs along the optical axis and is represented by M_z.

Back focal distance (BFD) The distance from the last surface of an optical system to the back focal plane.

Back focal length (f^*) Distance from the rear principal plane to the rear focal plane.

Baffles or glare stops Barriers to stray light that would otherwise bounce off the walls of a system and combine with the desirable light. These are placed at images of the aperture stop within the system.

Birefringent material Materials that have polarization-dependent indices of refraction, called the ordinary (n_o) and extraordinary (n_e) refractive indices, causing the x- and y-components of the electric field to experience different velocities as the wave is passing through the media.

Center of curvature The center point from which a spherical (curved) surface is drawn.

Chief ray A ray that passes through the center of the aperture stop in an optical system. Also called the principal ray.

Collimated light A wavefront which is a plane wave.
Collinear Points are collinear if they lie on the same line.
Concave A concave surface is indented relative to a flat surface.
Conjugate planes Planes that are in a one-to-one correspondence between any two spaces.
Convergence The approach of light to a fixed value or location, such as to a point.
Convex A convex surface bends outward and converges the curvature of a wavefront incident on it; it produces a spherical wavefront convergence (if spherical surface) to a point.
Crown Glass with a low index of refraction (< 1.6) and low dispersion (Abbe number greater than 55).
Critical angle (I_c) Angle at which total internal reflection (100% reflectance) occurs.
Depth of field How far the object can be moved longitudinally along the axis without creating more than an acceptable amount of blur.
Depth of focus The distance through which the focal plane may be moved from its ideal "best focus" position without seriously degrading the image.
Dielectric material A material such as glass in which no light is absorbed (the extinction coefficient $K_e = 0$).
Diffraction limit A limit on the maximum resolution of any optical system of a given *F/#* based on physical laws. An approximation to the diffraction limit for visible light is simply the *F/#* in micrometers.
Dispersion The angular spreading out of light upon refraction due to varying velocities in the media as the index of refraction varies with wavelength.
Divergence The spread of light rays propagating to infinity, with the corresponding wavefronts becoming increasingly larger.
Effective focal length (*efl*) The distance from a principal plane to the corresponding focal plane.
Electromagnetic (EM) wave A self-propagating wave consisting of electric and magnetic fields fluctuating together.
Entrance pupil The apparent size and location of the aperture stop when looking into the system from the object side.
Exact ray A real ray traced algebraically through a system; this tracing is usually done with computers because it is a difficult labor intensive process.
Exit pupil The apparent size and location of the aperture stop when looking back into the system from the image side.
***F*-number** (*F/#*) A measure of the light-gathering power of an optical system; it is related to the ultimate resolution capability. It is defined as the ratio of the effective focal length to the diameter of the entrance pupil.
Fermat's principle Light of a given frequency travels on a ray path of least time between two points.
Field-of-view (FOV) Expressed as ± some angle ($\pm\theta$). The outermost point of a scene capable of being transmitted through a lens to form an image, with the angle expressed with respect to the optical axis.
Field Stop The edge in the optical system that limits the extent of the image plane illuminated by light passing through the optical system. The field stop limits the size and shape of the image, but not its brightness. The field stop is located where the marginal ray crosses the axis. The field stop is located at the position of a real image that has a limited radial extent in the system. Placing a positive lens at the field stop can dramatically increase the field of view by allowing a real image formed within the system to accept light at higher angles that would otherwise hit the outer edge of the system.

Flint Glass with a relatively high index of refraction ($n_d > 1.6$) and a high dispersion (Abbe number less than 50).

Focal plane The plane at which an optical system brings parallel light to a point.

Focal point The intersection of the focal plane with the optical axis.

Frequency (υ) The number of cycles per unit of time, the reciprocal of the period, $1/T$, or c/λ.

Front focal length (*FFL*) The distance from the front principal plane of an optical system to the front focal plane.

Front focal distance The distance from the first lens surface to the first focal point.

Gaussian equation

$$\frac{1}{z'} = \frac{1}{z} + \frac{1}{f*} \quad \text{or} \quad \frac{n'}{z'} = \frac{n}{z} + \phi.$$

If the system is not in air. Used to find the focal length (or optical power $\phi = 1/f$), image distance (z'), or object distance (z) if two of these three values are known.

Group velocity The rate that the envelope of a waveform is propagating; i.e. the rate of variation of the amplitude of a waveform.

Gullstrand's equation $\phi = \phi_1 + \phi_2 - \phi_1\phi_2(t/n')$, used to find the total optical power of a two-element system separated by t in refractive index n'.

Handedness (parity) Image orientation after reflection. An image which undergoes an even number of reflections maintains its handedness (right-handed). However, an odd number of reflections change the handedness to odd (left-handed).

Hyperfocal distance The distance beyond which the object can be considered as if it were at infinity.

Incoherent light Light produced by a large number of individual radiators, producing a continuum of independent waves (incoherent radiation) with a spectrum of frequencies.

Index of refraction (n) The property of a material by which electromagnetic radiation is slowed down (relative to the velocity in a vacuum) when it travels inside the material.

Lagrange invariant A special case of the more general optical invariant in which the two rays are the chief and marginal rays. The product of the chief ray angle times the marginal ray height minus the marginal ray angle times the chief ray height with corresponding indices of refraction is a constant or invariant throughout an optical system. It determines the throughput of radiation from object to image. Bigger is better.

Lens maker's equation used to find the power of a thin lens.

$$\phi = (n_g - 1)\left(\frac{1}{R_1} - \frac{1}{R_2}\right).$$

Longitudinal waves Waves that need a medium in which to propagate, such as sound waves or water waves.

Marginal ray A ray that passes through an optical system near the edge of the aperture.

Meniscus In the lateral direction, the center of the lens is offset (± distance) from the edge location.

Meridional rays A ray that lies in the plane that contains the optical axis: typically, a ray in the y–z plane, also called a tangential ray.

Metamaterials Materials with a negative index of refraction.

Newtonian equation $zz' = ff^*$, used to find the image distance (z'), object distance (z), front focal distance (f), or back focal distance (f^*) if three of these four values are known. Eliminates the need to know lens thicknesses and principal point locations.

Nodal points Imaginary points about which an optical system may be rotated without changing the position of an image. For an optical system in air ($n = 1$), the nodal planes and principal planes coincide.

Numerical aperture (*NA*) The sine of the largest angle at which light enters the lens from a point near the front focal point. For first-order optics, the *NA* is equal to $(2F/\#)^{-1}$ or $n \sin u$.

Optical path length (*OPL*) The distance light travels through a medium is different from the width or thickness because the velocity of light slows in a medium denser than air; the product of the width of the medium and the refractive index.

Optical power The degree to which a lens or mirror converges or diverges light; equal to the reciprocal of the focal length of the lens or mirror.

Paraxial approximation A mathematical simplification used in ray-tracing calculations which requires rays to be near the optical axis and to have small angles relative to the axis such that $\sin u = u$.

Paraxial ray A ray representation of light that lies close to and almost parallel to the optical axis, which is usually the z axis in an x, y, z coordinate system.

Partial dispersion The amount a glass or optical material will spread light into its visible spectral components,

$$P = \frac{n_d - n_C}{n_F - n_C}.$$

Period (T) The time a wave takes to complete one cycle in amplitude.

Phase The relationship of the sinusoidal period of the electromagnetic wave, namely the term describing the internal argument of the sine function in an electromagnetic wave.

Phase velocity The rate at which the crests of the waveform propagate; or the rate at which the phase of the waveform is moving.

Plane waves Wavefronts which are flat and perpendicular to the rays.

Plano Flat surface with optical power equal to infinity

Point source A light source whose rays emanate in all directions from that source, such as starlight.

Polarization The shape in time formed by the electric field x and y vectors as seen from the negative z direction, such as linear, circular, and elliptical.

Principal planes Imaginary surfaces from which the focal lengths are measured in an optical system. For Gaussian optics the object and image distances are measured from these planes.

Principal points The intersection of the principal planes with the optical axis.

Radius of curvature The radius of a spherical surface.

Ray A geometrical optics representation of the electromagnetic wave as a straight line vector pointing in the direction of propagation with direction but no phase, and perpendicular to the propagating wavefronts.

Real image If an optical system produces a converging wavefront, the image will be a real image. The rays converge to a point for each point in the object and the image can be projected on a screen.

Reduced distance (L') The distance that an object appears to be from an observer, due to the refractive index when viewed in a denser medium than free space. Reduced thickness is L/n, where L is the thickness of the medium and n is the refractive index.

Reflectance (Fresnel reflectance, ρ) The percentage of light incident on a surface.

Reflection The phenomenon in which some of the light incident on a material travels at an equal but opposite angle to the incident angle upon incidence. The remaining light is refracted, or bent into the material depending on the index of refraction of the material.

Refraction The phenomenon in which some of the light incident on a material is bent into the material: it depends on the index of refraction of the material and is due to a change in velocity. The remaining light is reflected at an equal but opposite angle to the incidence angle.

Refractive index This is defined as the ratio of the speed of light in a vacuum to the speed of light in the material. It is always greater than 1 for standard materials and is limited to 6.

Resolution A figure of merit of the quality of a lens system; usually thought of as the smallest separation between two point sources capable of being distinguished in the image.

Roof When two mirrors make a 90° angle (dihedral angle $\theta = 90°$), both input and output rays are 180°, or anti-parallel, in the principal projection plane. A roof configuration is equivalent to a plane mirror, except that the handedness is even.

Sag The distance from a plane at the vertex to the actual curved surface of a lens where a parallel ray refracts.

Sagittal ray A ray that intersects the entrance pupil at points for which $\theta = \pm\pi/2$. It is a skew ray that originates in the x–z plane but intersects the pupil at $y_p = 0$.

Shape factor ($\mathcal{S}$) A numerical value that describes the shape of a lens, such as plano, equi-convex, concave-plano, etc. Convex lenses have negative values and concave lenses positive ones.

Sign convention The set of rules that govern the direction of light propagation and image formation in optical systems.

Skew ray A ray that originates from an object point in the y–z tangential (meridional) plane but does not propagate in this plane. Such a ray will intersect the entrance pupil at some arbitrary coordinates (x_p, y_p).

Snell's law The law used to describe how light refracts when encountering a material that changes its velocity, $n \sin I = n' \sin I'$.

Solid angle (Ω) A cone generated by a line that passes through the vertex and a point on a surface which is enclosed as this line is moved to contour this surface; a measure of how large that object appears at that point in three-dimensional space.

Speed of light (c) The constant velocity, approximately $3(10^8)$ m s^{-1}, in which light travels in a vacuum.

Spherical waves Waves emanating from a source or optical system as circles of energy, either converging (decreasing in diameter) or diverging (increasing in diameter).

Tangential ray A ray that is contained in the y–z plane and intersects the entrance pupil at $x_p = 0$.

Telecentricity If all chief rays on the object side or the image side of a system are going parallel to the optical axis, the system is said to be telecentric in object space or telecentric in image space. Telecentricity allows the object focal distance to vary without the magnification changing, so it is very useful for microscopes. Telecentricity is created by placing the aperture stop at the collective focal length of the lenses.

Thin lens A lens assumed to have no thickness.

Total internal reflection (TIR) A phenomenon in which all light is reflected upon interacting with a surface; it occurrs when the angle of incidence is equal to or greater than the critical angle, and the incident index of refraction is larger than the surface index of refraction.

Transmittance (Fresnel transmittance, τ) The percentage of light transmitted through a surface upon interaction, $\tau = [4n'n/(n'+n)^2]$, also equal to $1-\rho$, where ρ is the reflectance.

Transverse magnification The image magnification perpendicular to the optical axis, represented by M_t. If the magnification is negative the image is inverted. Conversely, a positive magnification indicates the image is upright.

Tunnel diagram The unfolding of a prism so that it lies flat along the z axis.

Vignetting The uneven truncation of ray fans in an optical system when approaching the system from off-axis object points. Off-axis object points have rays that approach too steeply, which are lost to the image due to shadowing.

Virtual image A representation of an actual object formed by a diverging wavefront, which seems to originate from a virtual image. The rays associated with this wavefront do not cross in real space – only in virtual space. There would not be an image on a screen placed at the virtual image location.

Visible spectrum The region of the electromagnetic spectrum which is generally divided into red, yellow, and blue light (600–700 nm, 500–600 nm, and 400–500 nm respectively) to represent human color sensitivity.

Wavefront An oscillation of energy which propagates from one point to another, carrying electromagnetic energy. It is an imaginary surface made up of a set of points with equal phase located at regular intervals from the source of light and perpendicular to the source rays.

Wavelength (λ) The length of a wave from one peak or valley to the next peak or valley. The length traveled by a wave in one period.

***ZZ'* diagram** The relationship between the object and image distances for a thin lens, represented graphically using a Cartesian coordinate system.

Index